murach's
HTML
and CSS

6TH EDITION

Zak Ruvalcaba

Anne Boehm

Mary Delamater

Editorial team

Authors:	Zak Ruvalcaba
	Anne Boehm
	Mary Delamater
Editors:	Joel Murach
	Scott McCoy
Production:	Juliette Baylon

Murach also has books on these subjects:

Web development

Modern JavaScript

JavaScript and jQuery

PHP and MySQL

ASP.NET Core MVC

Programming languages

Python

Java

C#

C++

Databases

MySQL

SQL Server

Oracle

Data science

Python for Data Science

R for Data Analysis

For more on Murach books, please visit us at www.murach.com

ISBN: 978-1-943873-21-0

Contents

Expanded contents

Section 1 Get started fast

Chapter 11   How to work with tables

Chapter 12   How to work with forms

Introduction

This 6th edition of our best-selling HTML and CSS book presents the skills that a web developer needs today. To do that, it uses the proven instructional approach that made the first five editions so popular. In addition, it updates and streamlines the previous edition so it works better than ever.

Who this book is for

We designed this book for anyone who wants to learn HTML and CSS, even if you have no programming experience. This book also works well for experienced web developers who want to expand and update their skills. Either way, when you finish this book, you'll have the HTML and CSS skills that you'll need on the job. In addition, you'll find that this book also makes it easy to quickly find answers to questions that come up on the job.

What this book does

To get you started right, the eight chapters in section 1 present a crash course that shows how to use HTML and CSS to develop web pages at a professional level. Chapter 3 shows how to use HTML to structure the contents of a web page. Then, chapters 4 through 7 show how to use CSS to format the page. Finally, chapter 8 shows how to use responsive web design to build web pages that look good and work right on devices with different screen sizes, from mobile phones to desktop computers.

When you finish section 1, you will have the perspective and skills you need for developing professional web pages. Then, you can add to those skills by reading any of the chapters in the next two sections, and you don't have to read those sections or chapters in sequence. In other words, you can skip to any of the chapters in the last two sections after you finish section 1.

The chapters in sections 2 and 3 present skills that you can learn whenever you need them. For example, if you want to learn how to create forms that get input from users, you can skip to chapter 12. If you want to learn how to add audio and video to your web pages, you can skip to chapter 13. If you want to learn the basic principles for designing a website, you can skip to chapter 16. And so on.

7 reasons why you'll love this book

- It presents modern best practices for using HTML and CSS and doesn't waste your time by explaining old practices that are no longer recommended.

- It teaches HTML and CSS in context so you can see how they work together to structure and format the pages of a website.

- It gets you started fast by presenting a complete subset of HTML and CSS in section 1 that allows you to begin doing productive work as soon as possible.

- It presents hundreds of examples that range from the simple to the complex. That way, you can quickly get the idea of how a feature works from the simple examples and also see how the feature is used in the real world from the more complex ones.

- The exercises at the end of each chapter provide a way for you to gain valuable hands-on experience without extra busywork.

- It uses *paired pages*, with the essential syntax, guidelines, and examples on the right page and clear explanations on the left page. This helps you learn faster by reading less.

- The paired-pages format is ideal for reference when you need to refresh your memory about how to do something.

Recommended software

To develop web pages with HTML and CSS, you can use any text editor. However, we recommend using Visual Studio Code (VS Code) because it provides many features that can help you develop web pages more quickly and with fewer errors.

In addition, we recommend using the Google Chrome browser to do the primary testing for a web page. That's because Chrome provides excellent developer tools for debugging your web pages.

This software can be downloaded for free from the internet. In addition, appendix A provides complete instructions for installing it.

The downloadable files

You can download all the files you need for getting the most from this book from our website. This includes the HTML and CSS files for:

- the complete web pages presented throughout this book

- the short examples presented throughout this book

- the starting points for the exercises at the end of each chapter

- the solutions to those exercises

These files let you test, review, and copy code. In addition, if you have any problems with the exercises, the solutions can help you over the learning blocks, which is an essential part of the learning process. Here again, appendix A provides complete instructions for downloading these files.

Support materials for trainers and instructors

If you're a college instructor or corporate trainer who would like to use this book for a course, we offer support materials that will help you set up and run your course as effectively as possible. These materials include instructional objectives, test banks, short exercises, case studies, and PowerPoint slides.

To learn more, please visit www.murachforinstructors.com if you're an instructor. If you're a trainer, please visit www.murach.com and click on the *Courseware for Trainers* link, or contact Kelly at 1-800-221-5528 or kelly@murach.com.

Please let us know how this book works for you

When we started writing the 1st edition of this book, our goal was to make it as easy as possible to master HTML and CSS. Based on all the positive feedback we've received on the first five editions, we seem to have succeeded. With this 6th edition, we worked hard to make sure this book continues to achieve our original goal.

Now, we hope we've succeeded. We thank you for buying this book. We wish you all the best with your web development. And if you have any comments, we would love to hear from you.

Zak Ruvalcaba
Author

Joel Murach
Editor

Section 1

Get started fast

The eight chapters in this section present the skills that you need to get started with writing HTML and CSS. These are the skills that you will use for almost every web page that you develop. And this is the minimum set of skills that every web developer should have.

When you complete this section, you'll be able to develop web pages at a professional level. Then, you can take your skills to the next level by reading the other chapters in this book.

After section 1, you don't have to read the chapters in sequence. Instead, you can skip to any chapter that presents the skills that you want to learn next. In other words, the eight chapters in this section present the prerequisites for the rest of the chapters in this book.

Chapter 1

Introduction to web development

This chapter introduces you to the concepts and terms that you need for working with HTML and CSS. When you finish this chapter, you'll have the background you need for learning how to build websites.

How web apps work

The *World Wide Web*, or *web*, consists of many components that work together to bring a web page to you over the *internet*. Before you start building web pages of your own, you should have a basic understanding of how these components work together.

The components of a web app

The first diagram in figure 1-1 shows that web applications, also known as web apps, consist of clients and a web server. The *clients* are the desktop computers and mobile devices, such as phones and tablets, that use the web app. They access the web pages through programs known as *web browsers*, such as Chrome, Edge, and Safari. The *web server* is a powerful computer that stores the files that make up a web app.

A *network* allows clients and servers to communicate. The internet is a large network that consists of many smaller networks. In a diagram, the "cloud" represents the network or internet that connects the clients and servers.

In general, you don't need to know how the cloud works. But it's nice to have a general idea of what's going on. That's why the second diagram in this figure gives you a conceptual view of the architecture of the internet.

To start, networks can be categorized by size. A *local area network* (*LAN*) is a small network of computers that are near each other and can communicate with each other over short distances. Computers on a LAN are typically in the same building or in adjacent buildings. This type of network is often called an *intranet*, and it can be used to run web apps for use by employees only.

By contrast, a *wide area network* (*WAN*) consists of multiple LANs that have been connected together over long distances using *routers*. To pass information from one client to another, a router determines which network is closest to the destination and sends the information over that network. A WAN can be owned privately by one company or it can be shared by multiple companies.

An *internet service provider* (*ISP*) is a company that owns a WAN that is connected to the internet. An ISP leases access to its network to other companies that need to be connected to the internet.

The internet is a global network consisting of multiple WANs that have been connected. ISPs connect their WANs at large routers called *internet exchange points* (*IXP*). This allows anyone connected to the internet to exchange information with anyone else.

The diagram in this figure shows how data is sent from the client in the top left to the server in the bottom right. First, the data leaves the client's LAN and enters the WAN owned by the client's ISP. Next, the data is routed through IXPs to the WAN owned by the server's ISP. Then, it enters the server's LAN and finally reaches the server. All of this can happen in a fraction of a second.

The components of a web application

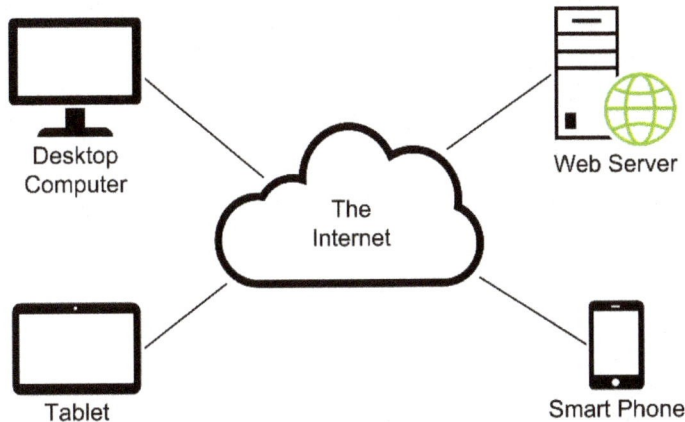

The architecture of the internet

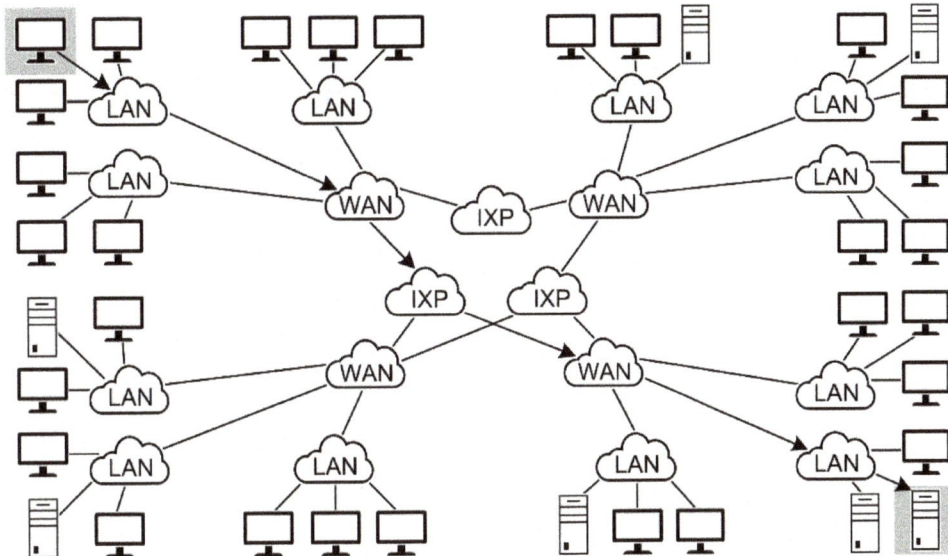

Description

- A web application consists of clients, a web server, and a network. The *clients* use programs known as *web browsers* to request web pages from the web server. The *web server* returns the pages that are requested to the browser.

- A *local area network* (*LAN*) directly connects computers that are near each other. This kind of network is often called an *intranet*.

- A *wide area network* (*WAN*) consists of two or more LANs that are connected by *routers*. The routers route information from one network to another.

- The *internet* consists of many WANs that have been connected at *internet exchange points* (*IXPs*). There are hundreds of IXPs located throughout the world.

- An *internet service provider* (*ISP*) owns a WAN and leases access to its network. It connects its WAN to the rest of the internet at one or more IXPs.

Figure 1-1   The components of a web application

The HTML for a web page

HyperText Markup Language (*HTML*) defines the content and structure of a web page. For example, figure 1-2 presents the HTML for a simple web page, which can be called an *HTML document*. A typical website sends this HTML from a web server to a web browser running on a client. Then, the browser renders the HTML into a web page that's displayed in the browser.

Although you're going to learn the details for coding HTML in chapter 3, here's a quick introduction to how the HTML works. This document uses *tags* to identify the *HTML elements* within the document. The *opening tag* for each element consists of the name for the element surrounded by angle brackets, as in <html> or <h1>. And the *closing tag* includes a forward slash before the name of the element, as in </html> or </h1>.

The basic structure of an HTML document consists of <head> and <body> elements that are coded within the <html> element. The <head> element contains elements that provide information about the document that isn't a part of the web page. In this figure, for example, the <title> element specifies a title of Mozart that the browser displays in the tab for the page.

On the other hand, the <body> element contains the elements that define the content that displays in the body of the web page. In this figure, for example, the <h1> element defines a level-1 heading, the <img> element defines an image, and the <p> elements define paragraphs.

Many elements can be coded with *attributes* in the opening tag. Each attribute consists of an attribute name, an equal sign, and an attribute value coded within a pair of quotes. In this figure, for example, the <img> element has two attributes named src and alt. In this case, the src attribute provides the name of the image file to display, and the alt attribute provides the text to display if the image can't be found.

An HTML file named index.html

```
<!DOCTYPE html>
<html lang="en">
<head>
    <meta charset="UTF-8">
    <meta name="viewport" content="width=device-width, initial-scale=1.0">
    <title>Mozart</title>
</head>
<body>
    <h1>Wolfgang Amadeus Mozart</h1>
    <img src="mozart.jpg" alt="Portrait of Wolfgang Amadeus Mozart">
    <p>Wolfgang Amadeus Mozart was a prolific and influential composer
        of the Classical period. He was born in 1756 and died in 1791.
        Despite his short life, his rapid pace of composition resulted in
        more than 800 works representing virtually every Western classical
        genre of his time.</p>
    <p>Born in Salzburg, Mozart showed prodigious ability from his earliest
        childhood. At age five, he was already competent on keyboard and
        violin, he had begun to compose, and he performed before European
        royalty.</p>
</body>
</html>
```

The HTML displayed in a browser

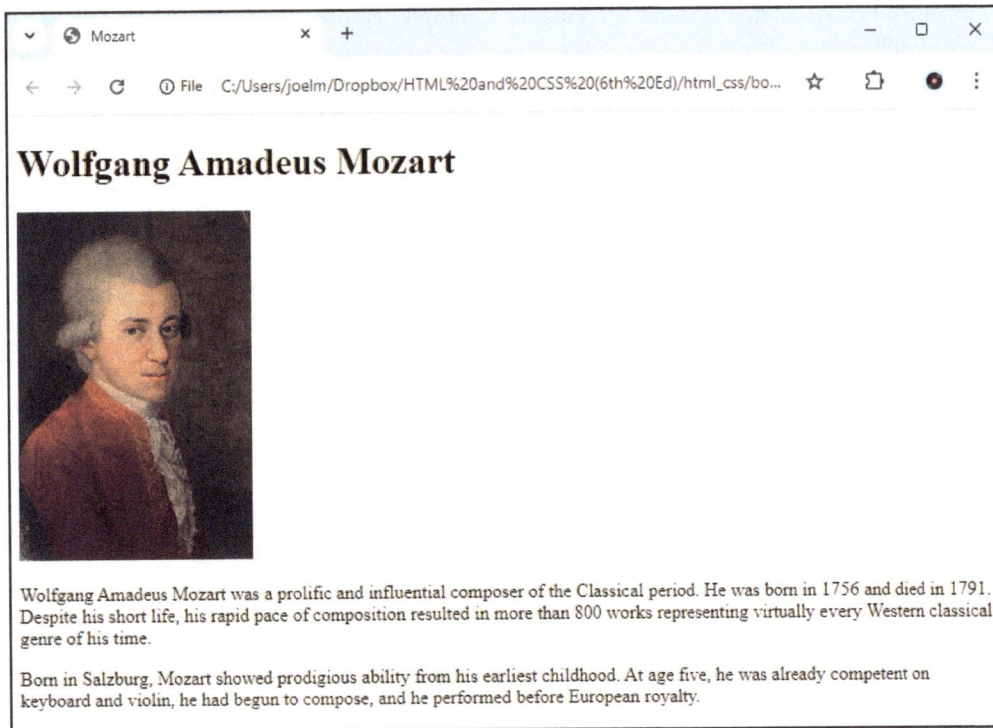

Description

- *HTML* (*HyperText Markup Language*) defines the structure and content of a web page.

Figure 1-2   The HTML for a web page

The CSS for a web page

You can use *Cascading Style Sheets* (*CSS*) to define the formatting for a web page. For example, figure 1-3 begins by showing some CSS that can be applied to the HTML presented in the previous figure. This CSS is stored in a file named main.css. Since it defines the style of a web page, this code is known as a *style sheet*.

After the CSS file, this figure shows a <link> element that you can use to apply this CSS to a web page. To do that, you can code a <link> element in the <head> section of the HTML file. Then, the href attribute of the <link> element specifies that the style sheet in the file named main.css should be applied to the HTML document.

If you compare the browser shown in this figure to the one in the previous figure, you can see that the formatting in the CSS file has been applied to the content in the HTML file. For example, the browser uses a different font for the text, centers the content in the browser window, displays the image to the right of the text, and so on. This gives you a quick idea of how much you can do with CSS.

Here's a brief introduction to how the CSS works. First, this CSS file consists of five *style rules*. Each of these style rules consists of a *selector* and one or more *declarations* enclosed in braces { }. The selector identifies one or more HTML elements, and the declarations specify the formatting for the elements.

The first style rule uses the * wildcard character to select all elements. Then, it sets the margins (space outside of an element) and padding (space within an element) to 0. This is often helpful since different browsers may provide different margin and padding values for different elements. As a result, resetting all of these values to 0 is often a good starting point for a style sheet.

The second style rule applies to the <body> element. Here, the first declaration specifies that the font family for the content should be Arial, Helvetica, or the default sans-serif type, in that order of preference. The second declaration sets the top and bottom margins to 0 and the left and right margins to auto, which centers the page in the browser window. The third declaration sets some padding (space) between the content in the body and the edge of the body. And the fourth declaration sets the maximum width to 600 pixels, which prevents the lines of text from becoming too long.

Similarly, the third style rule formats the <h1> element with a larger font size, the navy color, and a small amount of padding on its bottom. The fourth style rule formats the <img> element by displaying it on the right side of the body and by adding some padding to its left side. This causes the following <p> elements to display to the left of the image with a little space between the text and the image. And the fifth style rule adds some spacing to the bottom of each paragraph.

This should give you an idea of how HTML and CSS work together. In short, the HTML defines the content and structure of the document, and the CSS defines the formatting. This separates the content from the formatting, which is considered a best practice for creating and maintaining web pages.

A CSS file named main.css

```css
* {
    margin: 0;
    padding: 0;
}
body {
    font-family: Arial, Helvetica, sans-serif;
    margin: 0 auto;
    padding: 1em;
    max-width: 600px;
}
h1 {
    font-size: 200%;
    color: navy;
    padding-bottom: .3em;
}
img  {
    float: right;
    padding-left: 1em;
}
p {
    padding-bottom: .5em;
}
```

The element in a HTML file that links to a CSS file

```html
<link rel="stylesheet" href="main.css">
```

The formatted HTML displayed in a browser

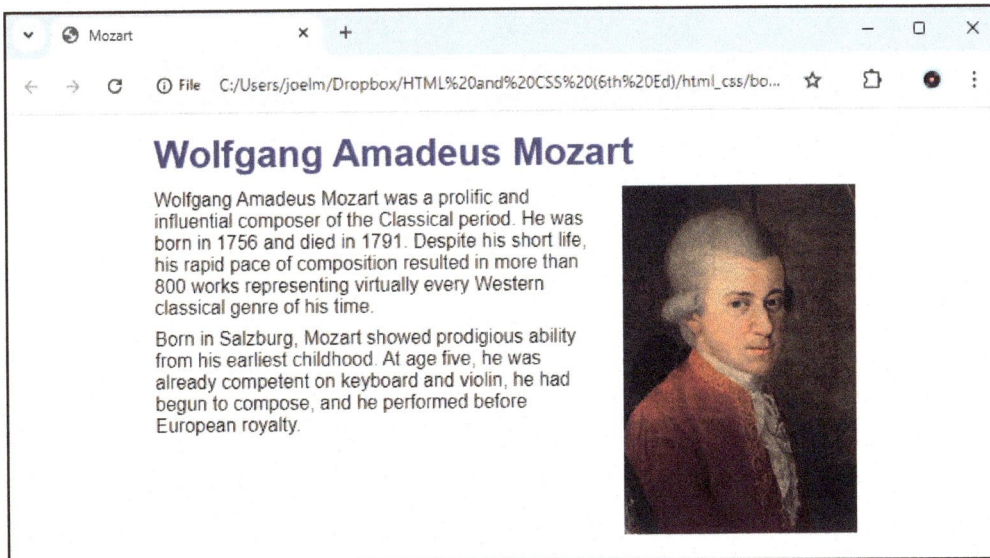

Description

- *CSS (Cascading Style Sheets)* defines the formatting for a web page by specifying the fonts, colors, borders, spacing, and layout of the pages.

Figure 1-3   The CSS for a web page

A short history of the HTML and CSS standards

Figure 1-4 presents a short history of the HTML and CSS standards. HTML standards have been around since 1995, but they didn't stabilize until version 5, which was adopted in 2014. This version is commonly referred to as HTML5.

Similarly, CSS standards have been around since 1996, but they didn't stabilize until version 3 and that version didn't get widespread use until 2010 or later. This version is commonly referred to as CSS3.

In recent years, new features were added to HTML5 and CSS3, but without new release numbers. As a result, there's no longer any reason to include the version numbers when referring to them.

This figure also describes two organizations that are involved in developing web standards. The first is the *World Wide Web Consortium*, which is commonly referred to as *W3C*. Until May of 2019, this was the group that developed the standards, and its website is a primary source for HTML and CSS information.

The second is the *Web Hypertext Application Technology Working Group* (*WHATWG*). This is a community of people interested in evolving HTML and related technologies, and its website is another primary source for HTML and CSS information. In May 2019, this group took over the development of the HTML standards, although the W3C continues to participate in the development process.

Highlights in the development of the HTML standards

Version	Description
HTML 1.0 to 4.01	HTML 2.0 was the first specification adopted as a standard by the W3C in November 1995. HTML 4.0 and 4.01 added new features and deprecated older features.
XHTML 1.0 to 1.1	XHTML 1.0 was adopted in January 2000 and reformulated HTML 4 using the syntax of XML. With XHTML 1.1, the control of the presentation of content was now done through CSS.
HTML 5 to 5.2	HTML 5 was adopted in October 2014 and replaced the current versions of both HTML and XHTML. HTML 5.1 and 5.2 were minor revisions of these standards.
HTML Living Standard	A standard that the WHATWG started developing as a split from the W3C in July of 2012. Because this standard is continually evolving, it doesn't use version numbers. Today, all modern browsers support this standard. In May 2019, the W3C ceded authority over the HTML standards to WHATWG. However, the W3C still participates in the development process.

Highlights in the development of the CSS standards

Version	Description
1.0	Adopted in December 1996.
2.0	Adopted in May 1998.
2.1	First released as a candidate standard in February 2004, it returned to working draft status in June 2005. It became a candidate standard again in July 2007.
3.0	A modularized version of CSS with the earliest drafts in June 1999. Some modules build on existing features of CSS 2.1, and others provide entirely new features. Each module is accepted as a standard independently.

Two important organizations for web standards

World Wide Web Consortium (W3C)

https://w3.org

An international community in which member organizations, a full-time staff, and the public work together to help develop web standards.

Web Hypertext Application Technology Working Group (WHATWG)

https://html.spec.whatwg.org

A community of people interested in evolving HTML and related technologies. It currently maintains the HTML standards.

Description

- Unlike the W3C standards, the WHATWG Living Standard is continually evolving.

Figure 1-4   A short history of the HTML and CSS standards

How web pages are processed

As described earlier, a web app consists of a client and a web server. In addition, a web app may use other kinds of *servers*, which are powerful computers that typically run at a remote location. These servers can store the files and databases that make up a web app, and they can run any code for the web app that's stored on the server.

The diagram in figure 1-5 shows how web apps typically work. To start, the web browser builds a request for a web page and sends it to the web server. This request is known as an *HTTP request* because it is formatted using *HyperText Transfer Protocol* (*HTTP*).

When the web server receives the HTTP request, it reads the request and returns an appropriate *HTTP response*. If the browser has requested a web page, the HTTP response typically contains the HTML, CSS, and JavaScript for the page.

When the browser receives the HTTP response, it *renders* (translates) the HTML, CSS, and JavaScript and displays the web page to the user. If the user requests another page, either by clicking a link or typing another web address into the browser, the process begins again. A process that begins with a client making a request and ends with a server returning a response is called a *round trip*.

What the server returns to the browser depends on various factors, including the parameters stored in the HTTP request. If, for example, the user has entered data into a form, the HTTP request includes that data as parameters. Then, the web server typically uses an *application server* to process the data in the HTTP request. This may include using *Structured Query Language* (*SQL*) to get data from a database that's running on a *database server*. Since this processing is done on the server, it's known as *server-side processing*.

In contrast to the server-side processing, *JavaScript* is a programming language that provides for *client-side processing*. Client-side processing works because all modern browsers have a *JavaScript engine* that can run JavaScript. This takes some of the processing burden off the server, which can make an app more efficient. In addition, this limits the number of round trips required by an app, which makes the app run faster.

JavaScript is commonly used to validate the data that the user enters into an HTML form before it is sent to the server for processing. In addition, JavaScript is commonly used to change the images that are displayed without using server-side processing. This includes image swaps, image rollovers, carousels, accordions, and slide shows.

An HTTP request and response for a web page

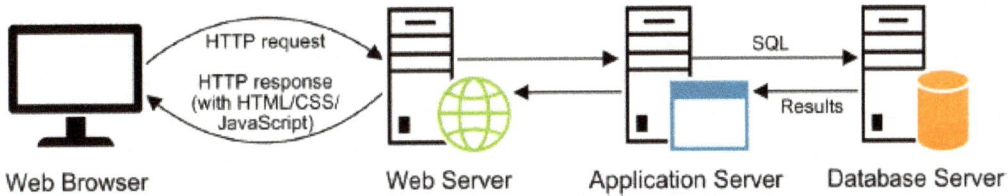

Some common uses of JavaScript

- Data validation
- Image swaps and rollovers
- Carousels and accordions
- Slide shows

Description

- The internet uses *HTTP* (*HyperText Transfer Protocol*) to send requests for web pages and to return responses that include the HTML, CSS, and JavaScript for a web page.
- When a user requests a web page, the web browser sends an *HTTP request* to the web server for the web page.
- When the web server receives the request, it retrieves the HTML, CSS, and JavaScript for the web page and sends it back to the browser as part of an *HTTP response*.
- When the browser receives the HTTP response, it *renders* the HTML, CSS, and JavaScript into a web page.
- The *JavaScript* language provides a way to add functionality to a web page.
- JavaScript runs within the web browser, not the web server. As a result, JavaScript processing is known as *client-side processing*. Client-side processing doesn't require a trip back to the server so it can help a web app run more efficiently.
- When a web server receives an HTTP request for a web page, it can pass the request on to an *application server* that runs code to process the request. Since this processing happens on a server, it's known as *server-side processing*.
- If an application server needs to work with a database, it can use a language known as *SQL* (*Structured Query Language*) to work with a database that's running on a *database server*.

Figure 1-5   How web pages are processed

Four web development issues

Whenever you develop a web app, you should be aware of the issues that are presented in the next four figures. As you progress through this book, you will learn how to handle each of them.

Responsive web design

Responsive web design (*RWD*) means that a website should work on all devices that access it. This may include desktop computers that have large screens or mobile phones that have small screens and may display content in portrait or landscape mode. For example, the screens shown in figure 1-6 show how a site that uses RWD adapts to the size and orientation of the screen. Here, you can see the home page of a website in a desktop browser with a relatively large screen and also on a mobile phone with a smaller screen in portrait mode. This shows that the look-and-feel of the page remains the same in both screen sizes. However, the design of the web page responds to the device that's accessing the web page to make the web page work well on all devices.

After the screens, this figure summarizes four benefits of responsive web design. First, it makes your site mobile-friendly. This is important because a high percent of people use mobile devices with small screens to access web pages. Second, it improves the way your site looks and works on devices with other screen sizes too, including large screens. Third, many search engines take RWD into consideration. As a result, using a responsive design can improve your site's ranking in search engines. Fourth, many users will leave a site if the site doesn't work well for their device. As a result, using a responsive design can reduce the bounce rate for your site.

Because RWD is such an important subject, this book covers the essential skills for working with it in section 1. In particular, chapter 7 shows how to implement a fluid design that works well on most devices. Then, chapter 8 shows how to use media queries to change the design of the web page depending on the screen size of the device.

A web page on a desktop computer and a mobile phone

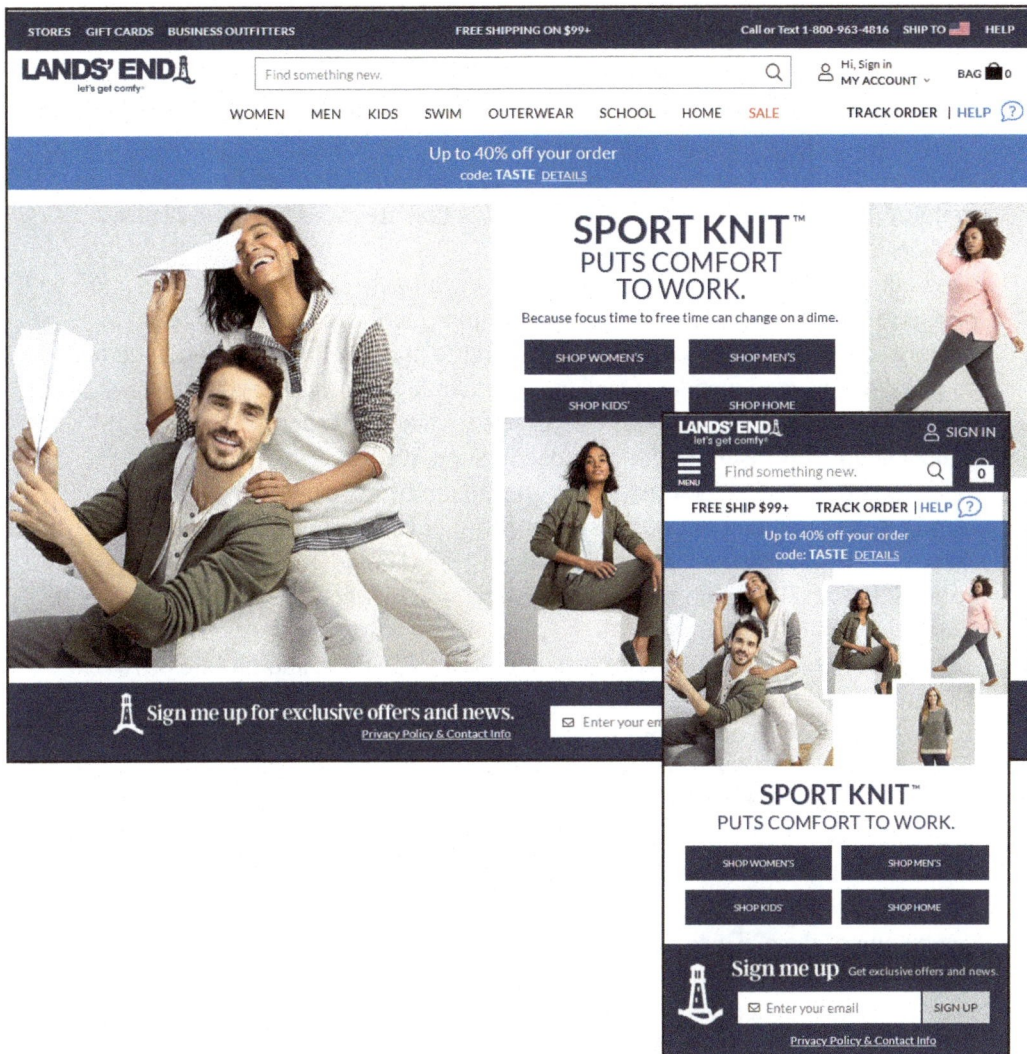

Four benefits of responsive web design

- Makes your site mobile-friendly
- Improves the way your site looks and works on devices with large and small screens
- Improves your site's rankings in search engines
- Reduces the bounce rate of your site

Description

- *Responsive web design* (*RWD*) means that a website should adapt to the screen size of the device that's accessing it.

Figure 1-6   Responsive web design

Cross-browser compatibility

If you want your website to be used by as many visitors as possible, you need to make sure that your web pages are compatible with all the browsers that people may use to access your website. That's known as *cross-browser compatibility*.

Not long ago, this was a significant development issue that required much attention. Today, however, most modern browsers including the five listed in figure 1-7 support almost all of the features that are provided by the latest versions of HTML and CSS.

In the past, Internet Explorer gave web developers the most problems because it was the least standard and didn't provide for automatic updates. In contrast, the five desktop browsers in this figure provide for automatic updates so you don't have to wonder whether they include the latest HTML and CSS features. Besides that, Internet Explorer has such a small share of the market today that most websites no longer need to support it.

In addition to desktop browsers, you also have to consider tablet and mobile browsers. So, how do you provide cross-browser compatibility in today's world? First, you should use the current versions of HTML and CSS as you develop your web pages. If you do that, almost everything that you learn in this book will run on all web browsers.

Second, you should test your pages on desktop, tablet, and mobile browsers. The good news is that the five desktop browsers in this figure provide Developer Tools that make that easy. For instance, the example at the top of this figure shows how you can use Chrome's Developer Tools to test a web page using the screen size for a mobile phone in landscape format. You'll learn more about this in chapter 8.

Eventually, you'll want to test your web pages on the actual browsers too. But since Chrome, Edge, and Opera are all Chromium-based, you only need to test on one of them. Then, you can test on Firefox and Safari just to be sure that they don't present any problems. And for good measure, you can test your web pages on actual tablets and mobile phones.

Chrome's Developer Tools with a page in mobile landscape format

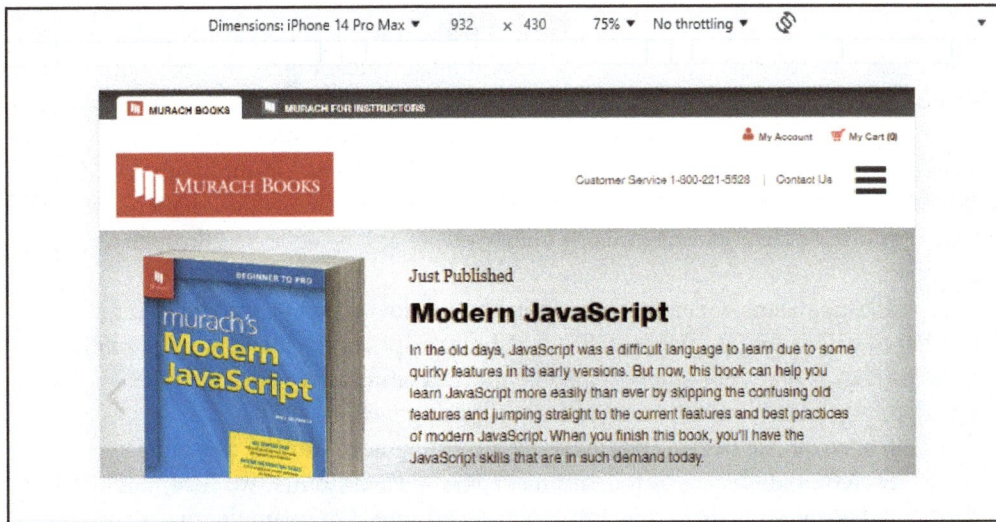

Five modern desktop browsers

- Chrome
- Opera
- Edge
- Firefox
- Safari

Testing guidelines

- Test your web pages on desktop browsers as well as tablet and mobile phone browsers.
- You can use Chrome's Developer Tools for the initial testing of your tablet and mobile layouts as described in chapter 8.

Description

- *Cross-browser compatibility* means that your web pages will work on any browser that accesses your website, including tablet and mobile phone browsers.

Figure 1-7   Cross-browser compatibility

Web accessibility

Web accessibility refers to the qualities that make a website accessible to as many users as possible, especially users with disabilities. Figure 1-8 presents some useful information about web accessibility.

For instance, visually-impaired users may not be able to read text that's in images, so you need to provide other alternatives for them. Similarly, users with motor disabilities may not be able to use the mouse, so you need to make sure that all of the content and features of your website can be accessed through the keyboard.

To a large extent, this means that you should develop your websites so the content is still usable if images, CSS, and JavaScript are disabled. A side benefit of doing that is that your site will also be more accessible to search engines, which rely primarily on the text portions of your pages.

This book presents the guidelines for providing accessibility as you learn the related HTML and CSS. And to learn more about accessibility, we recommend that you visit the sites that are identified in this figure. For example, the WebAim website is a good place to start because it presents an excellent introduction to accessibility.

However, you may also want to consult the website for the World Wide Web Consortium (W3C) for accessibility guidelines. For example, the second URL in this figure leads to the W3C accessibility guidelines for web content, and the third URL leads to accessibility guidelines for rich internet applications (RIAs).

The WebAIM website

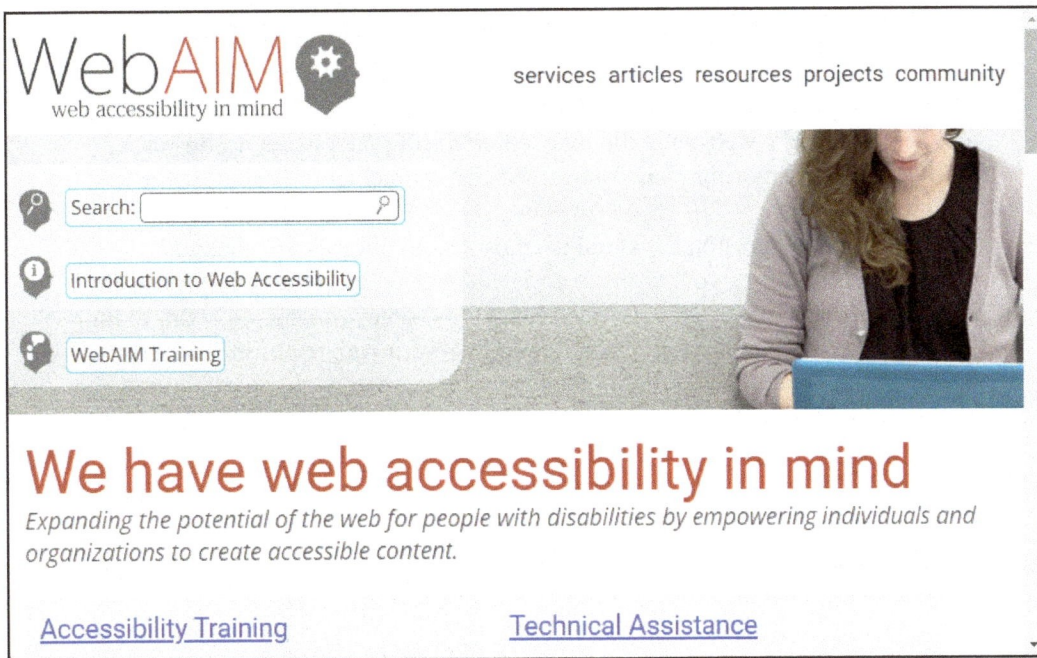

Accessibility laws that you should be aware of

- The Americans with Disabilities Act (ADA).
- Sections 504 and 508 of the federal Rehabilitation Act.
- Section 255 of the Telecommunications Act of 1996.

Types of disabilities

- Visual
- Hearing
- Motor
- Cognitive

URLs for more info

http://www.webaim.org

http://www.w3.org/TR/WCAG

http://www.w3.org/TR/wai-aria

Description

- *Web accessibility* refers to the qualities that make a website accessible to users, especially users with disabilities.

Figure 1-8   Web accessibility

Search engine optimization

Search engine optimization (*SEO*) refers to optimizing your website so your pages rank higher in search engines like Google and Bing. The example in figure 1-9 shows how important SEO can be. Here, the search term is "murach", and Murach Books website is the first search result that's listed. In this case, Google's search algorithm has determined that murach.com should be the first search result.

Because the algorithms that are used by search engines are changed frequently, optimizing your website and web pages requires a continual effort. That's why some developers specialize in SEO. To get you started right, though, this book presents the best practices for coding your web pages in a way that helps optimize them for search engines. In general, that means providing useful content, using HTML to clearly define the structure of that content, and using CSS to format that content in a way that works well on all devices.

A Google search for "murach"

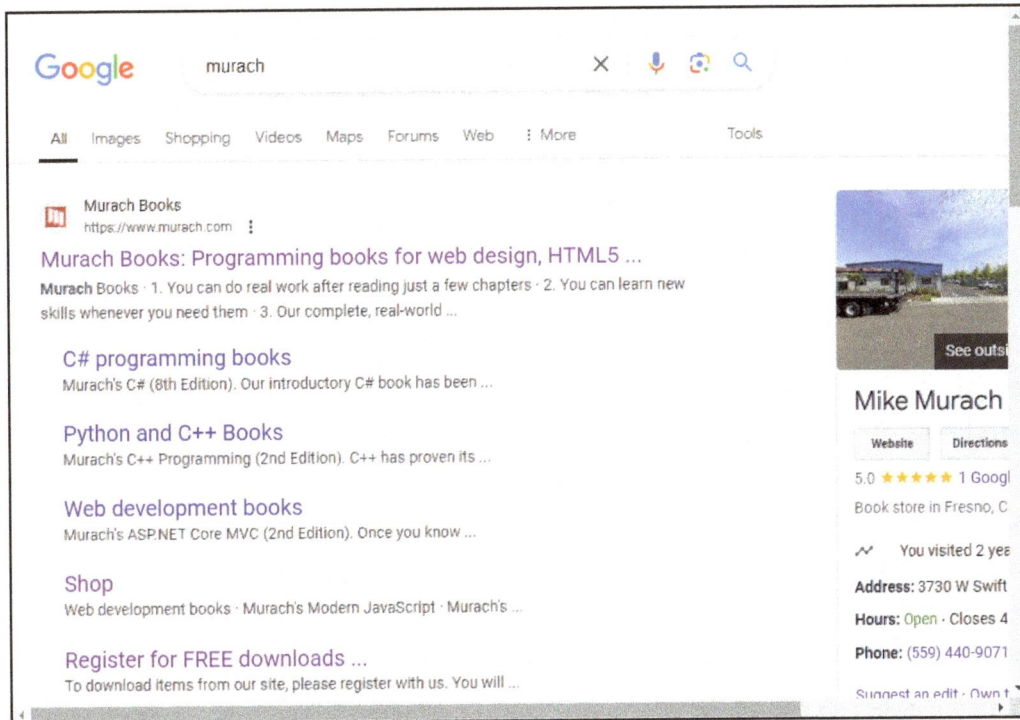

Description

- *Search engine optimization* (*SEO*) refers to the goal of optimizing your website so its pages rank high in the search engines.

Figure 1-9   Search engine optimization

Perspective

Now that you know the concepts and terms that you need for developing websites with HTML and CSS, you're ready to learn how to develop a web page. That's why the next chapter shows how to create, test, and validate a web page. After that, you'll be ready to learn all the details of HTML and CSS.

Terms

World Wide Web
web
internet
clients
web browser
web server
network
local area network (LAN)
intranet
wide area network (WAN)
router
internet service provider (ISP)
internet exchange points (IXP)
HyperText Markup Language (HTML)
HTML document
HTML element
opening tag
closing tag
attribute
CSS (Cascading Style Sheets)
style sheet
style rule

selector
declaration
World Wide Web Consortium (W3C)
Web Hypertext Application Technology
    Working Group (WHATWG)
HyperText Transfer Protocol (HTTP)
HTTP request
HTTP response
render a web page
round trip
application server
database server
server-side processing
JavaScript
JavaScript engine
client-side processing
responsive web design (RWD)
cross-browser compatibility
web accessibility
search engine optimization (SEO)

Chapter 2

How to code, test, and validate a web page

In this chapter, you'll learn how to create and edit the HTML and CSS files that define a web page. Then, you'll learn how to test those files to make sure they work correctly and how to validate them to make sure they conform to the latest HTML and CSS standards. When you're done with this chapter, you'll be ready to learn all the details of HTML and CSS coding.

Although you can use any text editor or IDE (integrated development environment) to work with HTML and CSS files, this chapter shows how to use a popular text editor called Visual Studio Code (or just VS Code). We recommend using VS Code because it's widely used by web developers, and we think it makes it easier to get started with developing web pages.

The HTML syntax

When you code an HTML document, you need to adhere to the rules for creating the HTML elements. These rules are referred to as the *syntax* of the language.

The basic structure of an HTML document

Figure 2-1 presents the basic structure of an *HTML document*. To start, every HTML document consists of two parts: the DOCTYPE declaration and the document tree.

When you start an HTML document, you can code the *DOCTYPE declaration* exactly as it's shown in this figure. It should be the first line of code in every HTML document that you create, and it tells the browser that the document is using a modern version of HTML.

The *document tree* begins after the DOCTYPE declaration. This tree consists of the *HTML elements* that define the web page. The first of these elements is the <html> element itself, which contains all of the other elements. This element can be referred to as the *root element* of the tree.

Within the <html> element, you should always code a <head> element and a <body> element. The <head> element contains elements that provide information about the page itself, while the <body> element contains the elements that provide the structure and content for the page.

The elements shown in this figure should be in every HTML document that you create. As a result, it's a good practice to start every HTML document from a template that contains this code or from another HTML document that's similar.

When you use HTML, you can code the names of the elements with lowercase, uppercase, or mixed case letters. For consistency, we recommend that you use lowercase letters unless uppercase is required. The one exception is the DOCTYPE declaration, which has historically been capitalized even though lowercase works too.

The basic structure of an HTML document

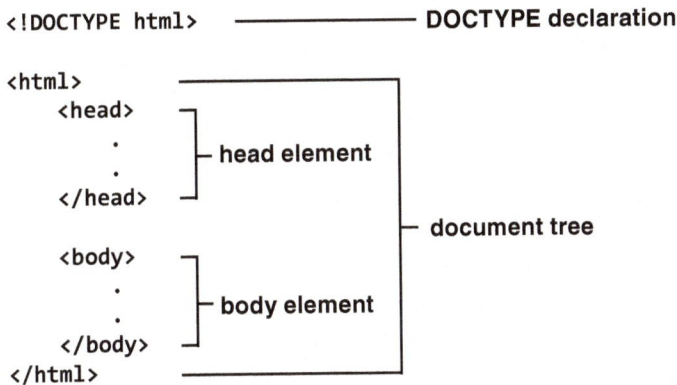

```
<!DOCTYPE html>  ———————— DOCTYPE declaration

<html>
    <head>
        .
        .          ┐— head element
    </head>        ┘
                              ┐— document tree
    <body>
        .
        .          ┐— body element
    </body>        ┘
</html>
```

A simple HTML document

```
<!DOCTYPE html>
<html lang="en">
<head>
    <meta charset="UTF-8">
    <meta name="viewport" content="width=device-width, initial-scale=1.0">
    <title>Mozart</title>
</head>
<body>
    <h1>Wolfgang Amadeus Mozart</h1>
    <p>Wolfgang Amadeus Mozart was a prolific and influential composer
        of the Classical period.</p>
    <p>Born in Salzburg, Mozart showed prodigious ability from his earliest
        childhood.</p>
</body>
</html>
```

General coding recommendation for HTML

- Although you can code the names of HTML elements using lowercase, uppercase, or mixed-case letters, we recommend using lowercase because it's easier to read.

Description

- An *HTML document* contains *HTML elements* that define the content and structure of a web page.
- Each HTML document consists of two parts: the DOCTYPE declaration and the document tree.
- The *DOCTYPE declaration* typically indicates that the document is using a modern version of HTML.
- The *document tree* starts with the <html> element that marks the beginning and end of the HTML code. This element can be referred to as the *root element* of the document.
- The <html> element typically contains one <head> element that provides information about the document and one <body> element that provides the structure and content of the document.

Figure 2-1   The basic structure of an HTML document

How to code elements and tags

Figure 2-2 shows how to code elements and tags. Most HTML elements start with an *opening tag* and end with a *closing tag* that is like the opening tag but has a slash within it. Thus, <h1> is the opening tag for a level-1 heading, and </h1> is the closing tag. Between those tags, you code the *content* of the element. In the first example, the content of the <h1> element is the text for the heading, and the content of the <p> element is the text for the paragraph.

However, some elements have no content or closing tag. These tags are referred to as *empty tags*. For instance, the <br> tag is an empty tag that starts a new line, and the <img> tag is an empty tag that identifies an image that should be displayed.

When you code HTML, it's common to code one element within another. This is known as *nesting* elements, and the third example shows the right and wrong way to code tags when one element is nested within another. In short, the tags for one element shouldn't overlap with the tags for another element. So, you should close the inner element before you close the outer element.

Two elements with opening and closing tags

```
<h1>Classical Music Genres</h1>
<p>Here is a list of links:</p>
```

Two empty tags

```
<br>
<img src="logo.jpg" alt="Music Logo">
```

Correct and incorrect nesting of tags

Correct nesting

```
<p>Order <i>today!</i></p>
```

Incorrect nesting

```
<p>Order <i>today!</p></i>
```

Description

- Most HTML elements have an opening tag, content, and a closing tag. Each tag is coded within a set of brackets (<>).

- An element's *opening tag* includes the tag name. The *closing tag* includes the tag name preceded by a slash. And the *content* includes everything that appears between the opening and closing tags.

- Some HTML elements have no content. For example, the <br> element, which forces a line break, consists of just one tag. This type of tag is called an *empty tag*.

- HTML elements are commonly *nested*. To nest elements correctly, though, you must close an inner set of tags before closing the outer set of tags.

- Technically, a tag is part of an element. However, in practice, many programmers use the terms *element* and *tag* interchangeably.

Figure 2-2   How to code elements and tags

How to code attributes

Figure 2-3 shows how to code the *attributes* for an HTML element. These attributes are coded within the opening tag of an element or within an empty tag. For each attribute, you code the attribute name, an equal sign, and the attribute value.

When you use modern HTML, the attribute value doesn't have to be coded within quotation marks unless the value contains a space. However, we recommend that you use quotation marks to enclose all values. Also, although you can use either double or single quotes, we recommend that you always use double quotes. That way, your code will have a consistent appearance that will help you avoid coding errors.

The examples in this figure show how one or more attributes can be coded. For instance, the first example shows two opening tags for an <a> element. The first has one attribute named href, and the second has three attributes named href, title, and class.

The second example shows an empty <img> element that has two attributes named src and alt. Here, the src attribute specifies the name of the image file to display, and the alt attribute specifies the text to display if the image file can't be found.

The third example shows how to use a *Boolean attribute*. A Boolean attribute can be on or off. To turn a Boolean attribute on, you code just the name of the attribute. In this example, the checked attribute turns that attribute on. This causes the related check box to be checked when it is rendered by the browser. If you want the attribute to be off when the page is rendered, you don't code the attribute.

The fourth example shows how to use two attributes that are commonly used to identify HTML elements. The id attribute uniquely identifies a single element on a web page. As a result, each id attribute on a web page should have a unique value. In this example, the id attribute has a value of page that uniquely identifies the <div> element.

By contrast, the class attribute can be used to mark one or more elements. As a result, a class attribute can use the same value for multiple elements. In this example, the class attribute has a value of nav_link that marks it as a member of the nav_link class.

How to code an opening tag with attributes

An opening tag with one attribute

```
<a href="contact.html">
```

An opening tag with three attributes

```
<a href="contact.html" title="Click to Contact Us" class="nav_link">
```

How to code an empty tag with attributes

```
<img src="logo.gif" alt="Music Logo">
```

How to code a Boolean attribute

```
<input type="checkbox" name="show_menu" checked>
```

Two common attributes for identifying HTML elements

An opening tag with an id attribute

```
<div id="page">
```

An opening tag with a class attribute

```
<a href="contact.html" title="Click to Contact Us" class="nav_link">
```

Coding rules

- An attribute consists of the attribute name, an equal sign (=), and the value for the attribute.
- Attribute values don't have to be enclosed in quotes if they don't contain spaces.
- Attribute values must be enclosed in single or double quotes if they contain one or more spaces, but you can't mix the type of quotation mark used for a single value.
- Boolean attributes can be coded as just the attribute name. They don't have to include the equal sign and a value.
- To code multiple attributes, separate each attribute with a space.

Our coding recommendation

- For consistency, enclose all attribute values in double quotes.

Description

- *Attributes* can be coded within opening or empty tags to supply optional values.
- A *Boolean attribute* can be on or off.
- The id attribute is used to identify a single HTML element, so its value can be used for just one HTML element.
- A class attribute is used to identify one or more HTML elements, so its value can be used for one or more HTML elements.

Figure 2-3   How to code attributes

How to code HTML comments

Figure 2-4 shows how to code comments in an HTML document. A *comment* provides a way to include text that describes or explains portions of code. To code an HTML comment, you can code the starting characters (<!--) for a comment, the text for a comment, and the ending characters (-->) for a comment. Then, the browser ignores the comment when it renders the page.

In this figure, for example, the first comment provides a brief description of the web page and the name of the author. Then, the second comment explains that the <ul> element displays an unordered list.

You can also use comments to *comment out* a portion of the code. This is useful when you're testing a web page and you want to temporarily disable a portion of code that you're having trouble with. Then, after you test the rest of the code, you can remove the comment and test that portion of the code. In this figure, the third comment comments out the last three items in the list. As a result, the browser doesn't display these items.

This figure also shows how to work with *whitespace*, which consists of characters like tab characters, blank lines, and extra spaces. Since the browser ignores whitespace when it renders an HTML document, you can use the whitespace characters to format your HTML so it's easier to read. In this figure, for example, the HTML document uses whitespace to indent and align the HTML elements and to put a blank line between the <head> and <body> elements. Since this makes your code easier to read, it's generally considered a good coding practice.

An HTML document with comments and whitespace

```html
<!DOCTYPE html>
<!--
    Description: Home page for website
    Author:      Joel Murach
-->
<html lang="en">
<head>
    <meta charset="UTF-8">
    <meta name="viewport" content="width=device-width, initial-scale=1.0">
    <title>Classical Music Info</title>
</head>

<body>
    <h1>Classical Music Info</h1>
    <h2>Genres</h2>
    <ul><!-- Display an unordered list -->
        <li>Symphony</li>
        <li>Concerto</li>
        <!-- Comment out three list items
        <li>Chamber Music</li>
        <li>Opera</li>
        <li>Choral Music</li>
        -->
    </ul>
</body>
</html>
```

Recommendation

- Use whitespace to indent lines of code and make them easier to read.

Description

- An HTML *comment* is text that appears between the <!-- and --> characters. Since web browsers ignore comments, you can use them to describe or explain portions of your HTML code that might otherwise be confusing.

- You can also use comments to *comment out* elements that you don't want the browser to display. This can be useful when you're testing a web page.

- An HTML comment can be coded on a single line or it can span two or more lines.

- *Whitespace* consists of characters like tab characters, line return characters, and extra spaces.

- Since whitespace is ignored by browsers, you can use it to indent lines of code and separate elements from one another by putting them on separate lines. This is a good coding practice because it makes your code easier to read.

Figure 2-4   How to code HTML comments

The CSS syntax

Like HTML, CSS has a syntax that you must follow when working with it.

How to code style rules

A CSS file consists of *style rules*. As the diagram in figure 2-5 shows, a style rule consists of a *selector* followed by a set of braces. Within the braces, you can code one or more *declarations* where each declaration consists of a *property* and a *value*. Here, the property is followed by a colon and the value is followed by a semicolon.

In this diagram, the selector is h1 so it applies to all <h1> elements in the HTML document. Then, the style rule consists of a single property named color that's set to the color named navy. As a result, the browser displays the content of all <h1> elements in navy blue.

The CSS that follows presents four other style rules. The style rule for the body selector contains three declarations. The first sets the font-family property, the second sets the padding property, and the third sets the max-width property. The style rule for the h1 and h2 selectors both have two declarations, and the style rule for the ul selector only has one declaration.

As with HTML, it's generally considered a good practice to use whitespace to make your CSS easier to read. For instance, this example codes the closing brace for each style rule on its own line and includes a blank line between each style rule. Similarly, this example codes each declaration on its own line and uses whitespace to indent each declaration.

The parts of a style rule

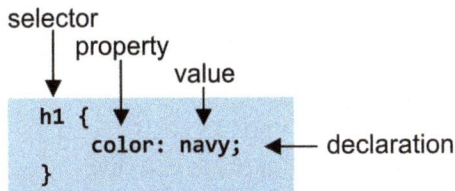

```
selector
    property
            value
h1 {
      color: navy;  ◄─── declaration
}
```

Four style rules

```
body {
    font-family: sans-serif;
    padding: 1em;
    max-width: 600px;
}

h1 {
    font-size: 180%;
    color: #0033cc;
}

h2 {
    font-size: 140%;
    border-bottom: 3px solid #0033cc;
}

ul {
    list-style-type: square;
}
```

Description

- A CSS *style rule* consists of a selector and zero or more declarations enclosed in braces.
- A CSS *selector* consists of the identifiers that are coded at the beginning of the style rule.
- A CSS *declaration* consists of a *property*, a colon, a *value*, and a semicolon.
- To make your code easier to read, you can use spaces, indentation, and blank lines within a style rule.

Figure 2-5   How to code style rules

How to code basic selectors

The selector of a style rule identifies the HTML element or elements that the declarations should be applied to. To give you a better idea of how this works, figure 2-6 shows how to use the three basic kinds of selectors.

The first kind of selector identifies HTML elements like <body>, <h1>, or <p> elements. For instance, the first two selectors in this figure apply to the <body> and <h1> elements. Since these selectors select elements by their type, they're known as *type selectors*.

The second kind of selector starts with a hash (#) and applies to the single HTML element that's identified by the id attribute. For instance, the #copyright selector applies to the HTML element that has an id attribute with a value of copyright. In this figure, that's the last <p> element in the HTML.

The third kind of selector starts with a dot (.) and applies to all of the HTML elements that are identified by the class attribute. For instance, the .base_color selector applies to all elements that have a class attribute with a value of base_color. In this figure, this includes the <h1> element and the last <p> element.

Starting with chapter 4, this book will present the details for coding style rules. But to give you an idea of what's going on in this example, here's a quick review of the code.

For the body selector, the first declaration sets the font-family either to Arial (if the browser has access to that font) or the sans-serif type that is the default for the browser. Then, the browser uses this font for all text that's displayed within the <body> element, unless it's overridden later by another style rule. So, in this example, the browser displays all of the text using the Arial font or its default sans-serif font.

The second declaration sets the maximum width of the body to 800 pixels. As a result, the body will be 800 pixels or less.

The third declaration sets the padding of the body to 1em, which is the height of the default font. As a result, the space between the body and the edge of the browser will be the height of the default font.

For the h1 selector, the first and only declaration sets the font-size to 160% of the default font size. As a result, the heading has a larger font size than the paragraph and list that follow it, which use the default font size of 100%.

For the #copyright selector, the first declaration sets the font size to 75% of the default font size. Then, the second declaration aligns the text with the right side of the container element, which is the <body> element in this figure.

For the .base_color selector, the first and only declaration sets the color to navy blue. As a result, the browser displays both of the HTML elements that have a class name of base_color in navy blue, not the default color of black.

This example shows three basic ways to identify the elements that you want to apply CSS formatting to. In addition, it shows how CSS separates the formatting from the content that's defined by the HTML.

HTML elements with id and class attributes

```
<body>
    <h1 class="base_color">Classical Music Genres</h1>
    <p>Here are some links:</p>
    <ul>
        <li><a href="symphony.html">Symphony</a></li>
        <li><a href="concerto.html">Concerto</a></li>
    </ul>
    <p id="copyright" class="base_color">All Rights Reserved</p>
</body>
```

CSS style rules that select elements by type, id, and class

Type

```
body {
    font-family: Arial, sans-serif;
    max-width: 800px;
    padding: 1em;
}
h1 {
    font-size: 160%;
}
```

ID

```
#copyright {
    font-size: 75%;
    text-align: right;
}
```

Class

```
.base_color {
    color: navy;
}
```

The elements displayed in a browser

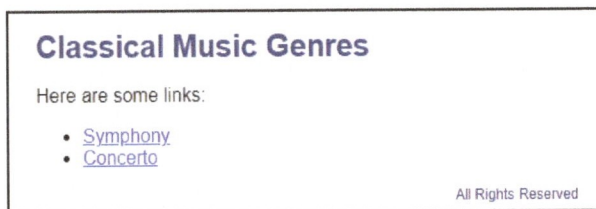

Classical Music Genres

Here are some links:

- Symphony
- Concerto

All Rights Reserved

Description

- To code a selector that selects all HTML elements of a specific type, you code the name of the element. This is referred to as a *type selector*.

- If an element is coded with an id attribute, you can code a selector for that id by coding a hash character (#) followed by the id value. This is referred to as an *id selector*.

- If an element is coded with a class attribute, you can code a selector for that class by coding a dot (.) followed by the class name. This is referred to as a *class selector*.

Figure 2-6   How to code basic selectors

How to code CSS comments

Within a CSS file, you can code comments that describe or explain what the CSS is doing as shown in figure 2-7. To code a comment, you start by coding the /* characters and end with the */ characters. Within those characters, you can code the text of the comment. Then, the browser ignores the comment.

In this figure, the first comment spans four lines. It provides a brief description of the style sheet and the name of the author. In addition, it uses additional * characters to make the comment stand out.

The next three comments describe the style rules in the file. These comments are coded on their own lines, just above the style rules that they describe.

The last comment describes a declaration in the style sheet. This comment is coded to the right the declaration, on the same line.

As with HTML, you can use comments to comment out portions of code that you want disabled. This can be useful when you're testing your CSS.

A CSS file with comments

```
/**********************************************************
* Description: Main style sheet
* Author:      Anne Boehm
**********************************************************/

/* Adjust the styles for the body */
body {
    font-family: sans-serif;
    padding: 1em;
    max-width: 600px;
}

/* Adjust the styles for the headings */
h1 {
    font-size: 180%;
    color: #0033cc;
}
h2 {
    font-size: 140%;
    border-bottom: 3px solid #0033cc;
}

/* Adjust the styles for the unordered list */
ul {
    list-style-type: square;        /* Change bullets to squares */
}
```

Description

- A CSS *comment* begins with the characters /* and ends with the characters */.
- A CSS comment can be coded on a single line, or it can span multiple lines.
- You can add additional * characters to make the comment stand out.

Figure 2-7   How to code CSS comments

How to use VS Code

HTML and CSS files are stored as text. As a result, you can use any text editor to work with the files for a web page. For this book, we recommend using Visual Studio Code (VS Code), a popular text editor that's available for free and runs on Windows, macOS, and Linux. This text editor provides all the features you need to work efficiently and productively. To learn how to install it, you can refer to appendix A.

For this chapter, we used the VS Code color theme named Light to give the text editor a white background. However, VS Code uses a color theme named Dark by default. As you might expect, it gives the text editor a dark background. If you want to change the default theme on your system, you can use the procedure shown in figure 2-8 to select the color theme you want.

How to work with folders

To work with a web page in VS Code, you start by opening a folder that contains all the subfolders and files for the web page. The first procedure in figure 2-8 shows how.

In this figure, the screen shows the dialog that you can use to open the html_css folder that contains all the web pages and examples presented in this book. In addition, this folder contains the files that you need to do the exercises at the end of each chapter.

In this screen, the Explorer on the left side of VS Code shows the subfolders and files for the folder that's opened by the Open Folder dialog. In other words, the Explorer shows what happens after the html_css folder is opened.

After you open a folder, you can hide or display the folders and files in the Explorer by clicking on the folder names. That makes it easy to find the folders and files that you're looking for. Then, you can add, rename, or delete a folder by using the procedures described in this figure.

Two of the procedures in this figure tell you to start by right-clicking on a folder. However, right-clicking isn't enabled by default on macOS. As a result, if you're using macOS, you may need to Ctrl-click (hold down the Ctrl key while you click) when this book instructs you to right-click. Or, if you prefer, you can modify your mouse options to enable right-clicking.

The dialog for choosing a top-level folder in VS Code

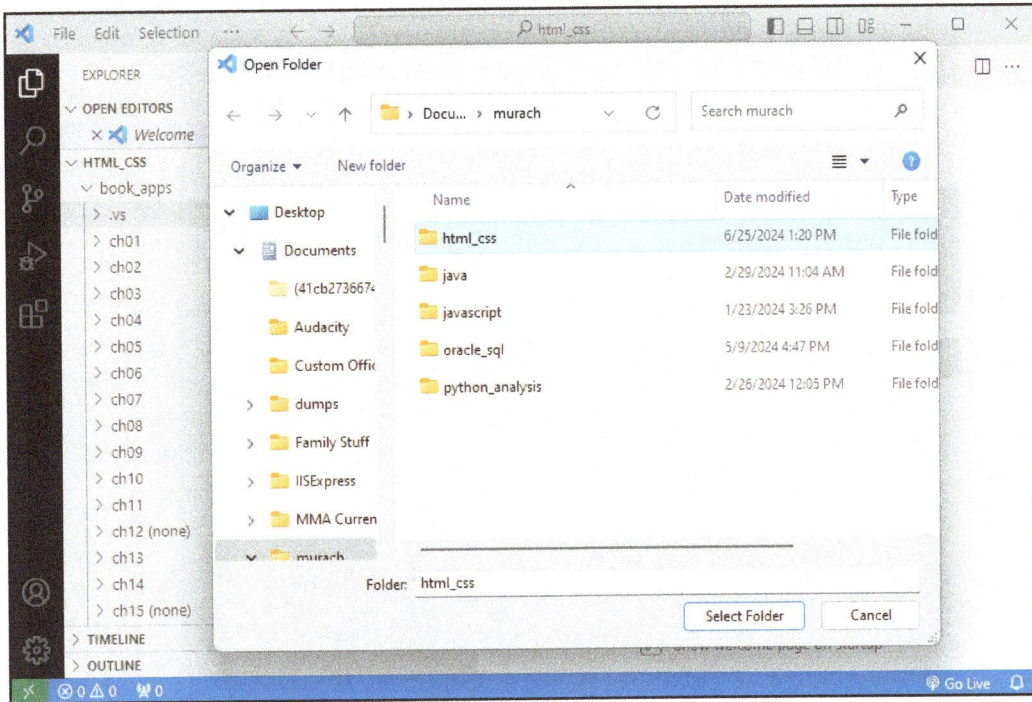

How to open the top-level folder

1. Select File→Open Folder from the menu system.
2. Use the resulting dialog to select the folder and click Select Folder. This sets up the folder and file tree that's shown in the Explorer.

How to hide or display the folders and files in the Explorer

- Click on the folder names.

How to add, rename, or delete a folder in the Explorer

- To add a subfolder to a folder, select the folder in the Explorer and click the New Folder icon that's displayed to the right of the top folder.
- To rename a folder, right-click on it and select Rename. Then, edit the name.
- To delete a folder, right-click on it and select Delete.

How to set the color theme

- Select File→Preferences→Color Theme from the menu bar.

Description

- Please refer to appendix A to learn how to install VS Code on Windows or macOS.
- If right-clicking isn't enabled on macOS, you can Ctrl-click instead.

Figure 2-8   How to work with folders

How to work with files

Figure 2-9 presents the basic skills for working with files in VS Code. To *open* a file, you find it in the Explorer and double-click on it. Then, it's displayed in a tab in the main window. This is called Standard Mode, and you use it when you're going to edit a file.

By contrast, to *preview* a file, you just click on it (not double-click). Then, the name of the file is displayed in italics on the tab, and the tab is re-used if you open or preview another file. This is called Preview Mode, and you use it when you want to take a quick look at a file.

To save, close, add, rename, and delete files, you can use the other procedures described in this figure. They work the way they do with most apps. However, when you add a new HTML or CSS file to a folder, you need to add the .html or .css extension to the name of the file. Then, VS Code will be able to tell if the file is an HTML or CSS file, and its editor will provide features that make it easier to work with that type of file.

VS Code with files in Standard and Preview mode

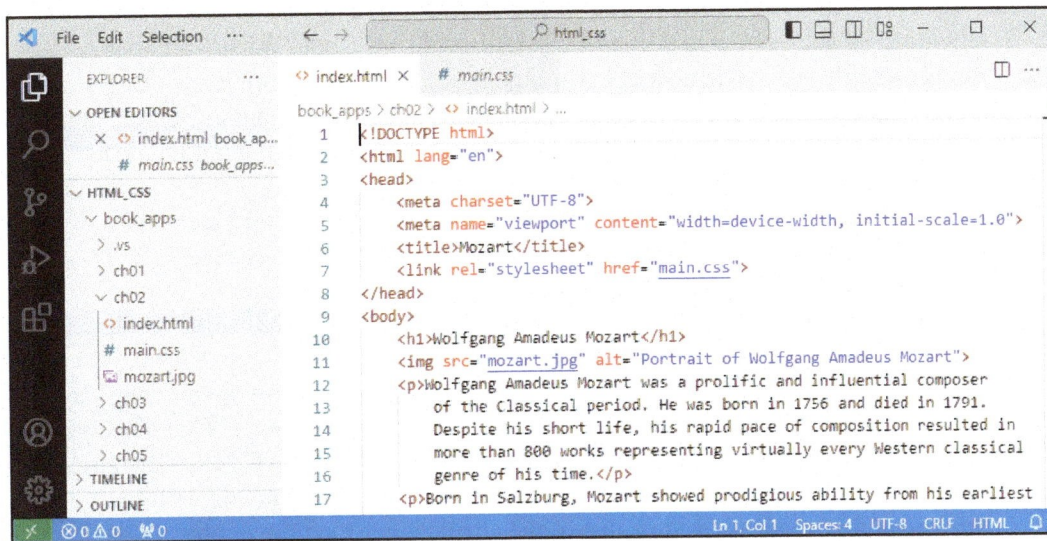

How to open or preview a file

- To *open* a file, double-click on it in the Explorer. This displays the file in a tab in the editor with the name of the file in normal font style, indicating that it's in Standard Mode.
- To *preview* a file, click on it in the Explorer. This displays the file in a tab in the editor with the name of the file in italics, indicating that it's in Preview Mode. If you open or preview another file, VS Code reuses the tab.

How to save or close a file

- If you want to save the changes to a file without closing it, select File→Save.
- To save the changes to more than one file, select File→Save All or click the Save All icon that shows when you point to Open Editors in the Explorer.
- To close a file, click the X in the upper right corner of the tab for the file, or click the X to the left of the filename in the Open Editors list.
- If you try to close a file that has been changed but not saved, you'll be asked if you want to save the changes.

How to add a new file to a folder

- Select the folder in the Explorer and click the New File icon that's displayed to the right of the top folder. Then, enter the name for the file and be sure to include the html or css extension.

How rename or delete a file

- To rename a file, right-click on it and select Rename. Then, edit the name.
- To delete a file, right-click on it and select Delete.

Figure 2-9   How to work with files

How to edit an HTML file

Figure 2-10 shows how to edit an HTML file with VS Code. When you open an HTML file, VS Code uses color to highlight the syntax components. It uses IntelliSense to provide *completion lists* that let you select the names of elements and attributes as you type. And when you enter an opening tag, it automatically adds the closing tag.

To display the completion list for an element, you type a left bracket (<). This displays a list of all the available elements. Then, if you type one or more letters, VS Code filters the list to just the elements that start with those characters. In this figure, for example, the letter *i* has been typed so VS Code only displays the elements that start with that letter.

You can use a similar technique to enter attributes within an opening tag. To start, enter a space and one or more letters after the element name. That displays a list of all the attributes that start with those letters. Then, when you select an attribute, VS Code inserts the attribute along with an equal sign and double quotes, and you can enter the value of the attribute between the quotes.

When you add a new file to a folder, VS Code doesn't generate any code for it. However, you can generate the starting code for an HTML file by entering an exclamation mark (!) and pressing the Tab key.

Because this starting code is minimal, another option is to copy code from another file and paste it into the new file. Or, you can open a similar file and then save it with a new name. Then, you can delete the code you don't need and add the code you do need.

As you become more familiar with HTML, you can take this to the next level by creating HTML files that contain the elements you need for different types of pages. Then, you can use those files as templates for creating new pages. To do that, you can open the template file and save it with a new name before you modify the file. That way, the original template file remains unchanged. In most cases, this works better than starting a new HTML file from scratch.

The completion list for selecting an HTML element

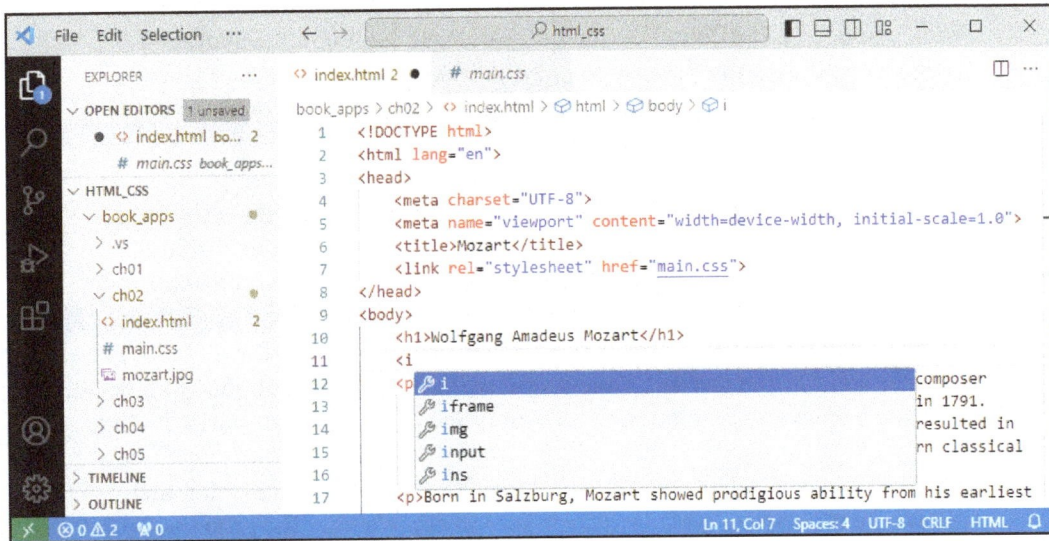

Common coding errors

- An opening tag without a closing tag.
- Misspelled element or attribute names.
- Quotation marks that aren't paired.
- Incorrect file references in <link>, <img>, or <a> elements.

How to use IntelliSense

- IntelliSense displays *completion lists* for elements and attributes. To insert an item from a completion list, click on it or highlight it and press the Tab or Enter key.
- When you enter a bracket at the end of an opening tag, VS Code enters the closing tag.

How to add starting code to a new file

- Enter an exclamation mark (!) into the file and press the Tab key.

Description

- VS Code provides features for editing HTML files like color coding and IntelliSense.
- VS Code makes it easy for you to add the starting code for an HTML file.

Figure 2-10   How to edit an HTML file

How to edit a CSS file

When you open a CSS file, VS Code uses color to highlight the syntax components. It uses IntelliSense to provide *completion lists* that let you select the names of elements, properties, and values. And when you enter an opening brace, it automatically adds the closing brace.

For example, if you start to enter the name of an element for a CSS style rule, VS Code displays a list of the elements with the letters you enter. If you enter one or more letters to start a property declaration, VS Code displays a list of the properties that start with those letters. And if you start an entry for a property value, VS Code displays a list of possible values.

The screen in figure 2-11 shows how completion lists work with CSS. Here, the letter *b* has been entered for the property in the third declaration. As a result, VS Code displays a completion list for all of the properties that start with that letter.

When you enter the name of a property, VS Code enters the colon that follows the property as well as the semi-colon at the end of the property declaration. Then, you just need to enter the value for the property.

The completion list for selecting a CSS property

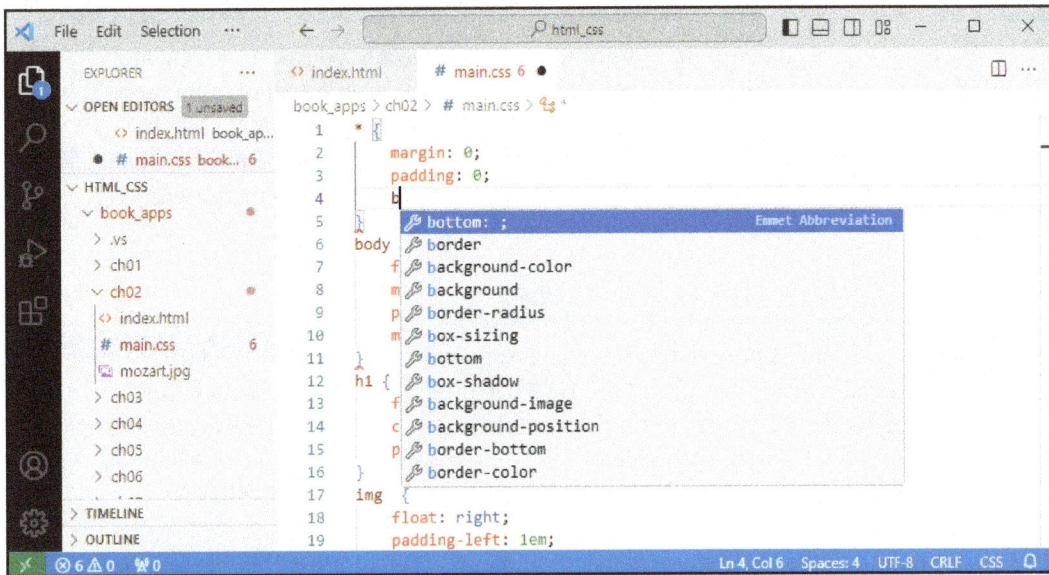

Common coding errors

- Braces that aren't paired correctly.
- Misspelled property names.
- Missing semicolons.
- Id or class names that don't match the names used in the HTML.

How to use IntelliSense

- IntelliSense displays *completion lists* for elements, properties, and values. To insert an item from a completion list, click on it or highlight it and press the Tab or Enter key.
- When you select a property from a completion list, VS Code adds the colon and the semicolon for the declaration. Then, you just need to enter or select the value.
- When you enter the left brace for a style rule, VS Code adds the right brace.

Description

- VS Code provides features for editing CSS files like color coding and IntelliSense.

Figure 2-11   How to edit a CSS file

How to use two VS Code extensions

By default, VS Code might not provide all the features you want. Fortunately, you can install extensions that provide features that aren't available by default. For example, since VS Code doesn't provide error checking for HTML by default, we recommend that you install an extension for that. Similarly, since VS Code doesn't provide an easy way to open an HTML file in a browser, we recommend that you install an extension for that.

How to install an extension

Figure 2-12 shows how to use VS Code to install an extension. To start, the first procedure shows how to install the HTMLHint extension that provides error checking for HTML files. To do that, you begin by clicking the Extensions icon in the left bar. Then, the left side of VS Code displays the Extensions window. This window lists any extensions that are already installed as well as any recommended extensions.

To find the HTMLHint extension, you can type "htmlhint" in the text box at the top of the Extensions window as shown in the first screen in this figure. When you do that, you'll notice that there are multiple extensions with that name. We recommend that you make sure to install the one developed by htmlhint.com.

The second procedure shows how to install the Open in Browser extension that makes it easy to open HTML files in a browser. To do that, you can type "open in browser" in the text box at the top of the Extensions window. When you do, VS Code displays several extensions with that name or a similar name. Because of that, you should make sure to install the one developed by TechER.

If you want to learn more about an extension, you can click it in the Extensions window. Then, VS Code displays more information about the extension in the main window, including buttons that let you disable or uninstall an installed extension.

VS Code after the HTMLHint extension has been installed

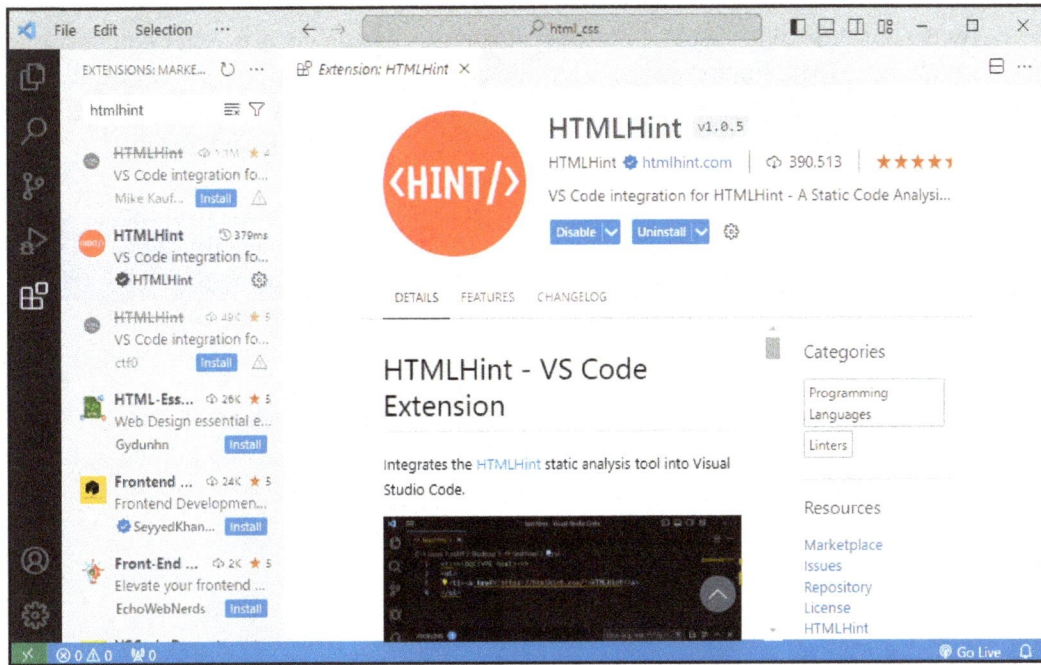

How to install HTMLHint

1. Click the Extensions icon (⊞) in the left sidebar.
2. Enter "htmlhint" in the text box at the top of the Extensions window to filter the available extensions.
3. Click the Install button for the HTMLHint extension from htmlhint.com.

How to install Open in Browser

1. Click the Extensions icon (⊞) in the left sidebar.
2. Enter "open in browser" in the text box at the top of the Extensions window to filter the available extensions.
3. Click the Install button for the Open in Browser extension from TechER.

Description

- By default, VS Code doesn't provide all the functionality that you might want for working with HTML and CSS. To add functionality, you can install an extension.

Figure 2-12   How to install an extension

How to find syntax errors

By default, VS Code checks CSS for syntax errors as you enter it. After you install the HTMLHint extension, VS Code also checks HTML for syntax errors as you enter it. Then, if VS Code detects an error, it displays a wavy line under the error as shown in the screen in figure 2-13. Here, "<link" is underlined because the tag doesn't end with a right bracket.

To display a description of this error, you can hover the mouse over the wavy underline. Or, you can display the Problems panel to view a list of all the errors. Then, you can click on an error to display it in the text editor. The Explorer also indicates whether a file contains errors by displaying the number of errors to the right of the filename. For example, the screen in this figure shows that the HTML file has one error.

Although HTMLHint can identify syntax errors, it can't detect other types of errors. As a result, you must discover those errors later when you test the web page.

The Problems panel with an error displayed

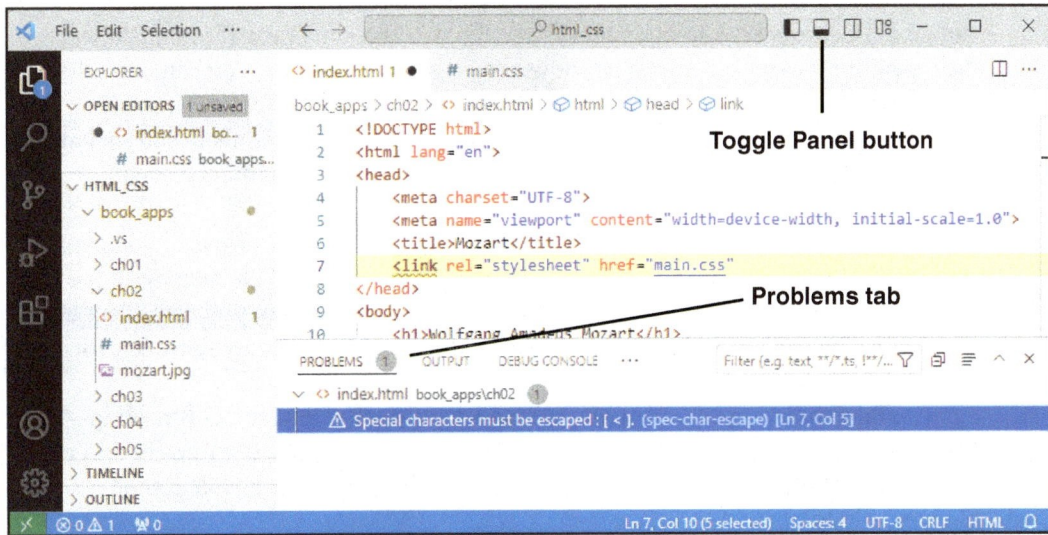

Two ways to display the Problems panel

- Click the Toggle Panel button that's displayed in VS Code.
- Select View→Problems from the menus.

Description

- By default, VS Code detects CSS syntax errors.
- If you install the HTMLHint extension, VS Code detects HTML syntax errors.
- If VS Code detects a syntax error, it underlines the error with a wavy line in the code editor and adds it to the Problems panel.
- In the code editor, you can get a description for an error by hovering the mouse over the wavy line.
- To view all the errors in a file, you can display the Problems panel. Then, you can click on an error to display it in the text editor.

Figure 2-13   How to find syntax errors

How to open an HTML file in a browser

Figure 2-14 shows how to open an HTML file in a web browser. If you installed the Open in Browser extension as shown earlier in this chapter, you can open the index.html file by right-clicking it and selecting the Open in Default Browser option. This opens the HTML file in your default browser, which is usually what you want. Or, if the index.html file is already open in the text editor, you can press Alt+B to open it in your default browser.

However, if you want to open the HTML file in a different browser, you can right-click the HTML file and select Open in Other Browser. This displays a list of the other browsers that are installed on your computer. Then, you can select a browser to display the file in that browser. Unfortunately, this may not work for the Edge browser.

Another way to open an HTML file is to right-click it and select Reveal in File Explorer (Windows) or Reveal in Finder (macOS). This starts File Explorer or Finder and displays the folder that contains the file. Then, you can double-click the file to open it.

Every time you open an HTML file in a browser, a new browser or browser tab is opened. This is OK, but it can clutter your computer with multiple open browsers or browser tabs. To avoid that, you can open the file in the browser just once. Then, after you use VS Code to fix the errors, you can save the changes, switch to the browser, and click on the Reload or Refresh icon to reload the file with the changes.

A menu with two items for opening an HTML file

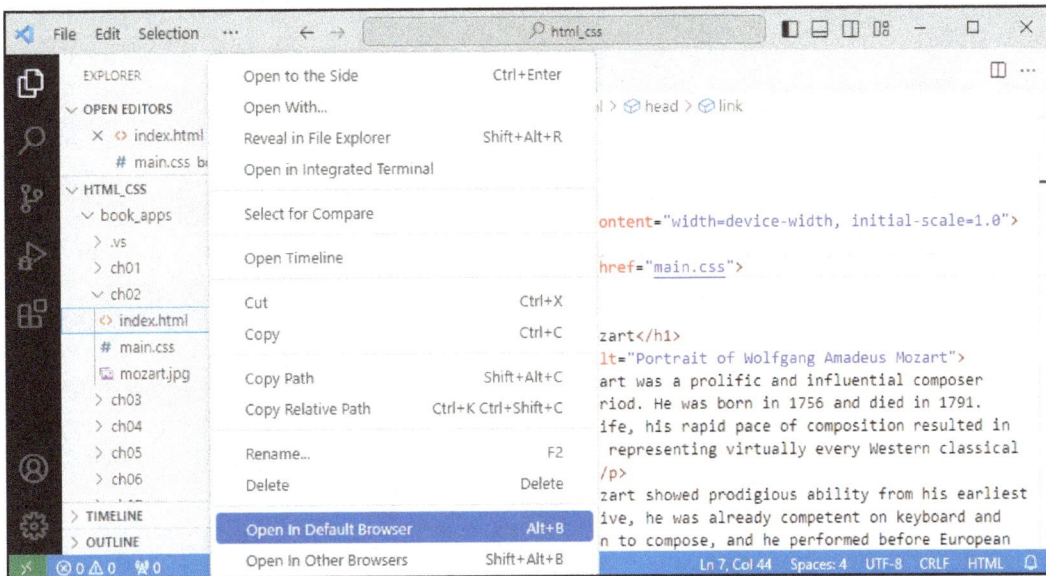

The web page in the Chrome browser

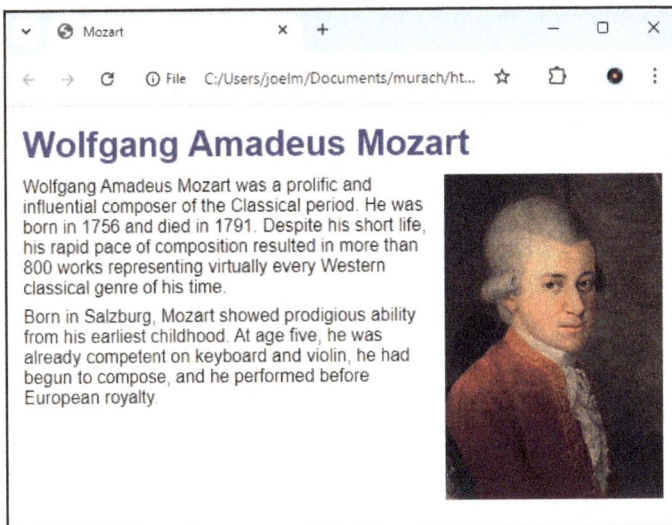

Description

- To open an HTML file with the Open in Browser extension, right-click the file in the Explorer window and select Open in Default Browser or Open in Other Browsers. However, this may not work for Edge. You can also use the extension's shortcut keys shown in the menu above to perform these operations.

- To open an HTML file without using the Open in Browser extension, right-click the file in the Explorer window and select Reveal in File Explorer (Windows) to display it in File Explorer or Reveal in Finder (macOS) to display it in Finder. Then, double-click the file.

Figure 2-14   How to open an HTML file in a browser

How to test and debug a web page

When you *test* a web page, your goal is to find all the errors. When you *debug* a web page, your goal is to fix all the errors.

How to test a web page

To test a web page, you start by running it. To do that, you can use VS Code to open the web page in a browser, or you can use one of the techniques shown in figure 2-15. Then, to test the web page, you check its content and appearance. You also check each link to make sure it goes where it's supposed to go. If you find errors, you should note them and then debug them.

When the web page works correctly in one browser, you'll want to test it in any other browsers that might access this page. An easy way to do that is to copy the URL from one browser, open another browser, and paste the URL into that browser.

When you test a web page in more than one browser, you will sometimes find that the page works on one browser, but doesn't work on another. That's usually because one of the browsers makes some assumptions that the other browser doesn't. If, for example, you have a slight coding error in an HTML file, one browser might make an assumption that fixes the problem, while the other might not.

How to debug a web page

To debug a web page, you find the causes of the errors and correct them. Often, the changes you make as you debug a web page are just minor adjustments or improvements. But sometimes, the web page doesn't look at all the way you expected it to. Often, these errors are caused by trivial coding problems like missing tags, quotation marks, and braces. However, finding these problems can be hard to do when your files consist of hundreds of lines of code. That's why chapter 4 shows how to use the Developer Tools for your browser to debug the HTML and CSS for a web page.

The web page in the Edge browser

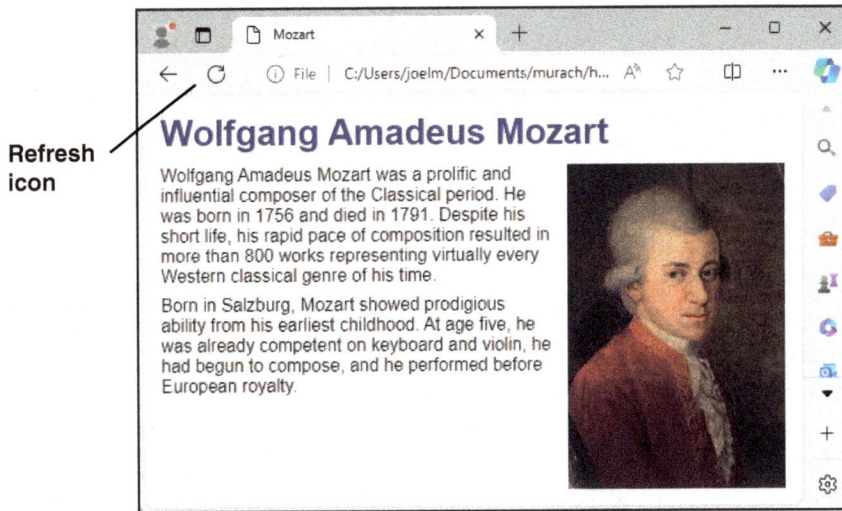

Refresh icon

Wolfgang Amadeus Mozart

Wolfgang Amadeus Mozart was a prolific and influential composer of the Classical period. He was born in 1756 and died in 1791. Despite his short life, his rapid pace of composition resulted in more than 800 works representing virtually every Western classical genre of his time.

Born in Salzburg, Mozart showed prodigious ability from his earliest childhood. At age five, he was already competent on keyboard and violin, he had begun to compose, and he performed before European royalty.

Two ways to run a web page

- If the HTML file for the web page is on your local file system, you can use File Explorer (Windows) or Finder (macOS) to find the HTML file. Then, you can double-click on the file to run the web page.
- If the web page is available from a server, you can enter the URL of the web page into the browser's address bar.

The components of a URL

```
https://www.murach.com/shop-books/index.html
```
protocol   domain name   path   filename

What happens if you omit parts of a URL

- If you omit the protocol, the browser uses the default of https:// or http://.
- If you omit the filename, the browser uses the default document name for the web server, typically index.html, default.htm, or some variation.

How to debug a web page

- Find the causes of the errors, fix the errors, and save the changes.
- After you fix the errors, test the page again. To do that, you can switch to the browser from your text editor or IDE and then click the Refresh or Reload icon.

Description

- When you *test* a web page, you try to find all the errors.
- When you *debug* a web page, you fix the errors and test again.

Figure 2-15   How to test and debug a web page

How to validate HTML and CSS

After you test and debug your web pages, the HTML and CSS is typically valid. However, you can take validation to the next level by using an official validation service to *validate* your files. These services are useful when a file is large, the HTML or CSS isn't working right, and you can't spot any errors. HTML and CSS validation also improves the likelihood that your code will work on other browsers besides the ones that you're using for testing.

How to validate HTML

One of the most popular ways to validate HTML is to use the W3C Markup Validation Service website that's shown in figure 2-16. When you use this website, you can provide the HTML that you want to validate in the three ways shown in this figure.

In this example, the Validate by File Upload tab is displayed. Then, you click the Choose File button to specify the file that you want to validate. Once that's done, you click the Check button to validate the document.

If the HTML is valid, the validator displays a message that says so. However, it may also display one or more warning messages. On the other hand, if the code contains errors, the validator displays a list of the errors. Then, you can use the error messages to help you correct your code.

The Markup Validation Service

`https://validator.w3.org/`

The web page

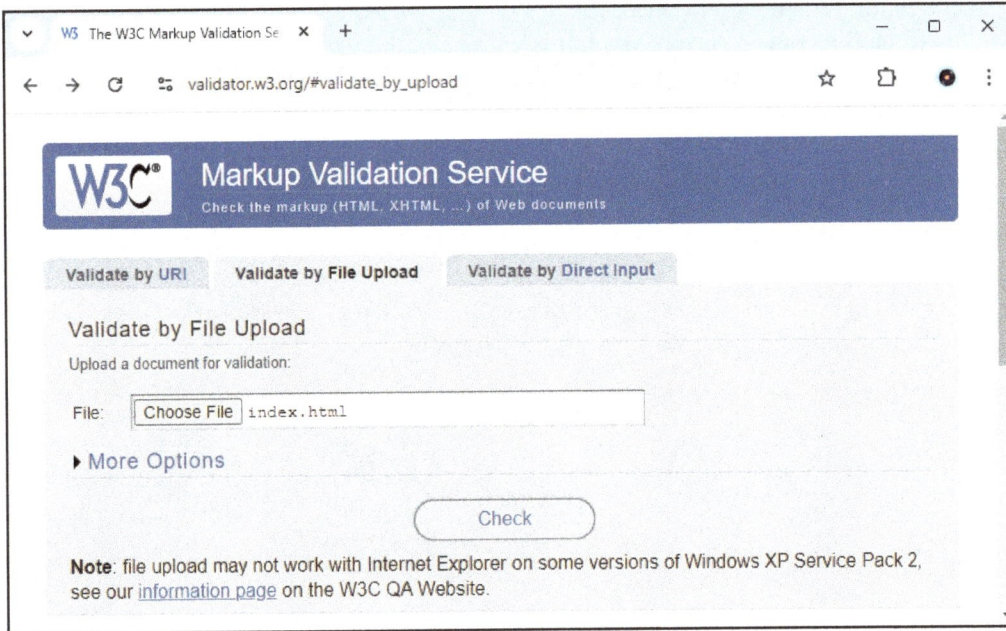

How to use the service

- Go to the URL for the service, use one of the tabs to specify the HTML to be validated, and click the Check button.

Three ways to specify the HTML to be validated

- If the HTML is available from a web server, use the Validate by URI tab to specify the URL for the web page.
- If the HTML file is on your local system, use the Validate by File Upload tab to choose the file.
- If you want to copy and paste some HTML, use the Validate by Direct Input tab.

Description

- *Validation* makes sure that your code is correct. It may find errors in your code that you weren't aware of.
- To validate HTML, you can use the W3C Markup Validation Service.
- If the HTML document is valid, the validator indicates that the document passed the validation. However, it may still display one or more warnings.
- If the HTML document isn't valid, the validator lists and describes each error.

Figure 2-16   How to validate HTML

How to validate CSS

One of the most popular ways to validate CSS is to use the W3C CSS Validation Service that's shown in figure 2-17. This service works much like the W3C service for validating an HTML file.

If the file contains errors when it is validated, the validation service displays a list of the errors. Then, you can use the error messages to identify and correct your coding errors.

The CSS Validation Service

`https://jigsaw.w3.org/css-validator/`

The web page

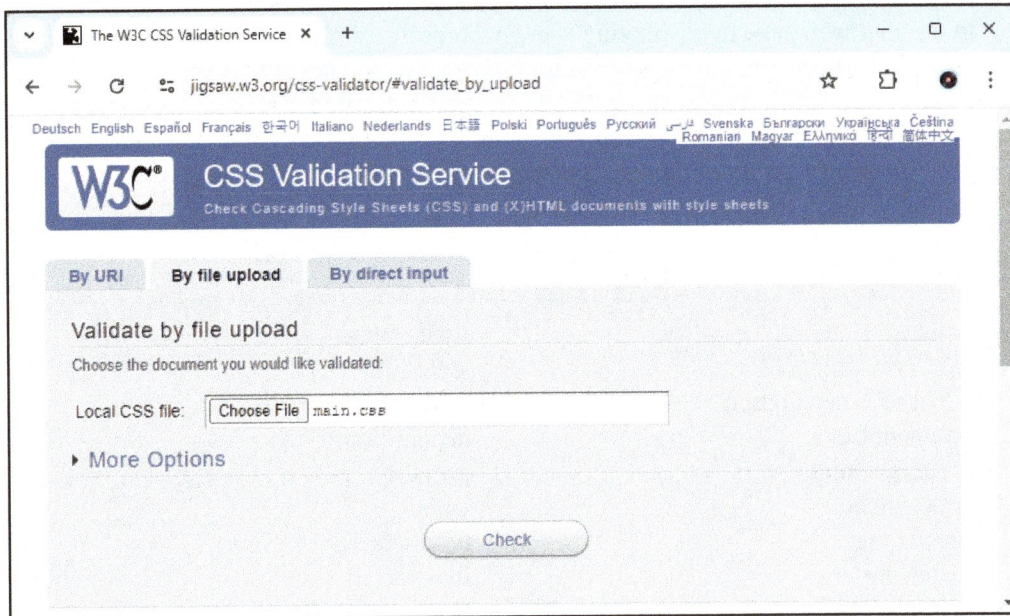

How to use the service

- Go to the URL for the service, use one of the three tabs to specify the CSS to be validated, and click the Check button.

Description

- To validate CSS, you can use the W3C CSS Validation Service.
- If the CSS is valid, the validator indicates that the CSS passed the validation. However, it may still display one or more warnings.
- If the CSS isn't valid, the validator lists and describes each error.

Figure 2-17   How to validate CSS

Perspective

Now that you've completed this chapter, you should be able to create and edit HTML and CSS files for a web page using VS Code. Then, you should be able to test that web page by displaying it in a web browser. You should also be able to validate the HTML and CSS files for a web page whenever that's necessary.

At this point, you're ready to learn the coding details for HTML and CSS. So, in the next chapter, you'll learn the details for coding the HTML that defines the structure and content for a web page. After that, you'll learn the details for coding the CSS that formats the content.

Terms

syntax	whitespace
HTML document	style rule
DOCTYPE declaration	selector
document tree	declaration
HTML element	property
root element	value
opening tag	type selector
closing tag	id selector
content of an element	class selector
empty tag	open a file
nested elements	preview a file
attribute	completion list
Boolean attribute	testing
comment	debugging
comment out	validation

Before you do the exercises for this book...

Before you do the exercises for this book, you should install VS Code and the Chrome browser and download the files for this book as described in appendix A (Windows) or B (macOS).

Exercise 2-1   Get started right with VS Code

This exercise gets you started with VS Code by having you open, run, and edit the HTML and CSS for the web page that's described in chapter 1.

Open the main folder for this book

1. Start VS Code.

2. Open this folder:

   `murach\html_css\book_apps`

   This displays the folders within the book_apps folder in the Explorer window.

3. Open this folder:

   `murach\html_css`

   This displays the book_apps, exercises, and solutions folders in the Explorer window, and you can expand all of these folders. This also closes the book_apps folder that you opened in the previous step.

View the code for a web page and run it

4. Install the HTMLHint extension.

5. Install the Open in Browser extension.

6. In the Explorer window, expand this folder:

   `exercises\ch02`

   This should display the files for a web page about Mozart.

7. Click the file named index.html to preview it in the code editor. Note that this also displays the filename in the list of files under Open Editors.

8. Click the file named main.css to preview it. Note that it replaces the HTML in the code editor with CSS and displays the filename in the list of files under Open Editors.

9. Right-click the file named index.html and select Open in Default Browser. This should open the index.html file in your default browser.

10. Switch back to VS Code. If you have another browser installed, right-click the index.html file again and select Open in Other Browsers. Then, click the name of the other browser to open the file in that browser. But remember, at the time of this writing, this may not work for Edge.

Edit the HTML

11. Double-click on the file named index.html to open that file.

12. In the index.html file, start a new line after the first <p> element but before the second one. Then, use code completion to add the opening and closing tags for an <h2> element.

13. Add "Early Life" between the <h2> tags. When you're done, the new HTML should look like this:

```
<h2>Early Life</h2>
```

14. Save this change.

15. Switch to the browser that's displaying this page. Note that it hasn't been updated.

16. Click the browser's Reload button to update the page. It should display the new <h2> element.

Edit the CSS

17. Double-click on the file named main.css to open that file. Note that both the HTML file and the CSS file are now displayed in separate tabs in the code editor and in the list of files under Open Editors.

18. In the main.css file, start a new line after the style rule for the <h1> element. Then, add a style rule for the <h2> element like this:

```
h2 {
    font-size: 160%;
    padding-bottom: .3em;
}
```

19. Save this change.

20. Switch to the browser that's displaying this page. Note that the format hasn't been updated.

21. Click the browser's Reload button to update the page. It should display the new formatting for the <h2> element.

Start a new HTML file

22. Right-click on the folder for the web page, select New File, and enter test1. html as the filename.

23. Enter an exclamation mark into the new file and press the Tab or Enter key. This should insert the starting code for an HTML document into the file.

24. Within the tags for the <body> element, enter your name.

25. Press Ctrl+S to save the changes to the new HTML file.

26. Right-click the new HTML file and select Open in Default Browser. This should open the new HTML file in your default browser and display your name in the browser's default font and font size.

Exercise 2-2   Validate HTML and CSS files

This exercise has you validate the HTML and CSS for a web page.

Validate an HTML file

1. Start VS Code and open the HTML file named index.html that's in this folder:

    `exercises\ch02`

2. Delete the ending > for the <img> element and save the file.

3. Go to the web page for the W3C Markup Validation Service.

4. Use the Validate by File Upload tab to validate the file. Then, scroll down the page to view the error messages and warnings that are displayed for this error.

5. Switch back to VS Code, fix the error, and save the file.

6. Validate the HTML file again. The service should display a message that indicates that the HTML is valid.

Validate a CSS file

7. Use VS Code to open the CSS file named main.css.

8. Delete the semicolon after the font-size property in the h1 style rule and save the file.

9. Go to the web page for the W3C CSS Validation Service.

10. Use the By File Upload tab to validate the file. Then, scroll down the page to see the error message.

11. Switch back to VS Code, fix the error, and save the file.

12. Validate the CSS file again. The service should display a message that indicates that the CSS is valid.

Chapter 3

How to use HTML to structure a web page

Chapter 2 presented the basic structure of an HTML document, and some basic techniques for coding the elements that make up a document. Now, you'll learn how to code the HTML elements that you'll use in most of your documents. In particular, you'll learn how to use these HTML elements to structure the content of a web page. Then, in the next three chapters, you'll learn how to use CSS to format those HTML elements.

How to code the head section

The head section of an HTML document contains elements that provide information about the web page rather than the content of the page. Figure 3-1 shows how to code it.

How to include metadata

For most web pages, the head section should include the <meta> elements shown in this figure to provide three items of information about the web page. This information is known as *metadata*.

The first <meta> element specifies the character encoding used for the page. In most cases, you can set the encoding to UTF-8 as shown in this example.

The second <meta> element specifies the viewport. This determines how a page scales when it's viewed on devices with different screen sizes. For now, you can code this element exactly as shown in this example.

The third <meta> element provides a description that can be used by search engines to index the page. To optimize your web page for search engines, this description should be a concise and high-quality summary of the content of the page, and it should be unique for each page. That's because it may be displayed in the search results of a search engine.

How to set a title and a favicon

The head section of every web page should also include a <title> element that describes the content of the page. In this figure, the <title> element specifies the name of the organization followed by two keywords, *speakers* and *luncheons*, that help to uniquely identify this page. Search engines may use this title to rank the page in its results, and they may display the title in the search results to help people decide whether they want to go to that page.

Although it can sometimes be helpful for the meta description or the title to include keywords, you should avoid repeating keywords. That's known as *keyword stuffing*, and it's a bad practice that may hurt the search engine optimization (SEO) for a page.

The content of the <title> element is displayed in the browser's tab for the page. In this figure, the tab isn't wide enough to display the entire title. As a result, it only displays the beginning of it. But if the user hovers the mouse over the tab, the browser should display the entire title in a tooltip.

The <link> element can be used in the head section to link a custom icon, called a *favicon*, to the web page. This causes the browser to display the icon to the left of the title in the tab for the page. But you don't need to code this element when you deploy the website to an web server. Instead, you can place the favicon in the root folder for the website and it will automatically appear in the tab for each page.

A web browser that shows the title and favicon for a page

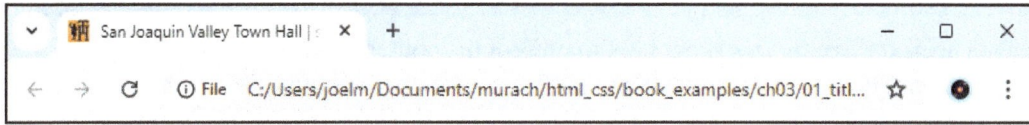

A head section that sets a title, a favicon, and three items of metadata

```
<head>
    <meta charset="utf-8">
    <meta name="viewport" content="width=device-width, initial-scale=1.0">
    <meta name="description" content="A yearly lecture series with speakers
        that present new information on a wide range of subjects">
    <title>San Joaquin Valley Town Hall | speakers and luncheons</title>
    <link rel="icon" href="favicon.ico">
</head>
```

SEO guidelines for the meta description and the title

- It should be a concise and high-quality summary of the content of the page that can be displayed in the search results for a search engine.
- It should be unique for each page.
- It's sometimes helpful to include one or two keywords, called *focus keywords*, but you should avoid repeating keywords.

Description

- The <meta> element provides information about the HTML document that's called *metadata*. It should be used to provide the charset, viewport, and description for each page.
- The <title> element specifies the text that's displayed in the browser's tab for the web page. It is also used as the name of a favorite or bookmark for the page.
- A *favicon* is an icon that appears to the left of the title in the browser's tab for the page. It may also appear to the left of the URL in the browser's address bar, and it may be used in a favorite or bookmark.
- To specify a favicon for a page, you use favicon.ico as the name of the favicon and you code a <link> element exactly as shown in this figure. Then, the favicon will be displayed when you test the page on your computer or local server.
- You don't need to code the <link> element for the favicon when you deploy the website to an web server. You just need to store the favicon in the root folder.

Figure 3-1   How to code the head section

How to structure the body section

The next three figures show how to present the content of a web page. That includes how to structure that content for search engine optimization and web accessibility.

How to code headings and paragraphs

Headings and paragraphs typically provide most of the text for a web page. They are defined by the HTML elements in the table in figure 3-2. These elements are *block elements*, which means that they start on a new line when they are displayed.

The example in this figure begins by coding the <html>, <head>, and <body> elements for the page. In the <html> element, the lang attribute identifies the language for the web page because that's a best practice. In this case, the HTML sets the lang attribute to en, which is the two-letter ISO code for English. However, if your web page uses another language, you should set the lang attribute to the two-letter code for that language.

Within the <body> element, the HTML uses the <h1>, <h2>, and <p> elements to structure the text for the document. When a browser displays these elements, it uses a default font and font size. In addition, it uses default settings to control the amount of space that's displayed above and below the elements.

When you use the heading elements, you should use the <h1> element for the most important heading on a page. In addition, you should only code one <h1> element on each page. For your next headings, you should go down one level at a time to indicate less import headings. In other words, the first heading level after <h1> should be <h2>, not <h3>. This structure improves search engine optimization for your site, and it makes your pages more accessible to devices like screen readers.

Incidentally, HTML provides for three more levels below <h3> with elements that range from <h4> through <h6>. However, most web pages don't need to use headings beyond <h3>.

Block elements for headings and paragraphs

Element	Description
h1	A level-1 heading
h2	A level-2 heading
h3	A level-3 heading
p	A paragraph

A page that uses headings and paragraphs

```html
<html lang="en">
<head>
    <!-- the elements for the head section go here -->
</head>
<body>
    <h1>San Joaquin Valley Town Hall Programs</h1>
    <h2>Pre-lecture coffee at the Saroyan</h2>
    <p>Join us for a complimentary coffee hour, 9:15 to 10:15 a.m. on the
        day of each lecture. The speakers usually attend this very special
        event.</p>
    <h2>Post-lecture luncheon at the Saroyan</h2>
    <p>Extend the excitement of Town Hall by purchasing tickets to the
        luncheons.</p>
</body>
</html>
```

How it looks in a browser

San Joaquin Valley Town Hall Programs

Pre-lecture coffee at the Saroyan

Join us for a complimentary coffee hour, 9:15 to 10:15 a.m. on the day of each lecture. The speakers usually attend this very special event.

Post-lecture luncheon at the Saroyan

Extend the excitement of Town Hall by purchasing tickets to the luncheons.

SEO guidelines

- Use the <h1> element to identify the most important information on the page.
- Code only one <h1> element per page.
- Use the <h2> and <h3> elements to identify lower levels of importance.

Description

- You should always use the lang attribute of the <html> element to identify the language for the web page.
- *Block elements* are the building blocks of a website. They always start on a new line and take up the full space of the area that they're in.

Figure 3-2   How to code headings and paragraphs

How to code the structural elements

Figure 3-3 presents the structural elements that were introduced with HTML5. To illustrate, the example in this figure shows how to use the <header>, <main>, and <footer> elements to divide a web page into three parts. Later, this structure makes it easy to use CSS to change the page layout and format the content.

All of these structural elements are block elements. But unlike the heading and paragraph elements, you can nest other block elements within these structural elements. In this figure, the HTML in the example nests an <h1> element within the <header> element, and it nests a <p> element within both the <main> and <footer> elements.

This example only uses three of the seven elements in the table. However, the web page that's presented at the end of this chapter uses the <nav> element. In addition, examples of the other elements are provided later in this book.

The HTML elements presented in figures 3-2 and 3-3 identify the type of content they contain. For example, an <h1> element identifies the level-1 heading for a page, and the <main> element identifies the main content for a page. Elements like these are called *semantic elements* because they give meaning to their content.

Semantic elements are good for the developer because they make the HTML easier to read and understand. In addition, they make it easier for search engines and browsers to identify the various types of content on a page. As a result, semantic elements improve search engine optimization. In addition, they improve accessibility by being more understandable to screen readers.

Structural elements

Element	Description
header	The header for a page
main	The main content for a page
section	A generic section that doesn't indicate the type of content
article	A written composition like an article, essay, or opinion piece
nav	A navigation section that contains links to other pages or placeholders
aside	A sidebar that is related to the content that's near it
footer	The footer for a page

A body section that uses three structural elements

```
<body>
    <header>
        <h1>San Joaquin Valley Town Hall</h1>
    </header>
    <main>
        <p>Welcome to San Joaquin Valley Town Hall. We have some
            fascinating speakers for you this season!</p>
    </main>
    <footer>
        <p>Copyright San Joaquin Valley Town Hall.</p>
    </footer>
</body>
```

How it looks in a browser

San Joaquin Valley Town Hall

Welcome to San Joaquin Valley Town Hall. We have some fascinating speakers for you this season!

© Copyright San Joaquin Valley Town Hall.

Accessibility and SEO guideline

- Use the structural elements to indicate the structure of your pages.

Description

- All of the structural elements are block elements that can contain other block elements.
- Elements that give meaning to their content are called *semantic elements*.

Figure 3-3   How to code the structural elements

When and how to use \<div\> elements

But what happens if you want to divide one of the structural elements into two or more parts and there isn't a semantic element for that? Then, you can use \<div\> elements as shown in figure 3-4. Here, the first example shows how two \<div\> elements can be used to divide a \<section\> element into two divisions. Then, you can use CSS to lay out and format these divisions separately.

The second example shows how \<div\> elements were used before HTML5 introduced the structural elements. Here, the \<div\> elements divide the page into header, main, and footer sections by using id attributes to identify them. But the \<div\> elements aren't semantic and don't help search engine optimization or web accessibility. That's why you should avoid using \<div\> elements if there's a semantic element that describes your content.

The <div> element

Element	Description
div	A block element that can be used to divide a structural element into groups

A <section> element that uses <div> elements to structure its content

```
<section>
    <div>
        <h2>The first priority</h2>
        <p>The elements for the first priority</p>
    </div>
    <div>
        <h2>The second priority</h2>
        <p>The elements for the second priority</p>
    </div>
</section>
```

How it looks in a browser

The first priority

The elements for the first priority

The second priority

The elements for the second priority

How <div> elements were used before HTML5

```
<body>
    <div id="header">
        <h1>San Joaquin Valley Town Hall</h1>
    </div>
    <div id="main">
        <p>Welcome to San Joaquin Valley Town Hall. We have some
            fascinating speakers for you this season!</p>
    </div>
    <div id="footer">
        <p>Copyright 2015, San Joaquin Valley Town Hall.</p>
    </div>
</body>
```

Accessibility and SEO guidelines

- Only use <div> elements when the HTML semantic elements don't apply.

Description

- Before HTML5, <div> elements were used to provide the structure for a web page. But now, <div> elements should only be used to provide structure within a semantic element.

Figure 3-4   When and how to use <div> elements

More skills for presenting text

Now that you know how to provide the overall structure for the body section, you're ready to learn some more skills for presenting the text of a web page.

How to code inline elements

In contrast to a block element, an *inline element* doesn't start on a new line. Instead, an inline element is coded within a block element. Figure 3-5 begins with a table that presents some inline elements that you can use to mark up and format the text for a web page.

Most of these elements are semantic elements that provide meaning. However, by default, most browsers provide formatting for these elements. For example, the <strong> and <em> elements provide two different levels of emphasis for text. By default, most browsers display <strong> elements in bold and <em> elements in italics.

For the record, you can also use the <b> and <i> elements to add bold and italics to text. However, since they only provide formatting, not meaning, it's a better practice to use the <strong> and <em> elements instead.

The <span> element is an inline element with no meaning. You can use this element if there isn't a semantic element that works for the text that you want to mark up. In other words, the <span> element is the inline equivalent of the <div> element. By default, browsers don't provide any formatting for a <span> element. However, you can use it to apply CSS formatting to a span of text within a block element. In addition, it's common to use JavaScript to work with <span> elements.

The example illustrates five of these elements. The first two lines show how to use the <strong> and <em> elements to emphasize text by displaying it in bold or italics. The third line shows how to use the <kbd> element to identify text as keyboard input, which formats it in a monospaced font. The fourth line uses the <q> element to add quotation marks to a quotation. And the fifth line shows how to use the <sub> element to identify text that should be displayed as a subscript.

Inline elements for text

Element	Description
strong	Strong emphasis, typically displayed in bold
em	Emphasis, typically displayed in italics
kbd	A keyboard entry, typically displayed in a monospaced font
code	Computer code, typically displayed in a monospaced font
var	A variable, typically displayed in a monospaced font
abbr	An abbreviation
dfn	A term that is defined elsewhere, typically displayed in italics
cite	A bibliographic citation like a book title, typically displayed in italics
q	A quotation, typically displayed within quotation marks
time	A date or time in a standard format
small	Small font size for "fine print" such as footnotes
sub	Subscript text
sup	Superscript text
span	A span of text

HTML that uses some of the inline elements

```
<p>Save a bundle at our <strong>big yearend sale</strong>.</p>
<p>If you don't get 78% or more on your final, <em>you won't pass.</em></p>
<p>When the dialog box is displayed, press <kbd>Enter</kbd>.</p>
<p>To quote Hamlet: <q>Conscience does make cowards of us all.</q></p>
<p>The chemical symbol for water is H<sub>2</sub>O.</p>
```

How it looks in a browser

Save a bundle at our **big yearend sale**.

If you don't get 78% or more on your final, *you won't pass.*

When the dialog box is displayed, press Enter.

To quote Hamlet: "Conscience does make cowards of us all."

The chemical symbol for water is H_2O.

Description

- An *inline element* is coded within a block element and doesn't begin on a new line.
- Most of the inline elements are semantic elements that provide meaning. The exception is the <span> element that you can use if a semantic element isn't available. The <span> element is in the inline equivalent of the <div> block element.
- Although you can use the <b> and <i> elements to apply bold and italics to text, it's a better practice to use the <strong> and <em> elements because they are semantic.

Figure 3-5   How to code inline elements for text

How to code character entities and special types of text

Figure 3-6 shows how to use *character entities* and three more block elements for special types of text. The first table presents some of the many character entities that provide a way to include characters like ampersands (&) and degree symbols in your text. As shown here, all character entities start with an ampersand (&) and end with a semicolon (;). Then, the rest of the entity identifies the character that the entity represents.

The second table presents three more block elements for coding special types of text. For instance, you can use the <blockquote> element to display a quotation, and the <address> element to present contact information. Like the structural elements, these elements are semantic and give meaning to their content. In addition, by default, browsers may apply formatting to these elements such as indenting block quotes or italicizing an address block.

The example shows how these block elements and character entities work. Here, the <blockquote> element identifies a quotation, and the <address> element identifies some contact information. This causes the browser to indent the quotation and italicize the address block.

The end of the example codes two character entities within a <p> element. The first provides the copyright symbol. The second provides an ampersand.

This shows that you can't just type an ampersand in your text. That's because the ampersand is the first character in a character entity. Similarly, because the left bracket (<) and right bracket (>) are used to identify HTML tags, you can't use those characters to represent less-than and greater-than signs. Instead, you need to use the < and > entities.

In some cases, the &, <, and > characters will work without coding them as character entities. However, if you want them to work for all purposes and pass all validation tests, it's a best practice to code them as character entities.

Common HTML character entities

Entity	Character	Entity	Character
&	&	°	°
<	<	±	±
>	>	¢	¢
©	©	‘	' (opening single quote)
®	®	’	' (closing single quote or apostrophe)
™	™		non-breaking space

Block elements for special types of text

Element	Description
blockquote	An indented block quotation
address	Contact information for the developer or owner of a website
pre	Preformatted text with preserved whitespace and a monospaced font

HTML that uses these block elements and character entities

```
<p>Ernest Hemingway wrote:</p>
<blockquote>
    Cowardice, as distinguished from panic, is almost always simply
    a lack of ability to suspend the functioning of the imagination.
</blockquote>

<address>
    1-800-221-5528<br>
    <a href="emailto:murachbooks@murach.com">murachbooks@murach.com</a>
</address>

<p>&copy; Mike Murach &amp; Associates, Inc.</p>
```

How it looks in a browser

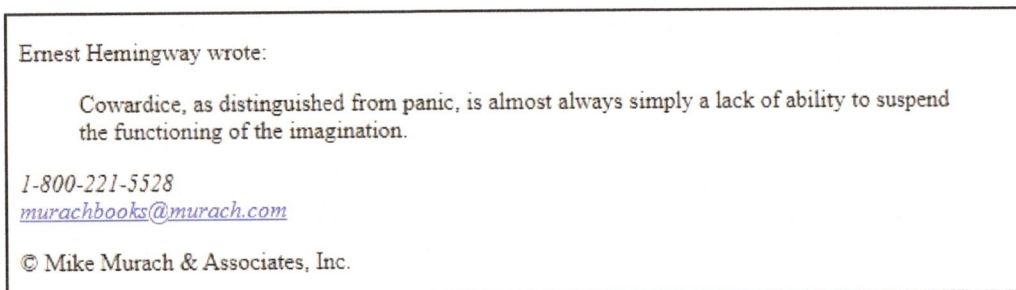

Ernest Hemingway wrote:

> Cowardice, as distinguished from panic, is almost always simply a lack of ability to suspend the functioning of the imagination.

1-800-221-5528
murachbooks@murach.com

© Mike Murach & Associates, Inc.

Description

- *Character entities* are used to display special characters in an HTML document.
- Block elements for special types of text are good semantically because they identify the type of content.

Figure 3-6   How to code character entities and special types of text

How to code links, lists, and images

Because you'll use links, lists, and images in most of the web pages that you develop, the figures that follow introduce you to these elements. But first, you need to know how to code absolute and relative URLs so you can use them with your links and images.

How to code URLs

Figure 3-7 presents some examples of absolute and relative URLs. But first, the diagram at the top of this figure shows the folder structure that's used by the examples. These folders are organized into three levels. The root folder for the site contains five subfolders, including the folders that contain the images and styles for the website. Then, the books folder contains subfolders of its own.

An *absolute URL* includes the domain name of the website. This is shown by the first example in this figure. Here, both URLs refer to pages at www.mywebsite.com. The first URL points to the index.html file in the root folder of this website, and the second URL points to the toc.html file in the books/php folder.

When used within a page on one website, an absolute URL typically refers to a file on another website. By contrast, a *relative URL* refers to a file within the same website. There are two types of relative URLs: a root-relative path and a document-relative path.

If a website is running on a web server, a *root-relative path* is a path that's relative to the root folder of the website. This is shown by the second example. Here, the leading slash indicates the root folder for the site. As a result, the first path refers to the login.html file in the root folder, and the second path refers to the logo.gif file in the images folder. However, if you run a web page from the file system as shown in the previous chapter, a root-relative path is relative to the root folder of the file system, not the root folder of the website. That's why the web pages presented in this book use document-relative paths.

A *document-relative path* is a path that's relative to the current document. This is shown by the third example. This example assumes that the paths are coded in a file that's in the root folder for the website. Then, the first path refers to a file in the images subfolder of the root folder, and the second path refers to a file in the books/php subfolder. This shows how to navigate down the levels of the folder structure.

But you can also navigate up the levels with a document-relative path. This is shown by the fourth example. This example assumes that the current document is in the books/java folder. Here, the first path navigates up one level to the index.html file in the books folder. The second path navigates up two levels to the index.html file in the root folder. And the third path navigates up two levels to the root folder and then down one level to the logo.gif file in the images folder.

This shows that there's more than one way to code the path for a file. For example, if you're coding an HTML file in the books folder, you can use either a root-relative path or a document-relative path to get to the images subfolder.

A folder structure for a website

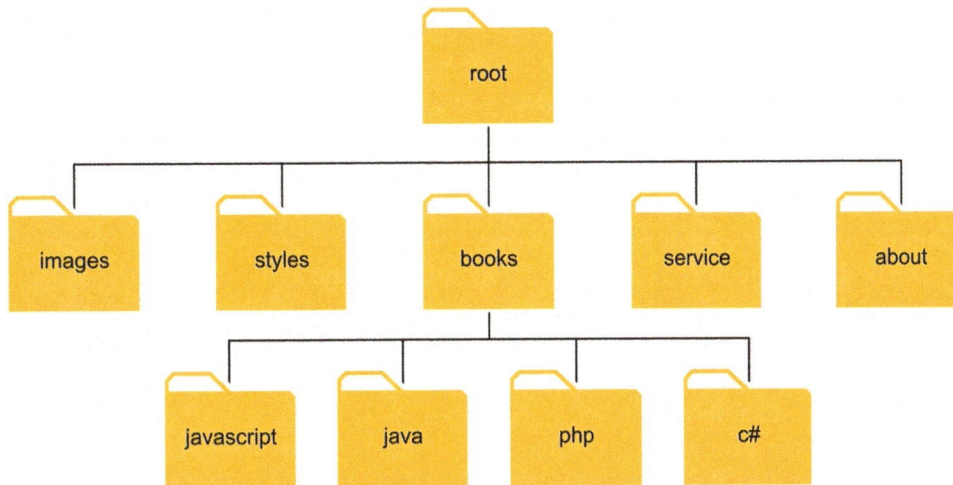

Absolute URLs

```
https://www.mywebsite.com/index.html
https://www.mywebsite.com/books/php/toc.html
```

Root-relative paths

```
/login.html              (refers to root/login.html)
/images/logo.gif         (refers to root/images/logo.gif)
```

Document-relative paths that navigate down

```
                         (current folder is root)
images/logo.gif          (refers to       root/images/logo.gif)
books/php/overview.html  (refers to       root/books/php/overview.html)
```

Document-relative paths that navigate up

```
                         (current folder is root/books/java)
../index.html            (refers to       root/books/index.html)
../../index.html         (refers to       root/index.html)
../../images/logo.gif    (refers to       root/images/logo.gif)
```

Description

- An *absolute URL* is a URL that provides the complete path to the resource including the domain name for the site. Absolute URLs let you display pages at other websites.

- A *relative URL* is relative to the folder that contains the current page.

- If a website is running on a web server, a *root-relative path* is relative to the root folder of the website. It always starts with a slash. If you run a web page from the file system, a root-relative path is relative to the root folder of the file system.

- A *document-relative path* is relative to the folder the current document is in. To go down one subfolder, you code the subfolder name followed by a slash. To go up one folder, you code two periods and a slash.

Figure 3-7   How to code URLs

How to code links

Most web pages contain *links* to other web pages or web resources. To code a link, you use the <a> element (or anchor element) as shown in figure 3-8. Because this element is an inline element, you usually code it within a block element like a <p> element.

In most cases, you only need to code one attribute for the <a> element, the href attribute. This attribute specifies the URL for the resource for the link. The examples in this figure show how this works.

The first example uses a relative URL to link to a page in the same folder as the current page. The second example uses a relative URL to link to a page that's relative to the current folder. The third example uses a relative URL to link to a page that's relative to the root folder. And the last example uses an absolute URL to link to a page at another website.

By default, most browsers underline links to indicate that they're clickable. As a result, most web users have been conditioned to associate underlined text with links. Because of that, you should avoid underlining any other text.

When a browser displays a link, the link has a default color depending on its state. For instance, most browsers display a link that hasn't been visited in blue and a link that has been visited in purple. However, you can use CSS as described in the next chapter to change these default colors.

When you create a link that contains text, it's a good practice for the text to clearly indicate the function of the link. In other words, you shouldn't use text like "click here" because it doesn't indicate what the link does. Instead, you should use text like the text shown in the examples in this figure. In short, if you can't tell where a link goes by reading its text, you should rewrite the text. This improves the search engine optimization and accessibility of your site.

The primary attribute of the <a> element

Attribute	Description
href	Specifies a relative or absolute URL for a link.

Link to a page...

In the same folder

```
<p>Go view our <a href="products.html">product list</a>.</p>
```

In a folder that's relative to the current folder

```
<p>Read about the <a href="../company/services.html">services we provide</a>.</p>
```

In a folder that's relative to the root folder

```
<p>View your <a href="/orders/cart.html">shopping cart</a>.</p>
```

At another web site

```
<p>To learn more about JavaScript, visit the
<a href="https://www.javascript.com/">official JavaScript web site</a>.</p>
```

How it looks in a browser

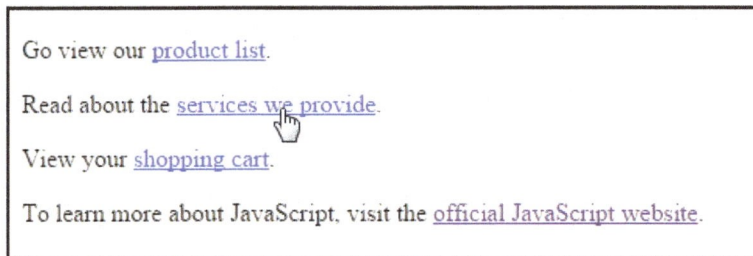

Go view our product list.

Read about the services we provide.

View your shopping cart.

To learn more about JavaScript, visit the official JavaScript website.

SEO and accessibility guideline

- The content of a link should be text that clearly indicates where the link is going.

Description

- The <a> element is an inline element that creates a *link* that loads another web page. The href attribute of this element identifies the page to load.
- The text content of a link is underlined by default to indicate that it's clickable.
- If a link hasn't been visited, it's displayed in blue. If it has been visited, it's displayed in purple. But you can change these values using CSS.
- If the mouse hovers over a link, by default, most browsers change the cursor to a hand with the finger pointer.

Figure 3-8   How to code links

How to code lists

Figure 3-9 shows how to code the two types of lists: ordered lists and unordered lists. To create an *unordered list*, you use the <ul> element. Then, within this element, you code one <li> (list item) element for each item in the list. The <li> element identifies the text that's displayed in the list. By default, when a browser displays an unordered list, it displays a bullet in front of each item. However, you can use CSS to change the type of bullet.

To create an *ordered list*, you use the <ol> element. Then, within this element, you code one <li> element for each item in the list. This works like the <ul> element, except that the browser displays numbers, not bullets, in front of the items. However, you can use CSS to change the type of numbers.

The example presented in this figure shows how this works. To start, an unordered list displays the names of several programming languages. That makes sense because the items in this list don't need to reflect any order. By contrast, the second list identifies three steps that need to be completed in a specific order. As a result, this example uses an ordered list for the second list.

Elements that create ordered and unordered lists

Element	Description
`<ul>`	An unordered list.
`<ol>`	An ordered list.
`<li>`	A list item.

HTML that creates two lists

```
<p>We have books on a variety of languages, including</p>
<ul>
    <li>JavaScript</li>
    <li>PHP</li>
    <li>Java</li>
    <li>C#</li>
</ul>

<p>You will need to complete the following steps:</p>
<ol>
    <li>Enter your billing information.</li>
    <li>Enter your shipping information.</li>
    <li>Confirm your order.</li>
</ol>
```

How it looks in a browser

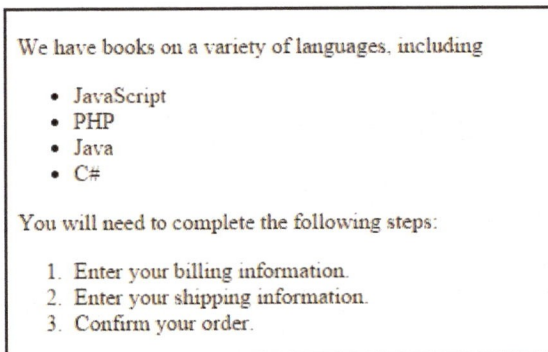

We have books on a variety of languages, including

- JavaScript
- PHP
- Java
- C#

You will need to complete the following steps:

1. Enter your billing information.
2. Enter your shipping information.
3. Confirm your order.

Description

- The two basic types of lists are *unordered lists* and *ordered lists*.
- By default, an unordered list is displayed as a bulleted list, and an ordered list is displayed as a numbered list.

Figure 3-9   How to code lists

How to include images

Images are an important part of most web pages. To display an image, you use the <img> element shown in figure 3-10. This is an inline element that's coded as an empty tag. When you code an <img> tag, two attributes are required: src and alt.

The src (source) attribute of an <img> element specifies the URL of the image that you want to display. In this figure, the example uses the src attribute to indicate that the image named murachlogo.gif can be found in the images subfolder of the current folder.

The alt attribute provides information about the image in case it can't be displayed or the page is being accessed by a screen reader. This is essential for visually-impaired users. The example in this figure uses an image for the logo of Murach Books. As a result, this example sets the value of the alt attribute to "Murach Logo". However, if an image doesn't provide any meaning, it's a good practice to code the value of the alt attribute as an empty string (""). You should do that, for example, when an image is only used for decoration.

You can use the height and width attributes of an <img> element to specify the size of an image. That can help the browser lay out the page as the image is being loaded. You can also use the height and width attributes to render an image larger (known as "stretching") or smaller than the original image. However, it's typically a better practice to use your image editor to make the image the right size for the page.

In this figure, the example codes an <img> element before an <h1> element. In this case, even though the <img> element is an inline element, it displays on its own line. That's because the <h1> element is a block element that starts a new line after the <img> element.

After the example, this figure lists three types of common image formats for web pages. To start, the JPEG format is typically used for photographs and scanned images, and the GIF format is typically used for simple illustrations or logos. The PNG format was developed specifically for the web as a replacement for the GIF format. As a result, for web development, you typically want to use PNG files.

There are other new image formats for web development such as WebP and AVIF. However, these formats aren't as widely used as the JPEG and PNG formats.

Attributes of the <img> element

Attribute	Description
src	The relative URL of the image to display. It is required.
alt	Alternate text that's displayed in place of the image. This text is read aloud by screen readers for users with disabilities. It is required.
height	The height of the image in pixels.
width	The width of the image in pixels.

HTML that displays an image

```
<img src="images/murachlogo.gif" alt="Murach Logo" height="75">
<h1>Mike Murach &amp; Associates, Inc.</h1>
```

How it looks in a browser

Mike Murach & Associates, Inc.

Common image formats

- JPEG (Joint Photographic Experts Group)
- GIF (Graphic Interchange Format)
- PNG (Portable Network Graphics)

Accessibility guidelines

- For images with useful content, always code an alt attribute that describes the image.
- For images that are used for decoration, code the alt attribute with no value ("").

Description

- The <img> element is an inline element that displays the image that's identified by the src attribute.
- The height and width attributes can be used to indicate the size of an image so the browser can allocate the correct amount of space on the page. These attributes can also be used to size an image, but it's usually better to use an image editor to do that.
- JPEG files have the JPG extension and are used for photographs. GIF files are used for small illustrations and logos. And PNG files combine aspects of JPEG and GIF files.

Figure 3-10　　How to include images

A structured web page

Figure 3-11 presents a simple web page that uses the HTML elements for structuring a web page. The purpose of this web page is to provide information about a series of lectures being presented by a non-profit organization.

The page displayed in a browser

The top of this figure shows how the Chrome browser formats the HTML for this page by default. However, in the next chapter, you'll learn how to use CSS to provide custom formatting for this page.

The HTML

The HTML for this page starts with the DOCTYPE declaration, the <html> element, and the <head> element. This example includes the elements and attributes that should be coded for every web page.

To start, the <html> element uses the lang attribute to identify the language for the web page as English. That's appropriate since the text for this page is written in English.

The <head> element includes the charset <meta> element, the viewport <meta> element, the <title> element, and a <link> element for a favicon. Although it isn't coded here, most pages should also include a <meta> element for the description. That's especially true for web pages where search engine optimization is a consideration.

The <body> element uses some of the structural elements presented in this chapter. Here, the <header> element contains an <h2> and an <h3> element. The <main> element contains an <h1> element, a <nav> element, and a <p> element. And the <footer> element contains one <p> element.

Within the <nav> element, an unordered list contains five items (although the HTML is shown for only two of them). Then, each list item contains an <a> element. This is a best practice because a <nav> element should contain a list of links. It's also a best practice to code the <a> elements within an unordered list, not a numbered list.

This page uses an <em> element to italicize the first four words in the <p> element within the <main> element. In addition, it uses a character entity that inserts the copyright symbol into the <p> element within the footer. But what's most important is that this page uses the structural and semantic elements. That's good for the developer as well as for search engine optimization and web accessibility.

A web page displayed in a browser

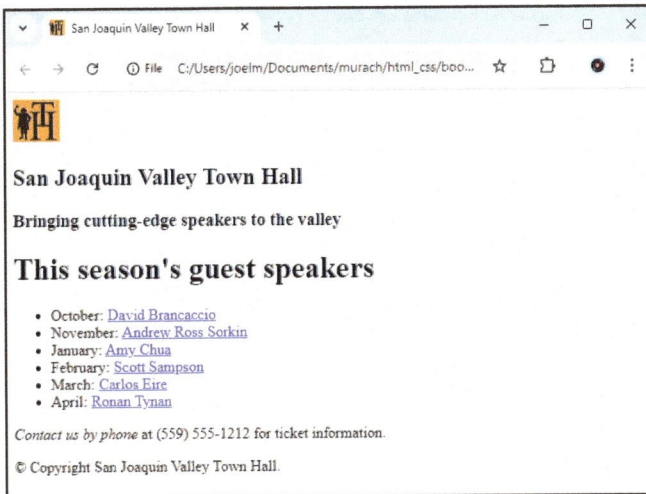

The HTML

```
<!DOCTYPE html>
<html lang="en">
<head>
    <meta charset="utf-8">
    <meta name="viewport" content="width=device-width, initial-scale=1.0">
    <title>San Joaquin Valley Town Hall</title>
    <link rel="icon" href="images/favicon.ico">
</head>
<body>
    <header>
        <img src="images/logo.jpg" alt="Town Hall Logo" width="50">
        <h2>San Joaquin Valley Town Hall</h2>
        <h3>Bringing cutting-edge speakers to the valley</h3>
    </header>
    <main>
        <h1>This season's guest speakers</h1>
        <nav>
            <ul>
                <li>October: <a href="speakers/brancaccio.html">
                    David Brancaccio</a></li>
                <li>November: <a href="speakers/sorkin.html">
                    Andrew Ross Sorkin</a></li>
                ...
            </ul>
        </nav>
        <p><em>Contact us by phone</em> at (559) 555-1212 for ticket
            information.</p>
    </main>
    <footer>
        <p>&copy; Copyright San Joaquin Valley Town Hall.</p>
    </footer>
</body>
</html>
```

Figure 3-11   A structured web page

Perspective

This chapter has presented most of the HTML elements that you'll need as you develop web pages. That includes both block and inline elements. With those skills, you can create web pages that use the default formatting that's provided by the browser. In the next three chapters, you'll learn how to use CSS to format your pages so they look just the way you want them.

As you might expect, there's a lot more to learn about HTML. That's why the chapters in section 2 of this book review some of the skills presented in this chapter and also present many other skills for working with HTML. In particular, they present more skills for working with lists, links, and images. However, the skills presented in this chapter should get you off to a good start.

Terms

metadata	absolute URL
keyword stuffing	relative URL
favicon	root-relative path
focus keywords	document-relative path
block element	link
semantic elements	ordered list
inline element	unordered list
character entity	

About the exercises

In the exercises for chapters 3 through 8, you'll develop a version of the Town Hall website that's similar to the one in the text. However, it will have different content, formatting, and page layouts.

As you develop this site, you'll use this folder structure:

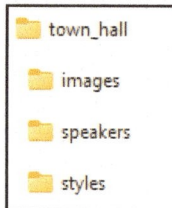

This is a realistic structure with images in the images folder, HTML pages for guest speakers in the speakers folder, and CSS files in the styles folder.

Exercise 3-1   Code the Town Hall home page

In this exercise, you'll code the HTML for the home page of the Town Hall website. When you're through, the page should look like this:

However, it should have three speakers, not just one.

Open the starting page and get the content for it

1. Use VS Code to open this HTML file:

    `exercises\ch03\town_hall\index.html`

    Note that it contains the head section for this web page as well as a body section that contains a <header> element with some text, a <main> element, and a <footer> element with some text.

2. Use your text editor to open this text file:

    `exercises\ch03\town_hall\index.txt`

    Note that it includes all of the text for the <main> element.

Code the <header> element

3. Code the <img> element that displays the image at the top of the page from the images folder. To locate the image file, use this path:

    `images/town_hall_logo.gif`

    Be sure to include the alt attribute, and set the height attribute to 80.

4. Code <h2> tags around the first line of text in the <header> element.

5. Code <h3> tags around the second line of text in the <header> element.

6. Test the home page in Chrome. If necessary, fix any problems and test again.

Code the <main> element

7. Copy all of the content for the <main> element from the index.txt file into the index.html file.

8. Code <h1> tags around "This season's guest speakers", and code <h2> tags around "Our Mission" and "Our Ticket Packages".

9. Code <p> tags around the first paragraph after the "Our Mission" heading, and code <blockquote> and <q> tags around the quote.

10. Code <ul> and <li> tags around the three items after the "Our Ticket Packages" heading.

11. Test the home page in Chrome. If necessary, fix any problems and test again.

12. After the "guest speakers" heading, format the month and name for the first speaker as one <h3> element with a <br> in the middle that rolls the speaker's name over to a second line.

13. Repeat the previous step for the next two speakers.

14. Code three links by adding <a> tags around the name for each speaker. The href attribute for each link should refer to an HTML file in the subfolder named speakers that uses the speaker's last name as the filename.

15. After the <h3> element for each speaker, code an <img> element that displays the image for the speaker. The JPG images are in the images subfolder with a filename that begins with the speaker's last name. Make sure to include the alt attribute for each element.

16. Test the home page in Chrome. If necessary, fix any problems and test again.

Add the rest of the HTML

17. In the footer, code <p> tags around the text. Also, use a character entity to add the copyright symbol to the start of the footer.

18. Add <sup> tags around the *th* in the second line of the header (as in 75th).

19. Test the home page in Chrome. If necessary, fix any problems and test again.

Test the links and add a link

20. Click on the link for the first speaker page. This should display a page that gives the speaker's name and says "This page is under construction". If this doesn't work, fix the href attribute in the link and test again. To return to the home page, you can click the browser's Back button.

21. Open the brancaccio.html file that's in the subfolder named speakers.

22. After the <h2> element, add a <p> element.

23. Within the <p> element, add a link that says "Return to home page". This link should refer to the index.html file. To do that, the <a> element should have an href attribute that navigates up one level in the folder structure like this:

```
../index.html
```

24. Test the home page in Chrome. If necessary, fix any problems and test again.

Chapter 4

How to use CSS
to format a web page

After you code the HTML that defines the structure and content of a web page, you're ready to code the CSS that formats the page. So that's what you'll start to learn in this chapter. Then, in chapter 5, you'll learn how to use the CSS box model for spacing, borders, and backgrounds. And in chapter 6, you'll learn how to use CSS for page layout.

An introduction to CSS

Before you can code the CSS for a web page, you need to know how to provide the styles for a web page. Then, you need to know how to apply the styles to specific HTML elements. In addition, for some styles, you need to know how to specify measurements and colors. That's what you'll learn in the next four figures.

How to provide the styles for a web page

The first example in figure 4-1 shows how to use an *external style sheet*, which is a CSS file that stores the styles for one or more web pages. To use an external style sheet, you code a <link> element that uses the href attribute to identify the file that contains the style sheet. In this example, the href attribute uses a relative URL to navigate down one folder to the styles folder and locate a file named main.css.

The second example shows that you can use two or more external style sheets for the same page. Then, the browser applies the styles in sequence from the first style sheet to the last. As a result, if two different styles apply to the same HTML element, the last one overrides the earlier one.

The third example shows two other ways to provide styles. First, you can embed styles in an HTML file by coding a <style> element that contains the styles that apply to that page. These styles are known as *embedded styles*. Second, you can code a style attribute for an HTML element with a value that contains all of the styles that apply to the element. These styles are known as *inline styles*.

Using an external style sheet separates the content from the formatting. This helps organize your code. In addition, it makes it easy to use the same styles for more than one HTML page. This provides a couple advantages over embedded styles or inline styles. First, it reduces code duplication, which makes your code easier to maintain. Second, it makes it easy to apply consistent formatting to all of the pages of a web site.

In general, it's considered a bad practice to use embedded styles or inline styles. However, they are useful in some situations. For instance, the download for this book uses embedded styles for some of the examples because coding the HTML and CSS in the same file makes it easier to see how the HTML relates to the CSS.

An external style sheet

```
<head>
    <title>San Joaquin Valley Town Hall</title>
    <link rel="stylesheet" href="../styles/main.css">
</head>
```

Two external style sheets

```
<head>
    <title>San Joaquin Valley Town Hall</title>
    <link rel="stylesheet" href="../styles/main.css">
    <link rel="stylesheet" href="../styles/speaker.css">
</head>
```

The sequence in which styles are applied

- From the first external style sheet to the last

Two other ways to provide styles (not recommended)

Embed the styles in the head section

```
<head>
    <title>San Joaquin Valley Town Hall</title>
    <style>
        body {
            font-family: Arial, Helvetica, sans-serif;
            font-size: 100%;
        }
        h1 {
            font-size: 250%;
        }
    </style>
</head>
```

Apply an inline style to a single element

```
<h1 style="font-size: 500%; color: red;">Valley Town Hall</h1>
```

The sequence in which styles are applied

- Styles from an external style sheet
- Embedded styles
- Inline styles

Description

- An *external style sheet* is a CSS file that stores one or more styles that can be applied to one or more HTML documents.
- When you specify a relative URL for an external CSS file, the URL is relative to the current file.
- It's generally considered a good practice to use an external style sheet.
- An *embedded style* is coded within the head section of an HTML document.
- An *inline style* is coded within the style attribute of an HTML element.
- It's generally considered a bad practice to use embedded and inline styles.

Figure 4-1   How to provide the styles for a web page

How to use the basic selectors

To apply styles to HTML elements, you code CSS *selectors* that identify the elements that you want to format. To get you started with selectors, figure 4-2 shows how to code and use four types of selectors.

To start, this figure shows the HTML for a <body> element that contains <main> and <footer> elements. Here, each of the <p> elements in the <main> element has a class attribute with a value of blue. This means that the elements are assigned to the same *class*. By contrast, the <p> element in the footer has an id attribute of copyright as well as a class attribute with two values: blue and right. This means that this element is assigned to two classes.

The CSS in this figure shows some style rules that format the HTML. To conserve vertical space, these examples use a single line for the style rules that contain a single declaration. However, for style rules that require more than one declaration, these examples use a separate line for each declaration. This makes the declarations easier to read.

The first style rule uses the *universal selector* (*). As a result, it applies to all HTML elements. This sets the margins for all elements to 0 and the padding for all elements to .25em (more about sizing in a moment).

The next three style rules use *type selectors* to select elements by type. To code a type selector, you just code the name of the element. As a result, the first type selector selects all <h1> elements, the second selects all <p> elements, and the third selects all <footer> elements. These style rules set the font family for the <h1> element, the left margin for all the <p> elements, and the background color for the <footer> element.

The next style rule uses an *id selector* to select an element by its id. To code an id selector, you code a hash character (#) followed by an id value that uniquely identifies an element. As a result, the first id selector selects the <p> element that has an id of copyright. Then, its declaration sets the font size for the paragraph to 80% of the font size for the page.

The next two style rules use *class selectors* to select HTML elements by class. To code a class selector, you code a dot (.) followed by the class name. As a result, the first class selector selects all elements that are assigned to the blue class, which are all three <p> elements. Then, the second class selector selects all elements that are assigned to the right class, which is the paragraph in the footer. The first style rule sets the color of the font to blue and the second style rule aligns the paragraph on the right.

A class selector has two advantages over an id selector. First, several elements can be in the same class. In this figure, for example, all three <p> elements are in the blue class. Second, an element can be a member of multiple classes. In this figure, for example, the third <p> element is in the blue class and the right class. By contrast, the id for an element must be unique. This means an id selector can only format a single element.

In this figure, the margin-left property for the <p> elements overrides the margin setting in the universal selector that applies to all elements. That's why the two paragraphs in the <main> element are indented. You'll learn more about that later in this chapter.

The HTML for the <body> element of a web page

```
<body>
    <main>
        <h1>This Season's Speaker Lineup</h1>
        <p class="blue">October: David Brancaccio</p>
        <p class="blue">November: Andrew Ross Sorkin</p>
    </main>
    <footer>
        <p id="copyright" class="blue right">&copy; Copyright</p>
    </footer>
</body>
```

CSS style rules that select elements

All elements

```
* {
    margin: 0;
    padding: .25em;
}
```

By type

```
h1 { font-family: Arial, sans-serif; }
p { margin-left: 3em; }
footer { background-color: lightgray; }
```

By ID

```
#copyright { font-size: 80%; }
```

By class

```
.blue { color: blue; }
.right { text-align: right; }
```

How it looks in a browser

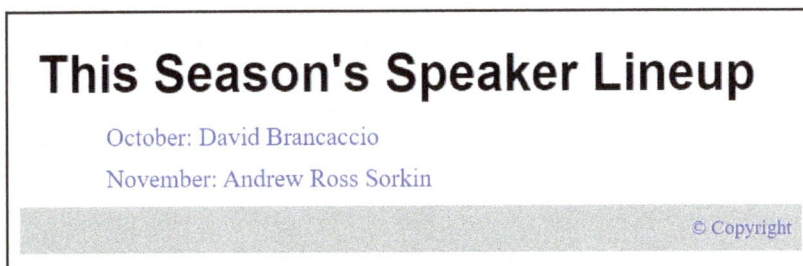

This Season's Speaker Lineup

October: David Brancaccio

November: Andrew Ross Sorkin

© Copyright

Description

- To select all elements, code the asterisk (*) character. This is the *universal selector*.
- To select all elements of a specific type, code an element name. This is a *type selector*.
- To select a single element with a specific id attribute value, code a hash (#) followed by the id value. This is an *id selector*.
- To select all elements with a specific class attribute value, code a dot (.) followed by the class name. This is a *class selector*.

Figure 4-2   How to use the basic selectors

How to specify measurements

Figure 4-3 shows five units of measure that are commonly used with CSS. Here, the first two are *absolute units of measure*. These units of measure are not relative to the size of other HTML elements. The next three units of measure are *relative units of measure*. These units of measure are relative to the size of other HTML elements.

Beyond the units of measure shown in this figure, CSS supports some other absolute units such as inches and picas. However, they aren't used as often as the units shown in this figure.

When you use a relative unit of measure to set font size, the measurement changes if the user changes the browser's font size. For example, if you set the size of a font to 80 percent of the browser's default font size, that element changes if the user changes the default font size in the browser. Because this lets the users adjust the font sizes to their own preferences, it's a best practice to use relative measurements for font sizes.

By contrast, when you use an absolute unit of measure like pixels or points, the measurement won't change even if the user changes the font size in the browser. For example, if you set the width of an element in pixels and the font size in points, the width and font size won't change even if the user changes the font size in the browser.

However, when you use pixels, the size changes if the screen resolution changes. That's because the screen resolution determines the number of pixels that the screen displays. For instance, the pixels on a screen with a resolution of 1280 x 1024 are closer together than the pixels on the same screen with a resolution of 1152 x 864. That means that a measurement of 10 pixels is smaller on the screen with the higher resolution. By contrast, a point is $1/72^{nd}$ of an inch no matter what the screen resolution is.

The examples in this figure show how you can use pixels, percentages, and ems in your CSS. Here, the bottom border for the header is set to 3 pixels. By contrast, the font sizes are set as percentages, and the margins and padding are set as ems.

The first style rule for the <body> element applies a font size of 100% to the body of the web page. This means that the font will be set to 100% of the default font size for the browser, which is usually 16 pixels. Although you can get the same result by omitting the font-size property, this property is typically included to make it clear that the default font size for the browser will be used.

Unlike ems, which are relative to the current font size, rems are relative to the font size of the <html> element, also known as the root element. Then, if you specify a font size for the <html> element, you can base all other font sizes on that size. You can also let the font size for the <html> element default to the browser's font size and base all other font sizes on that size.

Common units of measure

Symbol	Name	Type	Description
px	pixels	absolute	A pixel represents a single dot on a monitor. The number of dots per inch depends on the resolution of the monitor.
pt	points	absolute	A point is 1/72 of an inch.
em	ems	relative	One em is equal to the font size for the current font.
rem	rems	relative	One rem is equal to the font size for the root element.
%	percent	relative	A percent specifies a value relative to the current value.

The HTML for a <body> element

```
<body>
    <header>
        <h1>San Joaquin Valley Town Hall</h1>
    </header>
    <main>
        <p>Welcome to San Joaquin Valley Town Hall. We have some
            fascinating speakers for you this season!</p>
    </main>
</body>
```

CSS that uses relative units of measure and a fixed border

```
body {
    font-size: 100%;
    margin-left: 2em;
    margin-right: 2em;
}
header {
    padding-bottom: .75em;
    border-bottom: 3px solid black;
}
h1 {
    font-size: 200%;
}
```

How it looks in a browser

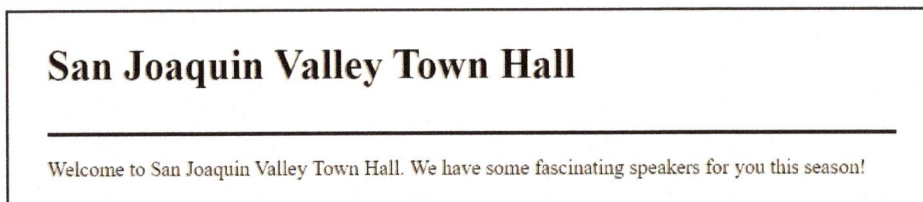

San Joaquin Valley Town Hall

Welcome to San Joaquin Valley Town Hall. We have some fascinating speakers for you this season!

Description

- You use the units of measure to specify a variety of CSS properties, including font-size, line-height, width, height, margin, and padding.
- An *absolute unit of measurement* is not relative to the size of another HTML element.
- A *relative unit of measurement* is relative to the size of another HTML element.

Figure 4-3   How to specify measurements

How to specify colors

Figure 4-4 shows three ways to specify colors. Specifically, these examples show three ways to code the color fuchsia. The easiest way is to specify the color by name. This figure lists the names for 16 basic colors, and it provides a URL for a web page that provides the CSS names for many more colors.

Another way to specify a color is to use an *RGB* (red, green, blue) *value*. One way to do that is to specify the percent of red, green, and blue that make up the color. For instance, the example in this figure specifies 100% red, 0% green, and 100% blue. When you use this method, you can also use any numbers from 0 through 255 instead of percentages. Then, 0 is equivalent to 0% and 255 is equivalent to 100%. This gives you more precision over the resulting colors.

The third way to specify a color is to use *hexadecimal*, or *hex*, *values* for the red, green, and blue values. This technique is used by many web designers. In hex, a value of 000000 is black, and a value of FFFFFF is white. The simple conversion of percentages to hex is 0% is 00, 20% is 33, 40% is 66, 60% is 99, 80% is CC, and 100% is FF. When you use this technique, the entire value must be preceded by the hash character (#).

When you use hex values for colors, you usually get the hex value that you want from a chart or palette that shows all of the colors along with their hex values. For instance, if you're using an IDE, you can often choose a color from a palette to insert the hex value for that color into your code.

To specify a color, you can use the color property to specify the color of an element's text, and you can use the background-color property to set the color of the element's background. In this figure, for example, the CSS sets the color of the <h1> element to blue (#00F or #0000FF), and it sets the background color of the body to a light yellow (#FFFFCC or #FFC).

You should also realize that the color property for an element is *inherited* by any of its *descendants*. If, for example, you set the color property of the <body> element to navy, that color is inherited by the <header> and <main> elements it contains. However, you can override an inherited property by coding a style rule with a different value for that property.

There are two more things to know about hex values. First, hex values aren't case sensitive. In other words, #0000FF and #0000ff are the same color. Second, you can only use a 3-digit hex value when both digits for red, green, and blue are the same. For instance, the hex value #0000FF specifies 00 for red, 00 for green, and FF for blue. As a result, you can shorten this hex value to #00F. By contrast, the hex value #931420 specifies 93 for red, 14 for green, and 20 for blue. As a result, you can't shorten this hex value.

Finally, whenever you're using colors, please keep the visually-impaired in mind by using colors with high contrast. For example, dark text on a light background is easy to read, and black on white is the easiest to read, even for the color blind. In general, if your pages are hard to read when they're displayed or printed in black and white, they aren't good for the visually-impaired.

The web page for CSS colors

www.w3.org/wiki/CSS/Properties/color/keywords

The 16 basic color names

black	silver	white	aqua	gray	fuchsia
red	lime	green	maroon	blue	navy
yellow	olive	purple	teal		

Three ways to specify a color

With a color name

```
color: fuchsia;
```

With an RGB (red-green-blue) value

```
color: rgb(100%, 0%, 100%);     /* using a percentage from 0% to 100% */
color: rgb(255, 0, 255);        /* using numbers from 0 to 255 */
```

With an RGB value that uses hexadecimal numbers

```
color: #FF00FF;                 /* could also be coded #F0F */
```

CSS that uses hexadecimal values to specify colors

```
body {
    background-color: #FFFFCC;     /* could also be coded as #FFC */
}

h1 {
    color: #00F;                   /* could also be coded as #0000FF */
}
```

How it looks in a browser

San Joaquin Valley Town Hall

Welcome to San Joaquin Valley Town Hall. We have some fascinating speakers for you this season!

Accessibility guideline

- Make sure there is enough contrast between the colors to make them accessible for people who are visually impaired or color blind.

Description

- Many graphic designers use *hexadecimal*, or *hex*, *values* to specify an *RGB value* because that lets them choose from over 16 million colors.
- With some text editors and IDEs, you can select a color from a palette of colors and have the color codes inserted into your style rules in either RGB or hex format.
- You can use a 3-digit hex value when both values for red, green, and blue are the same.

Figure 4-4   How to specify colors

How to work with text

Now that you have been introduced to CSS, you're ready to use it to work with text as shown in the next three figures.

How to set the font family and font size

Figure 4-5 shows how to set the font family and font size for text elements. The table at the top of this figure lists five of the most common generic font families. Below that, a screen shows one example of each generic font family.

Most websites use a sans-serif font as the primary font. That's because sans-serif fonts are easier to read in a browser than the other types of fonts. Conversely, serif fonts have long been considered the best for printed text. You can use serif and monospace fonts for special purposes, but it's generally considered a best practice to avoid the use of cursive and fantasy fonts.

When you code the values for the font-family property, you code a list of the fonts that you want to use. For instance, the first property in the first example in this figure lists Arial, Helvetica, and the generic sans-serif as the fonts. Then, the browser uses the first font in the list that is available to it. But if none of the fonts are available, the browser substitutes its default font for the generic font that's coded last in the list.

If the name of a font family contains spaces, like Times New Roman, you need to enclose the name in quotation marks when you code the list. This is shown by the second and third properties in the first example.

To set the font size for a font, you use the font-size property as shown by the second example. For this property, you should use relative measurements so the users can change the font sizes in their browsers. This is also essential for responsive web design.

When you use a relative unit of measurement, it's relative to the parent element. For example, the second property in the second example causes the font to be 150% larger than its parent element. So if the parent element is 16 points, this element is 24 points. Similarly, the third property specifies 1.5 ems so it's also 150% of the parent font.

The third example shows how the font family and font size can be set in the <body> element. Here, the font-family property changes the font to a sans-serif font. Then, the font-size property sets the font size to 100% of the browser's default size. Although this doesn't change the font size, it does make the size relative to the browser's default size, which is usually 16 pixels. Then, if the user changes the browser's font size, that change is reflected in the web page.

Like colors, the font properties that you set for an element are inherited by all of its descendants. For example, if the default font size is 16 pixels and you set the font-size property to 125%, the size of the element and any descendants will be 20 pixels.

Since all elements descend from the <body> element, setting the font family and size for the body element sets the default font for the web page. However, you can override these properties. For instance, the fourth example overrides the font family for the <body> element for all <p> elements.

Five common generic font families

Name	Description
serif	Fonts with tapered, flared, or slab stroke ends.
sans-serif	Fonts with plain stroke ends.
monospace	Fonts that use the same width for each character.
cursive	Fonts with connected, flowing letters that look like handwriting.
fantasy	Fonts with decorative styling.

One example of each generic font family

Times New Roman is a serif font. It is the default for most web browsers.

Arial is a sans-serif font that is widely used, and sans-serif fonts are best for web pages.

Courier New is a monospace font that is used for code examples.

Segoe Script is a cursive font that is not frequently used.

Impact is a fantasy font that is rarely used.

Specify a font family

```
font-family: Arial, Helvetica, sans-serif;
font-family: "Times New Roman", Times, serif;
font-family: "Courier New", Courier, monospace;
```

Specify the font size

```
font-size: 12pt;      /* in points - not recommended for most cases */
font-size: 150%;      /* as a percent of the parent element */
font-size: 1.5em;     /* same as 150% */
```

A font-family rule in the \<body\> element that's inherited by all descendants

```
body {
    font-family: Arial, Helvetica, sans-serif;
    font-size: 100%;
}
```

A font-family rule in a descendant that overrides the inherited font family

```
p { font-family: "Times New Roman", Times, serif; }
```

Description

- The fonts specified for the font-family property are searched in the order listed. If you include a font name that contains spaces, the name must be enclosed in quotes.
- If you specify a generic font last and the browser can't find any of the other fonts in the list, it uses its default font for the generic font that you specified.
- The font properties that you set for an element are inherited by all of its descendants.
- If you use relative font sizes, users can vary the sizes by using their browsers. If you use pixels, the font size varies based on the screen resolution.

Figure 4-5   How to set the font family and font size

How to style and format fonts

The first table in figure 4-6 summarizes the properties that you can use to style a font. Like the font-family and font-size properties, these properties are inherited by their descendants.

The first example in this figure shows how you can use some of these properties. There are several things to note about this example. First, if you want to remove the style from an element, you can set the font-style property to normal. Second, a font-weight value of bold is the same as a value of 700. Third, a font-weight value of lighter is relative to the parent element. And finally, a line-height value of 140% is the same as a value of 1.4em or 1.4.

The second and third examples show how to use a *shorthand property* named font to apply up to six font properties. To illustrate, the third example specifies bold and italic styling with a 14 pixel font that has 5 pixels between the lines (14/19). It also specifies that the Arial font or the default sans-serif font should be used. When you use the font property, the font size and family are required, but the rest of the properties are optional.

How to indent, align, and decorate text

The second table summarizes some properties for formatting text. For instance, you can use the text-indent property to indent the first line of text in a paragraph. You can use the text-align property to center text, align it on the right, or justify it. And you can use the vertical-align property to align text vertically within an element.

However, when you use the text-align property to justify text, the spacing between words is adjusted so the text is aligned on both the left and right sides of the element that contains it. Since that often makes the text more difficult to read, you should avoid using justified text in most cases.

If necessary, you can use the text-decoration property to display a line under, over, or through text. However, you'll rarely need to use this property because (1) you shouldn't underline words that aren't links, (2) you should use borders to put lines over or under text, and (3) you may never need to show that text has been crossed out.

However, you may want to use the text-decoration property to remove any text decoration that has been applied to an element. To do that, you specify a value of none for this property. For example, the text-decoration property of an <a> element is set to underline by default. If that's not what you want, you can set this property to none.

The fourth example in this figure shows how to use the text-align property. Here, the paragraph with an id of copyright is right-aligned.

Properties for styling fonts

Property	Description
font-style	How the font is slanted: normal, italic, and oblique.
font-weight	The boldness of the font: normal, bold, bolder, lighter, or multiples of 100 from 100 through 900, with 400 equivalent to normal. Bolder and lighter are relative to the parent element.
font-variant	Whether small caps should be used: normal and small-caps.
line-height	The amount of vertical space for each line. The excess space is divided equally above and below the font so it sets the spacing between lines.

Specify the properties for styling fonts

```
font-style: italic;
font-style: normal;        /* remove style */

font-weight: 700;
font-weight: bold;         /* same as 700 */
font-weight: lighter;      /* relative to the parent element */

line-height: 140%;
line-height: 1.4em;        /* same as 140% */
line-height: 1.4;          /* same as 140% and 1.4em */
```

The syntax for the shorthand font property

```
font: [style] [weight] [variant] size[/line-height] family;
```

Use the shorthand font property

```
font: italic bold 14px/19px Arial, sans-serif;
```

Properties for indenting, aligning, and decorating text

Property	Description
text-indent	The indentation for the first line of text. This property is inherited.
text-align	The horizontal alignment of text: left, center, right, and justify. Inherited.
vertical-align	The vertical alignment of text: baseline, bottom, middle, top, text-bottom, text-top, sub, and super.
text-decoration	A decoration: underline, overline, line-through, and none.

Right align the paragraph with an id of copyright

```
#copyright {
    text-align: right;
}
```

Description

- You can set the font-style, font-weight, and font-variant properties to a value of normal to remove any formatting that has been applied to these properties.
- The vertical-align property is often used with tables.
- The text-decoration property can be set to none to remove the underlines from links.

Figure 4-6   How to style fonts and text

How to float an image

By default, the block elements in an HTML document flow from the top of the page to the bottom of the page, and inline elements flow from the left side of the block elements that contain them to the right side. However, when you *float* an element, it's taken out of the flow of the document. Because of that, any elements that follow the floated element flow into the space that's left by the floated element.

Figure 4-7 presents the basic skills for floating an image so the text flows around it. To do that, you use the float property to specify whether you want the element floated to the left or to the right.

In this figure, the <header> element contains an <img> element followed by an <h1> element and an <h2> element. Then, the CSS for the <img> element sets the float property to left. This floats the image to the left, and the two headings float to the right of the image, as shown here.

The CSS for the <img> element also sets the margin-right property to 1em, which adds space between the image and the headings. You'll learn more about setting margins in chapter 5.

When you float an element, the subsequent elements flow around it depending on the height of the floated element. For example, if the width of the image in this figure is changed to 40 pixels, the height of the image is reduced, and the second heading flows below the image instead of beside it. So, you may need to adjust the height of a floated element or its top or bottom margins to get the effect you want.

If you want to stop the flow of elements into the space beside a floated element, you can use the clear property. In this figure, the <main> element sets the clear property to left. This keeps the <main> element from flowing into the space next to the image. In other words, this makes sure the <main> element always displays below the floated image, regardless of the height of the floated image. If the clear property wasn't set, the height of the floated image would cause the <main> element to flow next to the image rather than below it.

When you clear the float for an element that's floated to the right, you can set the value for the clear property to either right or both. Similarly, you can set the clear property to either left or both if the element ahead of it is floated to the left.

The properties for floating and clearing elements

Property	Description
float	Determines whether an element is floated. Possible values are left, right, and none. The default is none.
clear	Determines whether an element is cleared from flowing into the space left by a floated element. Possible values are left, right, both, and none. The default is none.

Some HTML that displays an image and three headings

```
<header>
    <img src="images/logo.gif" alt="Town Hall Logo" width="120">
    <h2>San Joaquin Valley Town Hall</h2>
    <h3>Bringing cutting-edge speakers to the valley</h3>
</header>
<main>
    <h1>This season's guest speakers</h1>
</main>
```

The CSS for floating the image

```
img {
    float: left;
    margin-right: 1em;
}
main {
    clear: left;
}
```

How it looks in a browser

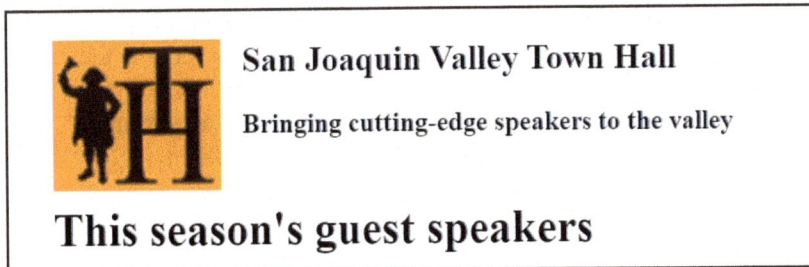

How it looks without the clear property for the <main> element

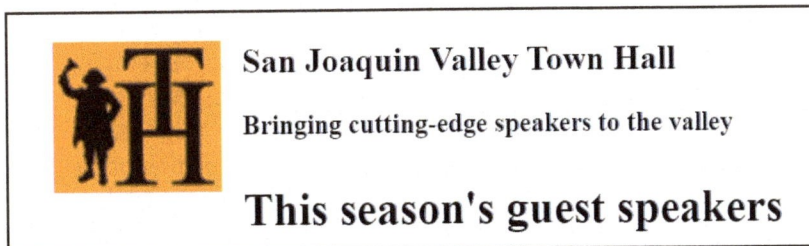

Description

- When you *float* an image to the right or left, the elements that follow flow around it.

Figure 4-7   How to float an image

How to use other selectors

So far, this chapter has shown how to use the universal, type, id, and class selectors. Now, you'll learn how to code other types of selectors. Once you understand how to code these types of selectors, you'll be able to apply CSS formatting to any elements in a web page.

How to code relational selectors

Figure 4-8 starts by presenting the *combinators* you use to code *relational selectors*. With these selectors, terms like *parent*, *child*, *sibling*, and *descendant* are used in the same way they are in a family tree. Child elements are at the first level below a parent element. Sibling elements are at the same level. And descendant elements include all the levels below a parent element.

A *descendant selector* selects all the elements contained within another element. In the HTML in this figure, all elements are descendants of the <main> element. The <li> elements are descendants of the <ul> and <main> elements. And the <a> elements are descendants of the <li>, <ul>, and <main> elements. To code a descendant selector, you code a selector for the parent element followed by a space and a selector for the descendant element.

If you want to select elements only when they're child elements of a parent element, you can code a *child selector*. To do that, you separate the parent and child selector with a greater than (>) sign.

An *adjacent sibling selector* selects an element that's coded at the same level as another element. For instance, in this figure, the <h1>, <p>, <ul>, and <section> elements are all siblings, and the <h1> and <p> elements are adjacent siblings. To code an adjacent sibling selector, you separate the selectors for the sibling elements with a plus (+) sign.

Unlike the adjacent sibling selector, a *general sibling selector* selects any sibling element whether or not the elements are adjacent. To code this type of selector, you separate the selectors for the sibling elements with a tilde (~).

The CSS in this figure shows how to code relational selectors. The first style rule selects all <p> elements that are descendants of the <main> element. This applies a left margin to the <p> element that's a child of the <main> element as well as the <p> elements that are children of the <section> element, since they are all descendants of <main>.

By contrast, the second style rule selects only <p> elements that are children of the <main> element. So, this rule applies the small caps font variant to the first paragraph in the HTML but none of the other paragraphs.

The third style rule selects any <p> element that's an adjacent sibling of an <h2> element. This applies bold to the first paragraph in the <section> element, since it comes right after the <h2> element.

Finally, the fourth style rule selects any <p> element that's a sibling of an <h2> element, adjacent or otherwise. This applies a cursive font to all the paragraphs in the <section> element, not just the first one.

Combinators that declare a relationship between two selectors

Name	Symbol	Relationship
Descendant	Space	Descendants of the specified element.
Child	>	Direct children of the specified element.
Adjacent sibling	+	The first next sibling of the specified element.
General sibling	~	All next siblings of the specified element.

The <main> element for a web page

```
<main>
    <h1>This Season's Town Hall speakers</h1>
    <p>We have some fascinating speakers for you this season!</p>
    <ul>
        <li>January: <a href="speakers/brancaccio.html">
            David Brancaccio</a></li>
        <li>February: <a href="speakers/fitzpatrick.html">
            Robert Fitzpatrick</a></li>
    </ul>
    <section>
        <h2>Post-lecture luncheons</h2>
        <p>Extend the excitement by going to the luncheons.</p>
        <p>A limited number of tickets are available.</p>
        <p>Contact us by phone at (559) 555-1212.</p>
    </section>
</main>
```

The relational selectors

```
main p { margin-left: 2em; }              /* descendent */
main > p { font-variant: small-caps; }    /* child */
h2 + p { font-weight: bold; }             /* adjacent sibling */
h2 ~ p { font-family: cursive; }          /* general sibling */
```

How it looks in a browser

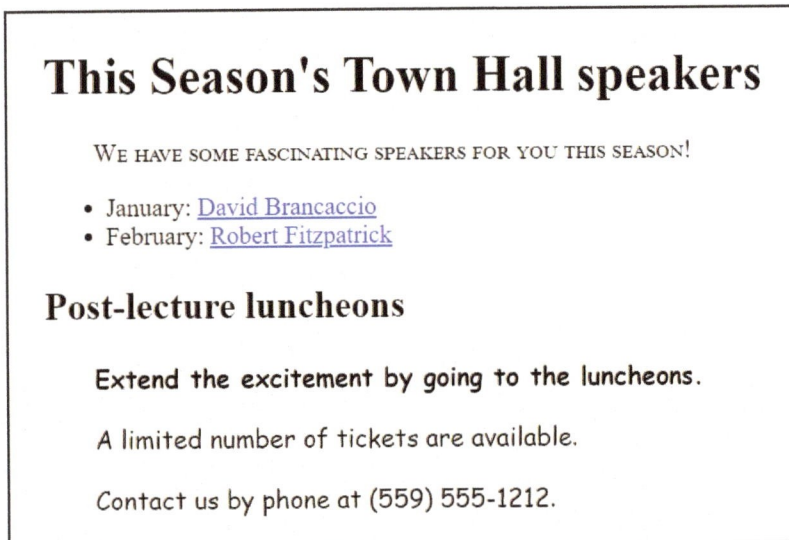

This Season's Town Hall speakers

WE HAVE SOME FASCINATING SPEAKERS FOR YOU THIS SEASON!

- January: David Brancaccio
- February: Robert Fitzpatrick

Post-lecture luncheons

Extend the excitement by going to the luncheons.

A limited number of tickets are available.

Contact us by phone at (559) 555-1212.

Figure 4-8   How to code relational selectors

How to code combination selectors

Figure 4-9 shows how to code combinations of selectors. To select an element type by class name, for example, you code the type name, followed by a dot (.) and the class name. So, the first style rule in this figure selects <ul> elements that have a class of speakers and changes the style of bullet that's used by the list items to square. As a result, the first list in this figure uses square bullets, not round ones.

You can also code multiple selectors for the same style rule. To do that, you separate the selectors with commas. So, the second style rule in this figure selects all <h1> and <h2> elements and applies the same color to them.

The color that the second style rule applies is often used in online tutorials and examples. It originated as a memorial to the young daughter of a designer who worked on CSS standards. If you want to learn more about it, you can search the internet for "rebeccapurple".

The <main> element for a web page

```
<main>
    <h1>This Season's Town Hall speakers</h1>
    <ul class="speakers">
        <li>January: <a href="speakers/brancaccio.html">
            David Brancaccio</a></li>
        <li>February: <a href="speakers/fitzpatrick.html">
            Robert Fitzpatrick</a></li>
    </ul>
    <h2>Post-lecture luncheons</h2>
    <ul>
        <li>Extend the excitement by going to the luncheons.</li>
        <li>A limited number of tickets are available.</li>
    </ul>
</main>
```

Combinations of selectors

A class within an element

```
ul.speakers { list-style-type: square; }
```

Multiple elements

```
h1, h2 { color: rebeccapurple; }
```

How it looks in a browser

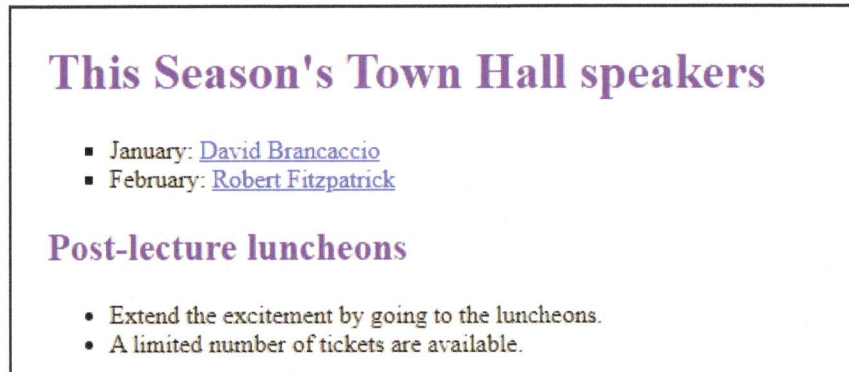

This Season's Town Hall speakers

- January: David Brancaccio
- February: Robert Fitzpatrick

Post-lecture luncheons

- Extend the excitement by going to the luncheons.
- A limited number of tickets are available.

Figure 4-9   How to code combination selectors

How to code attribute selectors

Figure 4-10 shows how to code *attribute selectors* that select elements based on an attribute or attribute value. This is illustrated by the style rules presented in this figure.

The first style rule uses the universal selector to select any element that has an href attribute, while the second style rule selects any <a> element that has an href attribute. These two rules apply regardless of the value of the href attribute. By contrast, the third style rule selects elements based on an attribute and that attribute's value. Specifically, this rule applies to any <input> element that has a type attribute whose value is button.

When you select elements based on an attribute value, it can be useful to select based on part of that value, rather than the entire value. For instance, let's say you have <a> elements assigned to the classes date-passed, date-today, and date-future. If you want to apply styles to all <a> elements assigned to these classes, you could use a combination selector, like this:

```
a.date-passed, a.date-today, a.date-future { }
```

But you could also use an attribute selector like the ones at the end of this figure. Specifically, you could select all <a> elements that have a class attribute whose value starts with "date", like this:

```
a[class^="date"] { }
```

The <main> element for a web page

```
<main>
    <h1>This Season's Town Hall speakers</h1>
    <ul>
        <li>January: <a href="speakers/brancaccio.html">
            David Brancaccio</a></li>
        <li>February: <a href="speakers/fitzpatrick.html">
            Robert Fitzpatrick</a></li>
    </ul>
    <input type="button" value="Contact us">
</main>
```

Attribute selectors

All elements with href attributes

```
*[href] { font-size: 125%; }
```

All <a> elements with href attributes

```
a[href] { font-family: Arial, sans-serif; }
```

All <input> elements with a type attribute that has a value of button

```
input[type="button"] { font-size: 1.5em; }
```

How it looks in a browser

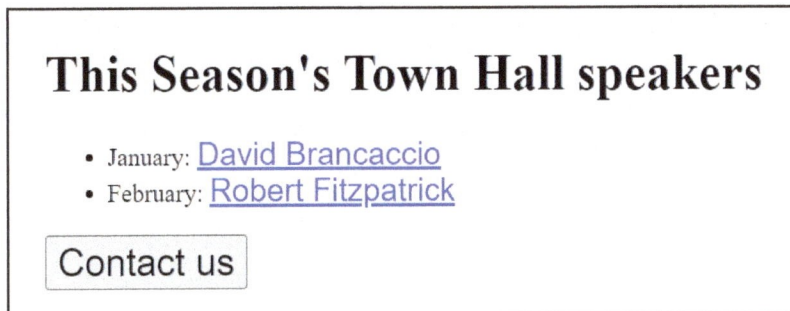

This Season's Town Hall speakers

- January: David Brancaccio
- February: Robert Fitzpatrick

Contact us

Attribute selectors that match part of the value of an attribute

```
[class^="date"]    /* attribute value starts with "date" */
[class$="date"]    /* attribute value ends with "date" */
[class*="date"]    /* attribute value contains "date" */
```

Figure 4-10   How to code attribute selectors

How to code pseudo-class selectors

Figure 4-11 shows how to code *pseudo-class selectors*, which are predefined classes that apply to specific conditions. To do that, you use the pseudo-classes in the table in this figure. For example, the :link pseudo-class refers to a link that hasn't been visited, and the :hover pseudo-class refers to the element that has the mouse hovering over it.

You can also use the :first-child, :last-child, and :only-child pseudo-classes to refer to specific relationships. For example, :first-child refers to an element that's the first of its siblings, and :only-child refers to an element that has no siblings.

The first pseudo-class selector shown in this figure causes all links that haven't been visited to be displayed in green. By default, most browsers display these links in blue.

The second selector is a combination selector that applies to any link that has the mouse hovering over it or the focus on it. As the accessibility guideline in this figure indicates, you should always code the :hover and :focus pseudo-classes for links in combination. That way, the formatting is the same whether the user hovers the mouse over a link or presses Tab to move the focus to it.

The third pseudo-class selector causes the text in the first <li> element in the <ul> element to be bold. However, it doesn't cause any other sibling <li> elements to be bold.

In addition to the pseudo-classes presented in this figure, there are others that can be useful. For example, the :root pseudo-class is presented later in this chapter, and some other pseudo-classes are presented later in this book.

Common CSS pseudo-classes

Pseudo-class	Description
:link	A link that hasn't been visited. By default, blue, underlined text.
:visited	A link that has been visited. By default, purple, underlined text.
:hover	An element with the mouse hovering over it.
:focus	An element like a link or form control that has the focus.
:first-child	An element that's the first of its siblings.
:last-child	An element that's the last of its siblings.
:only-child	An element that has no siblings.

The <main> element for a web page

```html
<main>
    <p>Welcome to San Joaquin Valley Town Hall.</p>
    <p>We have some fascinating speakers for you this season!</p>
    <ul>
        <li><a href="brancaccio.html">David Brancaccio</a></li>
        <li><a href="sorkin.html">Andrew Ross Sorkin</a></li>
        <li><a href="chua.html">Amy Chua</a></li>
    </ul>
</main>
```

The pseudo-class selectors

```css
a:link { color: green; }
a:hover, a:focus { color: fuchsia; }
li:first-child { font-weight: bold; }
```

How it looks in a browser

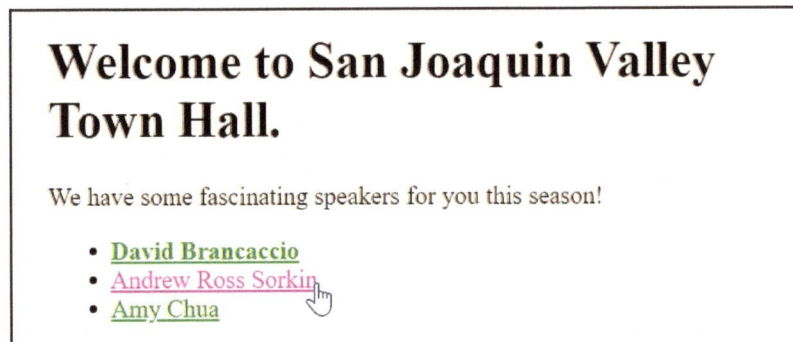

Accessibility guideline

- Apply the same formatting to the :hover and :focus pseudo-classes for an element. That way, those who can't use a mouse will have the same experience as those who can.

Description

- *Pseudo-classes* are predefined classes that apply to specific conditions.
- For a full list of pseudo-classes, see the documentation at this URL:
  https://developer.mozilla.org/en-US/docs/Web/CSS/Pseudo-classes

Figure 4-11   How to code pseudo-class selectors

How to code pseudo-element selectors

Figure 4-12 shows how to code *pseudo-elements selectors*. These selectors use the pseudo-elements shown in this figure to select part of the content of an element or to add new content before or after the element. Note that the double colons (::) distinguish pseudo-elements from pseudo-classes.

For instance, you can use the ::first-line selector to select the first line of content in an element that contains more than one line. And you can use ::before or ::after selectors to insert content before or after the content of an element. To do that, you set the content property with the content you want to insert.

The first pseudo-element selector shown in this figure adds bold to the first line in the <main> element. This applies to the first line displayed in the browser, not the first line in the HTML file.

The second pseudo-element selector increases the font size of the first letter in every <p> element. This creates an effect that's sometimes used in the first paragraph in a block of text that contains many paragraphs.

The third selector combines a pseudo-class and a pseudo-element to add an image after the content of the last <li> element. This works because the pseudo-element is coded after the pseudo-class, and there's only one pseudo-element in the selector. By the way, adding images like this is a common use of the ::before and ::after pseudo-elements.

In addition to the pseudo-elements presented here, there are others that can be useful. For a complete list, you can visit the URL shown in this figure.

Common CSS pseudo-elements

Pseudo-element	Description
::first-letter	Selects the first letter in the content of an element. Only works with block level elements.
::first-line	Selects the first line of text in the content of an element. Only works with block level elements.
::before	Inserts content before the content of the selected element.
::after	Inserts content after the content of the selected element.
::selection	Selects the portion of an element that's selected by the user.

The <main> element for a web page

```
<main>
    <p>Welcome to San Joaquin Valley Town Hall.</p>
    <p>We have some fascinating speakers for you this season!</p>
    <ul>
        <li><a href="brancaccio.html">David Brancaccio</a></li>
        <li><a href="sorkin.html">Andrew Ross Sorkin</a></li>
        <li><a href="chua.html">Amy Chua</a></li>
    </ul>
</main>
```

The pseudo-element selectors

```
main::first-line { font-weight: bold; }
p::first-letter { font-size: 150%; }
li:last-child::after { content: url(images/new.png); }
```

How it looks in a browser

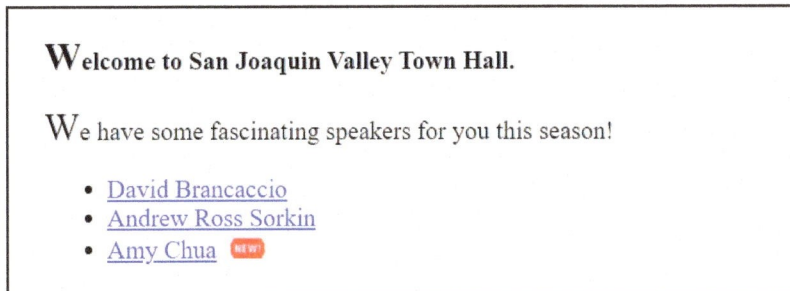

Welcome to San Joaquin Valley Town Hall.

We have some fascinating speakers for you this season!

- David Brancaccio
- Andrew Ross Sorkin
- Amy Chua NEW!

Description

- *Pseudo-elements* let you select a specified part of an element or insert content before or after an element.

- Pseudo-elements use double colons (::), while pseudo-classes use single colons (:).

- You can only use one pseudo-element per selector, and it must come after any pseudo-classes or relational selectors.

- For a full list of pseudo-elements, see the documentation at this URL:

  https://developer.mozilla.org/en-US/docs/Web/CSS/Pseudo-elements

Figure 4-12   How to code pseudo-element selectors

How the cascade rules work

The term *Cascading Style Sheets* (*CSS*) refers to the fact that more than one style sheet can be applied to a single web page. But that means that more than one style rule may be applied to the same element. And you've already seen that more than one style in the same style sheet can be applied to the same element.

So what happens when the styles that are applied to an element conflict? When that happens, CSS applies the *specificity* rules presented in figure 4-13 to determine which style rule takes effect. In other words, it applies the style rule with the highest specificity. For example, the #highlight selector is more specific than the .highlight selector so its style rules are applied. And if the specificity is the same for two or more styles, the style rule that appears later in the code is applied.

This works the same if you provide two or more style sheets for an HTML document. Then, the specificity rules still apply, but if the specificity is the same, the style rules in each style sheet override the style rules in the preceding style sheets. The same idea applies if you accidentally code two style rules for the same element in a single style sheet. The one that's coded last takes precedence if the specificity is the same.

It's possible to identify one or more of your declarations as important so they take precedence over other declarations. To do that, you code !important as part of the declaration, as shown by the example in this figure. This is known as an *important declaration*. However, important declarations are considered a poor practice because they can make it harder to understand when the style rules of your CSS are applied. Although you may see important declarations in other CSS, it's a good practice to avoid using them in your own CSS.

To understand the cascade rules for CSS, you also need to know that users can create *user style sheets* for their browsers that provide default style rules. For example, users with poor vision can create user style sheets that provide for large font sizes. In addition, some of the preferences a user can choose in the settings of the browser work like a user style sheet. For instance, most browsers provide a checkbox for users to say whether the site is able to override their preferences. This works like an important declaration in a user style sheet. In addition, the web browsers themselves provide some default styles, and some of these styles may have important declarations. That's why you need to be aware of how user style sheets and web browser styles can affect your web pages.

With that as background, this figure lists the levels of the *cascade order* from highest to lowest. This shows that the important declarations in the web browser override the important declarations in a user style sheet, which override the important declarations in the style sheets for a web page. However, the normal declarations in the style sheets for a web page override the normal declarations in a user style sheet. And all of these override the default declarations in the web browser. Within each of these levels of the cascade order, the specificity rules apply.

What happens when more than one style is applied to an element

- The style rule with the highest specificity is applied.
- If the specificity is the same for two or more style rules in a group, the style rule that appears later in the code is applied.

How to determine the specificity of a selector

- An id is the most specific.
- A class, attribute selector, or pseudo-class selector is less specific.
- An element or pseudo-element selector is least specific.

How to identify a declaration as important (not recommended)

```
.highlight {
    font-weight: bold !important;
}
```

The cascade order for applying style rules

Apply the style rule from the first group in which it's found:

- Important declarations in the browser
- Important declarations in a user style sheet
- Important declarations in the style sheet for a web page
- Normal declarations in the style sheet for a web page
- Normal declarations in a user style sheet
- Default declarations in the browser

Description

- When two or more style rules are applied to an HTML element, CSS uses the *cascade order* and *specificity* rules to determine which style rule to apply.
- It's possible to code an *important declaration* that takes precedence over other declarations. In general, this is considered a bad practice.
- A user can create a *user style sheet* that provides a default set of style rules for web pages. Users with poor vision often do this so the type for a page is displayed in a large font.
- Some of the preferences a user can choose in the settings of the browser work like a user style sheet.
- If you want to create or remove a user style sheet for a browser, you can search the internet for the procedures that the browser requires.

Figure 4-13   How the cascade rules work

How to organize your CSS

Now that you've learned some of the basics for working with CSS, you're ready to learn two techniques that can help you organize your code. These techniques help make your CSS easier to understand and maintain.

How to create and use custom properties

Figure 4-14 shows how to use a CSS feature called *custom properties*, or *CSS variables*. This feature lets you define a property value, or variable, that you can use throughout a CSS file.

This figure starts with an example that defines a custom property named primary-color that has a hex value of #0000ff. To do that, you code two dashes, the name of the custom property, a colon, the value of the custom property, and a semicolon.

Once the custom property is defined, other style rules can refer to it as primary-color instead of coding the hex value for the color. To do that, you code the var keyword followed by parentheses. Within the parentheses, you code two dashes and the name of the variable. In the first example in this figure, both the <h1> and <a> elements use the custom property named primary-color to set their color properties.

When a property like a color is used in many places in a CSS file, a custom property can prevent coding errors and make the code easier to understand. For example, it's easy to make an entry error when entering the same hex value in multiple places. In addition, entering the same hex value in multiple places makes it harder to change that color later. As a result, if you find yourself entering a hex value in multiple places, you should consider using a custom property instead. Then, if you want to change the color later, you only have to change the hex value in one place, not in multiple places.

When declaring a custom property, it's important to specify an appropriate selector. That's because the selector determines where the property can be used. In the first example, the selector is the :root pseudo-class. This pseudo-class refers to the <html> element, but it has a higher specificity than that element. That means that you can apply this custom property anywhere in the HTML document. And that's how you'll typically use custom properties.

However, it's also possible to create a custom property with a more specific selector as shown in the second example. Here, the selector is for a <section> element. As a result, the custom property can only be applied to HTML elements within a <section> element. This can be helpful if the property value is used several times within the HTML element, but you don't want the property value to be used outside that element.

A custom property that's available to the entire web page

Declare the custom property

```
:root {
    --primary-color: #0000ff;
}
```

Use the custom property

```
h1 {
    color: var(--primary-color);
}
a:link {
    color: var(--primary-color);
}
```

A custom property that's only available within a <section> element

Declare the custom property

```
section {
    --section-color: #f2972e;
}
```

Use the custom property

```
section h1 {
    color: var(--section-color);
}
section a:link {
    color: var(--section-color);
}
```

Description

- *Custom properties*, also called *CSS variables*, are properties that can be used throughout an entire document or within specific parts of a document.

- To create a custom property, you code it within a style rule that has a selector just like any other style rule. Then, that selector determines where the custom property can be used.

- To create a custom property that can be used throughout a document, you can declare it in the :root pseudo-class. This is the same as coding it in the <html> element except that the :root pseudo-class has a higher specificity.

- When a property like a color is used in many places in a CSS file, a custom property can help prevent coding errors and make the code easier to understand.

Figure 4-14   How to create and use custom properties

How to nest CSS

It's common to apply styles to a structural HTML element like a <header>, <footer>, or <aside> and also apply styles to elements within that element. For instance, the first example in figure 4-15 has a style rule that aligns the text in a <header> element. Then, it has three relational selectors that float images and format headings that are within that <header> element.

In the first example, the header selector is coded four times, which clutters the code. In addition, because all the selectors are coded on the same level, it can take a moment to determine which are the relational selectors.

Fortunately, you can address both of these issues by *nesting* the relational selectors. To do that, you start by coding a parent selector. Then, within the braces of the parent, you code the *nesting selector* (&), which refers to the parent selector, followed by the relational selector.

The second example rewrites the first example to nest the relational selectors. It starts by selecting the <header> element and applying a style to the header. Then, it uses the nesting selector followed by the img selector to select all images in the header. After that, it uses the nesting selector and the h2 and h3 selectors to select the headings in the header.

In the second example, the header selector is only coded once. This makes the code cleaner. In addition, the nested relational selectors are indented, which makes the code easier to read.

The third example shows that you can also nest pseudo-classes and pseudo-elements. Here, the nested CSS for a <main> element adds bold to the first line, enlarges the first letter, and formats the last <li> element with small caps.

When nesting first became available in CSS, the nesting selector was required. Now, it's optional in some cases but required in others. For instance, in the second example, you can remove all three nesting selectors and the CSS works fine in Chrome. By contrast, if you remove the three nesting selectors in the third example, the first two style rules don't work properly in Chrome.

Because of this, we recommend that you always use the nesting selector. That way, you don't have to worry about whether it's required. In addition, it makes your CSS more consistent and provides backward compatibility with older browsers that require the nesting selector.

A selector and some relational selectors

```
header {
    text-align: center;
}
header img {
    float: left;
}
header h2 {
    font-size: 240%;
}
header h3 {
    font-size: 130%;
    font-style: italic;
}
```

A selector with nested relational selectors

```
header {
    text-align: center;
    & img {
        float: left;
    }
    & h2 {
        font-size: 240%;
    }
    & h3 {
        font-size: 130%;
        font-style: italic;
    }
}
```

Nested pseudo-classes and pseudo-elements

```
main {
    &::first-line {
        font-weight: bold;
    }
    &::first-letter {
        font-size: 150%;
    }
    & li:last-child {
        font-variant: small-caps;
    }
}
```

Description

- Nesting can make your CSS easier to read by grouping related styles and eliminating repeating selectors.
- The *nesting selector* (&) refers to the parent selector. The nesting selector isn't required in all cases, but we recommend using it.

Figure 4-15   How to nest CSS

A formatted web page

So far, this chapter has shown how to code selectors, format text, and organize your CSS. Now, it shows how to put these skills together to provide the formatting for a web page.

The page displayed in a browser

Figure 4-16 presents a web page that uses an external style sheet. This web page uses a sans-serif font, applies colors to some of the headings and text, centers the headings in the header, adds line height to the items in the unordered list, applies italics and bold to portions of text, and right-aligns the footer.

Overall, this formatting improves the graphic design of the page. However, there's still plenty of room for improvement. For instance, there should probably be less space after "San Joaquin Valley Town Hall" and more space at the left and right sides of the page. To make these improvements, you need to know how to work with the CSS box model as shown in the next chapter.

A formatted web page displayed in a browser

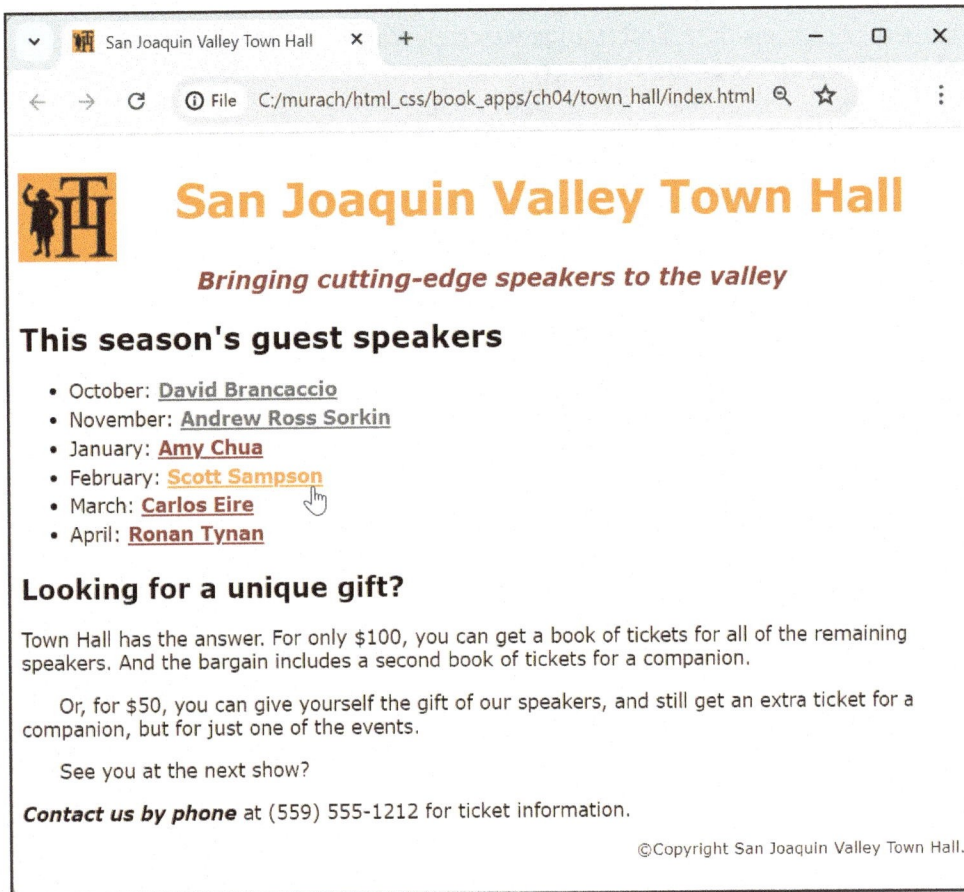

San Joaquin Valley Town Hall

Bringing cutting-edge speakers to the valley

This season's guest speakers

- October: David Brancaccio
- November: Andrew Ross Sorkin
- January: Amy Chua
- February: Scott Sampson
- March: Carlos Eire
- April: Ronan Tynan

Looking for a unique gift?

Town Hall has the answer. For only $100, you can get a book of tickets for all of the remaining speakers. And the bargain includes a second book of tickets for a companion.

Or, for $50, you can give yourself the gift of our speakers, and still get an extra ticket for a companion, but for just one of the events.

See you at the next show?

Contact us by phone at (559) 555-1212 for ticket information.

©Copyright San Joaquin Valley Town Hall.

Description

- All text uses a sans-serif font.
- The first two headings use two different colors.
- Some of the text uses italics and bold.
- The links in the list use colors.
  - The first two links are gray because the lecture dates have passed.
  - The remaining links use the same color as the second heading. However, they use the same color as the first heading when they have focus or are hovered over.
- The copyright information is right-aligned.

What's wrong with the graphic design of this web page

- The spacing above and below the block elements and around the body of the page could be improved. In the next chapter, you'll learn how to do that.

Figure 4-16   The page displayed in a browser

The HTML

Figure 4-17 presents the HTML for the web page. In the head section, the second <link> element refers to an external style sheet that's stored in the styles folder.

In the body section, the <header> and <main> elements define the two main sections of the body. Within the <main> element, the <a> elements in the first and second <li> elements have class attributes that assign them to the date_passed class to indicate that the dates for those events have already passed.

After the <nav> element, the HTML includes a second <h2> element and four <p> elements. And the second and third <p> elements have their class attributes set to indent. Within the third <p> element, this HTML uses the <em> element to emphasize the text that says, "Contact us by phone". By default, this causes the browser to italicize the text. However, the CSS that's presented in the next figure also adds bold to this element.

At the bottom of the HTML, the <footer> element includes a <p> element that has its class attribute set to copyright. This makes it possible for the CSS that's presented in the next figure to format this <p> element differently than the other <p> elements on this page.

The HTML

```
<!DOCTYPE HTML>

<html lang="en">
<head>
    <title>San Joaquin Valley Town Hall</title>
    <meta charset="utf-8">
    <link rel="icon" href="images/favicon.ico">
    <link rel="stylesheet" href="styles/main.css">
</head>

<body>
    <header>
        <img src="images/logo.gif" alt="Town Hall Logo" width="80">
        <h2>San Joaquin Valley Town Hall</h2>
        <h3>Bringing cutting-edge speakers to the valley</h3>
    </header>
    <main>
        <h1>This season's guest speakers</h1>
        <nav>
            <ul>
                <li>October: <a class="date_passed"
                    href="speakers/brancaccio.html">David Brancaccio</a></li>
                <li>November: <a class="date_passed"
                    href="speakers/sorkin.html">Andrew Ross Sorkin</a></li>
                <li>January: <a href="speakers/chua.html">
                    Amy Chua</a></li>
                <li>February: <a href="speakers/sampson.html">
                    Scott Sampson</a></li>
                <li>March: <a href="speakers/eire.html">
                    Carlos Eire</a></li>
                <li>April: <a href="speakers/tynan.html">
                    Ronan Tynan</a></li>
            </ul>
        </nav>

        <h2>Looking for a unique gift?</h2>
        <p>Town Hall has the answer. For only $100, you can get a book of
            tickets for all of the remaining speakers. And the bargain includes
            a second book of tickets for a companion.</p>
        <p class="indent">Or, for $50, you can give yourself the gift of our
            speakers, and still get an extra ticket for a companion, but for
            just one of the events.</p>
        <p class="indent">See you at the next show?</p>
        <p><em>Contact us by phone</em> at (559) 555-1212 for ticket
            information.</p>
    </main>
    <footer>
        <p class="copyright">Copyright San Joaquin Valley Town Hall.</p>
    </footer>
</body>
</html>
```

Figure 4-17   The HTML for the web page

The CSS

Figure 4-18 shows the CSS for the web page. To start, notice how the code in this file is structured. It begins by declaring two custom properties, or variables, that contain the hex values for the light and dark accent colors that the web page uses. Since the selector for these variables is the :root pseudo-class, all of the subsequent style rules can use them.

After the variables for the colors, the CSS defines the style rules for specific elements. In effect, this sets the default formatting for these elements.

After the default styles, the CSS defines style rules that are specific to the <header>, <main>, and <footer> elements. The declarations in these style rules override the default ones for elements because they're more specific. For instance, the font size for the footer p.copyright selector overrides the font size for the body selector.

In the style rule for the <body> element, the two properties specify the font family and font size. Because these properties are inherited, they become the defaults for the document. Also, because the font size is specified as 100 percent, the actual font size is determined by the default font size for the browser. The font sizes for the headings are also specified as percents, so they are based on the size that's calculated for the <body> element.

In the style rules for the <a> element, the color for the :link and :visited pseudo-classes is set to the variable for the dark color. Similarly, the color for the :hover and :focus pseudo-classes is set to the variable for the light color.

In the style rule for the unordered list, the line-height property is specified to increase the spacing between the items. Also, a font weight of bold is specified for the <em> element. However, this element is also italicized by default. As a result, the <em> element is displayed in bold and italic.

The style rules for the <header> and <main> elements nest their respective relational selectors. This groups the styles for these elements and removes duplicate selectors, which makes the CSS easier to read.

In the style rule for the <header> element, the image is floated to the left so the elements that follow it flow to the right. In addition, the h2 selector overrides the font size set in the h2 selector, while the h3 selector sets the font style to italic. Both of these selectors use the variables to set the colors of the elements.

In the style rule for the <main> element, the clear property stops the flow of the text around the image that is floated left. In addition, <p> elements assigned to the indent class are indented 2 ems, and the color of <a> elements assigned to the date_passed class is set to gray.

In the style rule for the <footer> element, <p> elements assigned to the copyright class are aligned to the right and have a copyright symbol inserted before the content of the paragraph. In addition, this selector overrides the font size set in the p selector.

The CSS

```
/* variables for the colors */
:root {
    --light: #f2972e;
    --dark: #931420;
}

/* default styles */
body {
    font-family: Verdana, Arial, Helvetica, sans-serif;
    font-size: 100%;
}
h1 { font-size: 150%; }
h2 { font-size: 135%; }
h3 { font-size: 120%; }
p, li { font-size: 95%; }
ul { line-height: 1.5; }
a, em { font-weight: bold; }
a:link, a:visited { color: var(--dark); }
a:hover, a:focus { color: var(--light); }

/* the styles for the header */
header {
    text-align: center;
    & img {
        float: left;
    }
    & h2 {
        font-size: 225%;
        color: var(--light);
    }
    & h3 {
        font-style: italic;
        color: var(--dark);
    }
}

/* the styles for the main content */
main {
    clear: left;
    & p.indent {
        text-indent: 2em;
    }
    & a.date_passed {
        color: gray;
    }
}

/* the styles for the footer */
footer p.copyright {
    font-size: 80%;
    text-align: right;
    &::before {
        content: "\00A9"; /* copyright symbol */
    }
}
```

Figure 4-18   The CSS for the web page

How to use Developer Tools to inspect styles

In chapter 2, you learned about the Developer Tools that are provided by a browser. If you have problems with cascading styles when you test a web page, you can use the Developer Tools to determine what's happening, as shown in figure 4-19. In most browsers, you can access these tools by pressing the F12 key. If that doesn't work, you can right-click (Windows) or Ctrl-click (macOS) on a page and select Inspect or Inspect Element.

This figure shows a Chrome browser displaying the Developer Tools in a panel below the web page from figure 4-16. Here, the left pane of the tools panel displays all of the HTML elements of the page. Then, you can expand or collapse a group of elements by clicking on the triangle symbol before the element. In this figure, the <body> and <main> elements are expanded.

To inspect the styles for an element, you click on the element in the Elements pane. In this figure, the user has clicked on the <h1> element.

Another way to inspect styles is to click on the inspect icon in the upper left of the toolbar for the Developer Tools (the icon with the arrow pointing to a square). Then, when you click an element on the page, it's selected in the Elements pane.

When an element is selected, its styles display in the Styles pane of the Developer Tools panel. In this figure, the Styles pane displays on the right side of the panel. However, some browsers may display the Styles pane below the Elements pane. Either way, the Styles pane shows all of the styles that apply to the element, from the first styles at the bottom of the pane to the last styles at the top.

In this example, the bottom of the pane shows the styles that the element inherits from the <body> element. Here, the pane crosses out the font-size style to show that it is overridden. The next group up shows the styles that the user agent style sheet applies, which is the style sheet provided by the browser. Here again, the font-size style is overridden. Above that, you can see that the style rule for the <h1> element in the <main> element in the main.css style sheet overrides the lower values with a font size of 170%.

If you experiment with this feature of the Developer Tools, you'll discover how useful it can be. Just select an element to see how the styles are applied.

Chrome's Developer Tools displaying the styles for an element

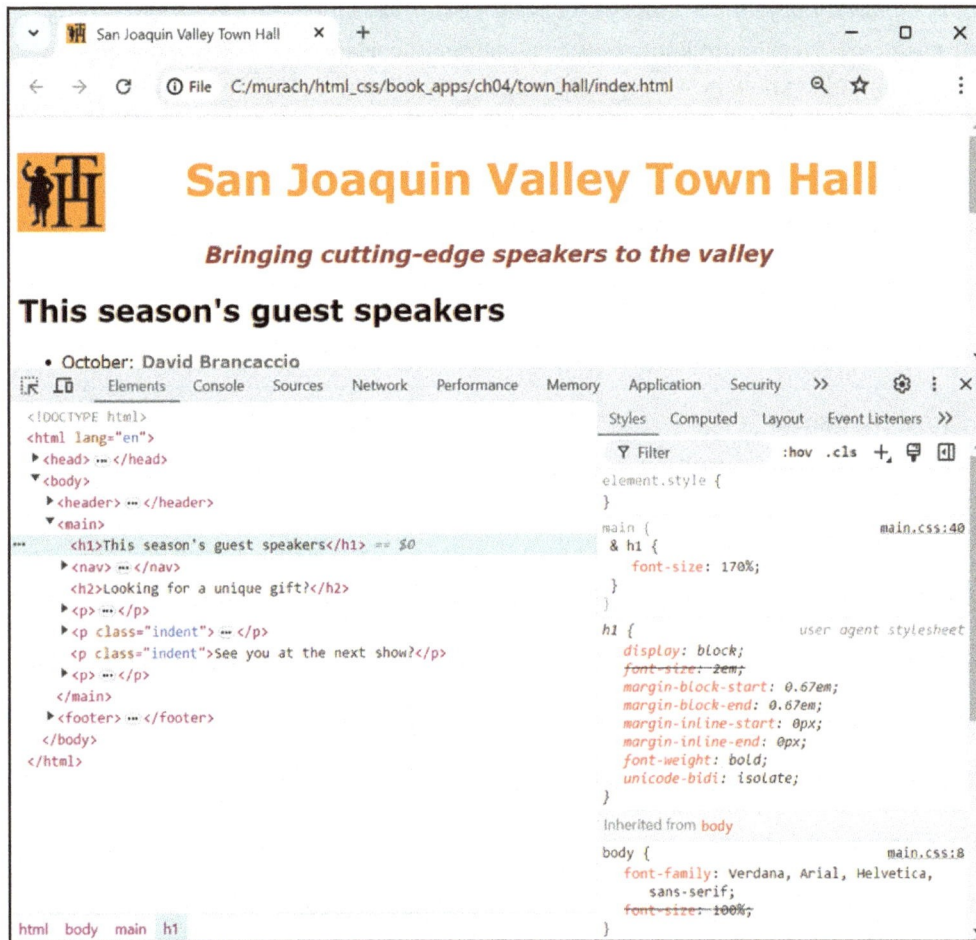

How to use Chrome's Developer Tools

- To display the panel for the tools, press F12. If that doesn't work, right-click (Windows) or Ctrl-click (macOS) on the web page and select Inspect or Inspect Element.

- To inspect the styles that have been applied to an element, click on the element in the Elements pane of the Developer Tools panel. Or, click on the inspect icon at the left of the toolbar for this panel, and then click on an element in the web page.

- The styles that apply to the selected element display in the Styles pane of the Developer Tools panel.

Description

- To help debug your web pages, you can use a browser's Developer Tools to inspect the styles that apply to an element, including how the styles in one style sheet override the styles in another style sheet.

Figure 4-19   How to use Developer Tools to inspect styles

Perspective

At this point, you should know how to code many variations of CSS selectors. You should also know how to specify measurements and colors and how to apply the CSS properties for formatting text. That gets you off to a good start with CSS, but there's still much more to learn.

In the next chapter, you'll learn how to use the CSS box model to set the margins, padding, and borders for the elements in your pages. That way, you can improve the graphic design for your web pages. Then, in chapter 6, you'll learn how to use CSS for page layout.

Terms

external style sheet	relational selector
embedded styles	parent
inline styles	child
selector	sibling
class	descendant selector
universal selector	child selector
type selector	adjacent sibling selector
id selector	general sibling selector
class selector	attribute selector
absolute unit of measure	pseudo-class selector
relative unit of measure	pseudo-element selector
RGB value	Cascading Style Sheets (CSS)
hexadecimal (hex) value	cascade order
inherit a property	important declaration
descendant	user style sheet
font family	custom property
shorthand property	CSS variable
float	nesting selector
combinator	

Exercise 4-1   Format the Town Hall home page

In this exercise, you'll format the Town Hall home page by using the skills that you've learned in this chapter. When you're through, the page should look like this.

Open the HTML file and update the head section

1. Use your text editor to open this HTML file:

    ```
    exercises\ch04\town_hall\index.html
    ```

2. Add the following link element to the end of the head section:

    ```
    <link rel="stylesheet" href="styles/main.css">
    ```

3. Run the web page in Chrome to view it. Note that no styles are applied yet.

Open the CSS file and add the default styles

4. Use your text editor to open this CSS file:

    `exercises\ch04\town_hall\styles\main.css`

5. Add a style rule for the body that sets the font family to Arial, Helvetica, sans-serif and sets the font size to 100%.

6. Add a style rule for any link that has the focus or has the mouse hovering over it that italicizes the text and also sets the color to the hex value #800000.

7. Add a style rule for the <h1> element that sets the font size to 150%.

8. Add a style rule for the <h2> element that sets the font size to 130% and sets the color to #800000.

9. Add a style rule for the <h3> element that sets the font size to 105%.

10. Refresh the page to view your changes.

Format the header

11. Add a style rule for the <header> element.

12. Add a nested style rule for the <h2> element that sets the font size to 170% and indents the heading 30 pixels.

13. Add a nested style rule for the <h3> element that sets the font size to 130%, sets the font style to italic, and indents the heading 30 pixels.

14. Add a nested style rule for the <img> element that floats the image to the left.

15. Refresh the page to view your changes.

16. In the HTML file, find the <img> element in the header and change the value of the height attribute to 120.

17. Refresh the page to view your changes. Note that the Our Mission heading is now next to the image instead of below it.

Format the main content

18. Add a style rule for the <main> element and set the clear attribute to left.

19. Refresh the page to view your changes. Note that the Our Mission heading is now under the image again.

20. In the HTML, change the height attribute of the image back to 80.

Format the footer

21. Add a style rule for the <footer> element and a nested style rule for the <p> element that centers the element text.

22. Refresh the page to view your changes.

Use the Developer Tools to review the styles for the page

23. In Chrome, display the Developer Tools. Next, expand the <main> element in the Elements pane and click on one of the <h2> elements.

24. Review the styles for the <h2> element in the Styles pane, and notice how the font-size style for the <body> element in the main style sheet and the <h2> element in the user agent style sheet are overridden by the font-size style for the default <h2> element in the main style sheet.

25. Click the icon in the Developer Tools toolbar that has a square with an arrow pointing to it, and then click on the <h2> element in the header to make sure that it's selected in the Elements pane.

26. Review the styles for this <h2> element to see that they're similar to the styles for the default <h2> element. However, the font size for this element is larger and it has a text indent.

27. When you're done with the Developer Tools, close the panel by clicking the "X" in the upper right corner.

Use a CSS class as a selector

28. In the HTML, find the <h3> element in the header and enclose the text "75" in an <em> element. Then, add a class attribute with a value of large to the <em> element.

29. In the CSS, create a style rule that sets the font size to 125% for the class named large.

30. Refresh the page to view your changes.

Use a CSS variable

31. Create a custom property named accent-color and assign it a hex value of #800000. The selector should be for the root element.

32. Replace every occurrence of the hex value #800000 with this custom property.

33. Refresh the page. It should look the same as before.

34. Change the value of the custom property to the hex value #4d0080.

35. Refresh the page to view your changes. When you're done, change the custom property back to its previous value.

Chapter 5

How to use the CSS box model

The last chapter showed how to use some basic CSS properties to format the text on a web page. Now, this chapter shows how to control the spacing between elements and how to display borders and backgrounds. To do that, you need to learn how to use the CSS box model.

An introduction to the box model

When a browser displays a web page, it places each HTML block element in a box. That makes it easy to control the spacing, borders, and other formatting for elements like headers, sections, footers, headings, and paragraphs. Some inline elements like images are placed in a box as well. To work with boxes, you use the CSS *box model*.

How the box model works

Figure 5-1 presents a diagram that shows how the box model works. By default, the box for a block element is as wide as the block that contains it and as tall as it needs to be based on its content. However, you can explicitly specify the size of the content area for a block element by using the height and width properties. You can also use properties to set the borders, margins, and padding for a block element.

The diagram in this figure shows that *padding* is the space between the content area and a border. On the other hand, a *margin* is the space between the border and the outside of the box.

If you need to calculate the overall height of a box, you can use the formula in this figure. To start, you add the values for the margin, border width, and padding for the top of the box. Then, you add the height of the content area. Last, you add the values for the padding, border width, and margin for the bottom of the box. The formula for calculating the overall width of a box is similar.

When you set the height and width properties for a block element, you can use any of the units that you learned about in the last chapter. However, it's generally considered a best practice to use relative units of measure. That way, the size of the page changes according to the size of the browser window. This is called a *fluid layout*, and it's an important part of responsive web design.

The CSS box model

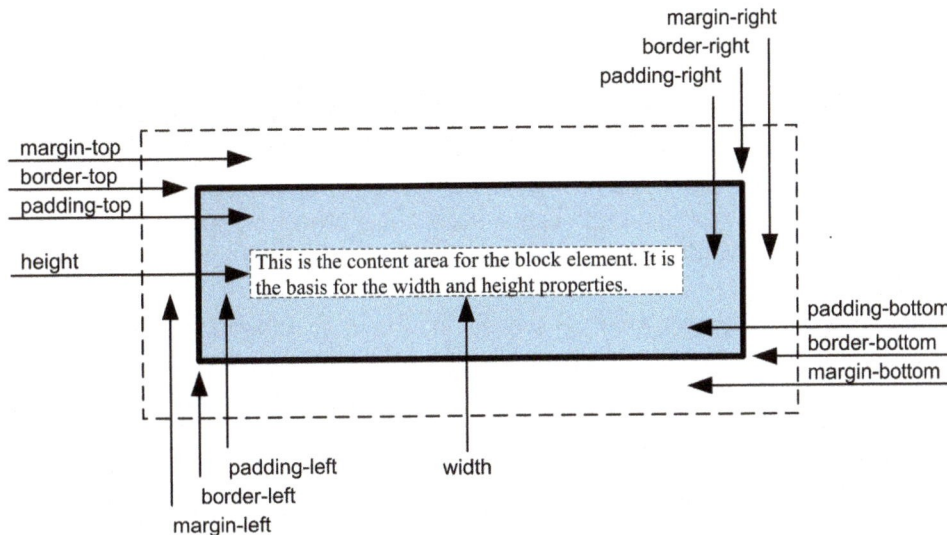

The formula for calculating the height of a box

```
top margin + top border + top padding +
height +
bottom padding + bottom border + bottom margin
```

The formula for calculating the width of a box

```
left margin + left border + left padding +
width +
right padding + right border + right margin
```

Description

- The CSS *box model* lets you work with the boxes that a browser places around each block element as well as some inline elements. This lets you add formatting such as margins, padding, and borders.
- *Padding* is the space between an element and its border.
- A *margin* is the space between the border of an element and either its containing block or the element next to it.
- By default, the box for a block element is as wide as the block that contains it and as tall as it needs to be based on its content.
- You can use the height and width properties to specify the size of the content area for a block element.
- You can use other properties to control the margins, padding, and borders for a block element. Then, these properties are added to the height and width of the content area to determine the height and width of the box.

Figure 5-1   How the box model works

A web page that shows how the box model works

Figure 5-2 presents a simple web page that's designed to show how the box model works. The HTML for the body of this page only defines four elements. Then, the CSS adds borders to these four HTML elements: a dotted 3-pixel border to the body, a solid 2-pixel border to the <main> element, and dashed 1-pixel borders to the <h1> and <p> elements. When a browser displays this web page, these four borders show how the margins and padding for these boxes work.

For the <body> element, the CSS sets the margin to 1em for all four sides. As a result, the browser uses this measurement for the margin on the left, top, and right of the body border, but not on the bottom. That's because the browser window determines the bottom margin for the body.

For the <main> element, the CSS sets the width to 80% and the margins to 2em on all four sides of the box. As a result, the browser uses these margins on the left, top, and bottom of the main box, but not on the right. That's because the width of this element is less than 100%.

The style rule for both the <h1> and <p> elements begins by setting the border and the padding for all four sides of these elements. Then, it sets the padding on the left side of the element to 20 pixels. Since the padding on the left side of these elements is set to a fixed unit of measure, the text for these elements is aligned. If you used relative units of measure here instead, the text wouldn't be aligned. That's because the font sizes for these elements are different.

Note that this style rule sets the padding for all four sides of both elements to 1em. Then, it sets the left padding to 20 pixels. Because it's coded after the padding property, the padding-left property overrides the left padding specified by the padding property.

The style rule for the <h1> element sets the top margin to .5em, the right and left margins to 0, and the bottom margin to .25em. As a result, there is more space above the <h1> element than below it.

The style rule for the <p> element sets all the margins to 0. As a result, all of the space between the <h1> and <p> elements is due to the padding of the <p> element and the bottom margin of the <h1> element.

A web page that shows how the box model works

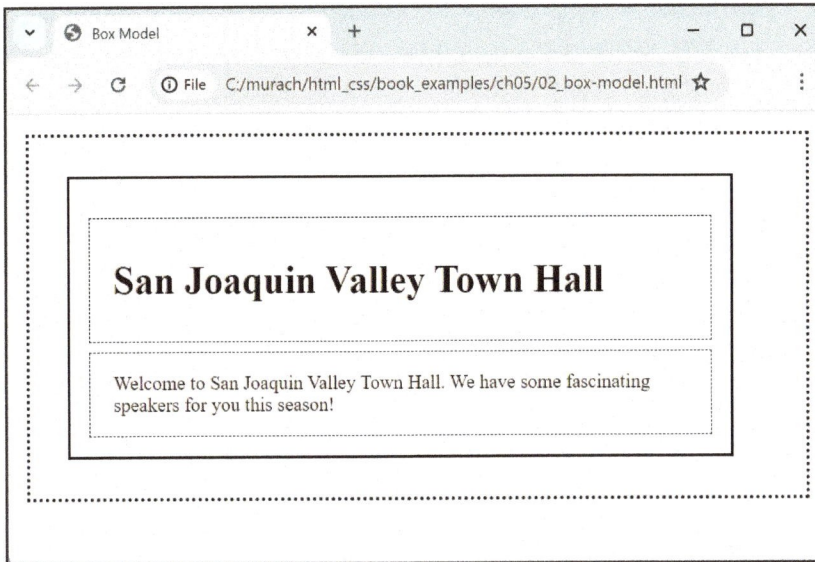

The HTML for the <body> element

```
<body>
    <main>
        <h1>San Joaquin Valley Town Hall</h1>
        <p>Welcome to San Joaquin Valley Town Hall.
        We have some fascinating speakers for you this season!</p>
    </main>
</body>
```

The CSS

```
body {
    border: 3px dotted black;
    margin: 1em;
}
main {
    border:  2px solid black;
    width:   80%;
    margin:  2em;            /* all four sides */
    padding: 1em;            /* all four sides */
}
h1, p {
    border: 1px dashed black;
    padding: 1em;            /* all four sides */
    padding-left: 20px;      /* use absolute measurement for left side */


}
h1 {
    margin: .5em 0 .25em;    /* .5em top, 0 right and left, .25em bottom */
}
p {
    margin: 0;               /* all four sides */
}
```

Figure 5-2   A web page that shows how the box model works

How to size and space elements

The previous figure should give you a general idea of how the box model works. Now, the figures that follow present the details of using properties to set the size of an element and the spacing between the elements on a page.

How to set widths and heights

Figure 5-3 presents the properties for setting heights and widths. By default, the width and height properties are set to a value of auto. As a result, the size of the content area for the element is automatically adjusted so it's as wide as the element that contains it and as tall as the content it contains. To change that, you can use the height and width properties.

The first two examples in this figure show how this works. Here, the first declaration in each example specifies an absolute value using pixels. Then, the second declaration specifies a relative value using percents. As a result, the width in the second declaration is set to 75% of the *containing block*. Next, the third declaration uses the auto keyword. That sets the width based on the size of the containing element and the height based on the content of the element. Because the default width and height is auto, you usually won't need to use that value.

In addition to the width and height properties, you can use the min-width, max-width, min-height, and max-height properties to specify the minimum and maximum width and height of the content area. These properties are often helpful for coding pages that implement responsive web design.

Properties for setting widths and heights

Property	Description
width	The width of the content area for a block element. Or auto (the default) to calculate the width of the box based on the width of its containing block.
height	The height of the content area for a block element. Or auto (the default) to calculate the height of the area based on its content.
min-width	The minimum width of the content area for a block element regardless of its content.
max-width	The maximum width of the content area for a block element. Or none to indicate that there is no maximum width.
min-height	The minimum height of the content area for a block element regardless of its content.
max-height	The maximum height of the content area for a block element. Or none to indicate that there is no maximum height.

Set the width of the content area

```
width: 450px;          /* an absolute width */
width: 75%;            /* a relative width */
width: auto;           /* width based on its containing block (the default) */
```

Set the height of the content area

```
height: 125px;
height: 50%;
height: auto;          /* height based on its content (the default) */
```

Set the minimum and maximum width and height

```
min-width: 450px;
max-width: 600px;
min-height: 120px;
max-height: 160px;
```

Description

- The width and height properties can be set to absolute or relative values.
- If you specify a percent for the width property, the width of the content area for the block element is based on the width of the block that contains it, called the *containing block*.
- If you specify a percent for the height property, the height of the content area for the block element is based on the height of the containing block.
- The min-height property can be used to be sure that an element has the specified height even if the content doesn't fill the element.
- The min-width, max-width, min-height, and max-height properties are often used with responsive web design.

Figure 5-3   How to set widths and heights

How to set margins

Figure 5-4 presents the properties for setting margins. When you set margins, you can use individual properties like margin-top or margin-left to set individual margins as shown in the first example.

However, instead of setting individual margins, you can use the margin property to set the margins for all four sides of a box. When you use this *shorthand property*, you can specify one, two, three, or four values. If you specify all four values, they are applied to the sides of the box in a clockwise order: top, right, bottom, and left. If you have trouble remembering this order, you can think of the word *trouble*.

If you specify fewer than four values, this property still sets the margins for all four sides of the box. For example, if you only specify one value, each margin is set to that value. If you specify two values, the top and bottom margins are set to the first value, and the left and right margins are set to the second value. And if you specify three values, the top margin is set to the first value, the left and right margins are set to the second value, and the bottom margin is set to the third value. This is illustrated by the second example in this figure.

Although it isn't shown here, you can also specify the auto keyword for any margin. In most cases, you'll use this keyword to center the <body> element in the browser window or to center a block element within its containing block. To do that, you specify auto for both the left and right margins. For this to work, you must also set the width of the element as shown later in this chapter.

Different browsers may have different default margins for the block elements. Because of that, it's a good practice to set the top and bottom margins of the elements that you're using. That way, you can control the space between elements like headings and paragraphs. Later in this chapter, you'll learn more about handling different browser default styles.

Finally, if you specify a bottom margin for one element and a top margin for the element that follows it, the margins are *collapsed*. That means the smaller margin is ignored and only the larger margin is applied. One way to get around this is to set the margins to zero and use padding for the spacing.

How to set padding

The properties for setting padding are also presented in figure 5-4. These properties work much like the properties for setting margins. For example, you can set the padding for the sides of a box individually, or you can set the padding for all four sides of a box at once by using the shorthand padding property.

Properties for setting margins

Property	Description
margin-top	The top margin
margin-right	The right margin
margin-bottom	The bottom margin
margin-left	The left margin
margin	A shorthand property for setting the margin for all four sides of a box. One value sets all four margins. Two values set the (1) top and bottom and (2) right and left margins. Three values set the (1) top, (2) right and left, and (3) bottom margins. Four values set the top, right, bottom, and left margins (think *trouble*).

Set the margin on a single side of an element

```
margin-top: .5em;
margin-left: 1em;
```

Set the margins on multiple sides of an element

```
margin: 1em;               /* 1em for all four sides */
margin: 0 1em;             /* top and bottom 0, right and left 1em */
margin: .5em 1em 2em;      /* top .5em, right and left 1em, bottom 2em */
margin: .5em 1em 2em 1em;  /* top .5em, right 1em, bottom 2em, left 1em */
```

Properties for setting padding

Property	Description
padding-top	The top padding
padding-right	The right padding
padding-bottom	The bottom padding
padding-left	The left paddding
padding	A shorthand property for setting the padding for all four sides of a box. Works the same as the margin property.

Set the padding on a single side of an element

```
padding-top: 0;
padding-right: 1em;
```

Set the padding on multiple sides of an element

```
padding: 1em;              /* all four sides */
padding: 0 1em;            /* top and bottom 0, right and left 1em */
padding: 0 1em .5em;       /* top 0, right and left 1em, bottom .5em */
padding: 0 1em .5em 1em;   /* top 0, right 1em, bottom .5em, left 1em */
```

Description

- If you specify a bottom margin for one element and a top margin for the element that follows it, the margins are *collapsed*, which means that only the larger margin is applied.
- If you set the top and bottom margins for elements to zero, you can use padding to set the spacing between the elements.

Figure 5-4   How to set margins and padding

How to work with box sizing

The first table in figure 5-5 presents the box-sizing property that determines how the browser calculates the width and height of an element. Then, the second table presents the two possible values for this property, content-box and border-box. The default is content-box.

When the box-sizing property of an element is set to content-box, the browser adds any values for border and padding to the width and height of the element. By contrast, when it's set to border-box, the browser adjusts the height and width so they include the border and padding values. When the box-sizing value is context-box, you can have unexpected results, as the examples in this figure illustrate.

In the first example, the box-sizing property of the <p> element is set to content-box by default. Then, the CSS sets its width to 100% of the containing block, sets its border to 2 pixels, and sets its padding to 1em for all four sides.

When the browser displays this paragraph, it calculates its width as 100% of the containing block plus 2 pixels for the left border plus 1em for the left padding plus 1em for the right padding plus 2 pixels for the right border. This makes the width of the paragraph greater than the width of the element that contains it. This excess width causes the paragraph to overflow its containing block to the right.

In the second example, the CSS sets the box-sizing property of the <p> element to border-box. As a result, when the browser displays this paragraph, it calculates its width by adjusting the value for 100% of the containing block to include the values for the border and padding. This keeps the paragraph within its containing block.

Property for sizing an element

Property	Description
box-sizing	Controls how the width and height of an element are calculated. Possible values are content-box and border-box. Content-box is the default.

The values for the box-sizing property

Value	Description
content-box	Adds the border and padding values to the height and width.
border-box	Includes the border and padding values in the height and width.

HTML for a \<p\> element

```
<p>Welcome to San Joaquin Valley Town Hall.
   We have some fascinating speakers for you this season!</p>
```

CSS for the \<p\> element

```
p {
    background-color: lightgray;
    width: 100%;
    border: 2px dotted black;         /* border and padding added to width */
    padding: 1em;
}
```

How it looks in a browser

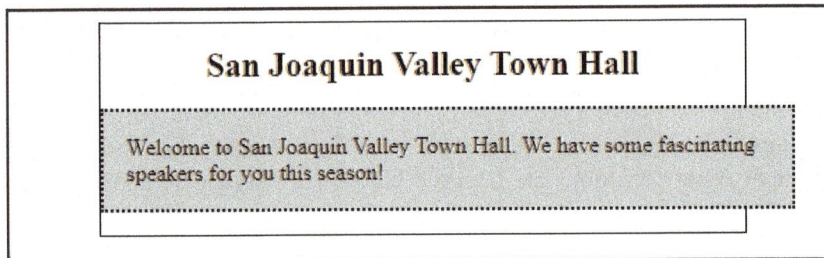

Set the box-sizing property to border-box

```
p {
    ...
    box-sizing: border-box;
}
```

How it looks in a browser

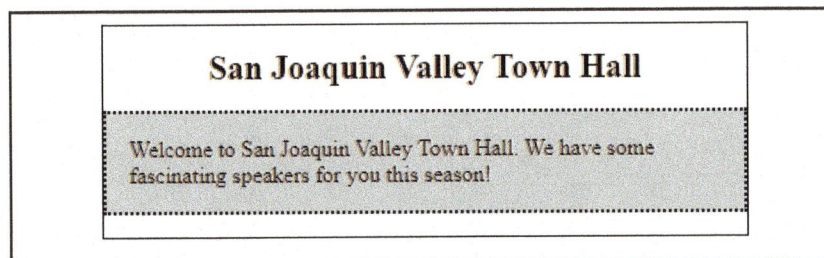

Figure 5-5   How to work with box sizing

How to use a reset selector

Web browsers provide default styles for many HTML elements. In the past, there was enough variation between default browser styles to make the same web page render differently in different browsers.

To correct for this, many web developers added style sheets to their web pages that caused the web page to render the same in all browsers. These style sheets became known as normalize or reset style sheets. A *normalize style sheet* applies CSS style rules that make minor adjustments to the default browser styles. By contrast, a *reset style sheet* removes all browser defaults by setting them to zero. Then, you set the styles you want.

Today, the modern browsers render most HTML and CSS exactly the same, and even older browsers render code in a way that's acceptable even if it's not exactly the same. As a result, there's no longer much need for normalize or reset style sheets. In addition, you can use the universal selector to set defaults to resolve minor formatting differences or to code a simple *reset selector*.

Figure 5-6 shows a reset selector that sets the margins and padding for all of the elements to zero. That way, you can get the spacing that you want by applying margins and padding that overrides the margins and paddings specified by the reset selector.

This reset selector also sets the box-sizing property to border-box. That way, the height and width of all of the elements includes any border and padding values, so you don't get the unexpected results shown in the last figure.

If you find you need more than this, you can use a third-party normalize or reset style sheet. This figure provides the URLs for three of the most popular ones. To use them, you download the CSS file for the style sheet and code a <link> element to that file in each page. When you code this link, make sure to code it before any other style sheets. Then, the normalize or reset styles are applied first so the browser variations are handled before your styles are applied.

A reset selector

```css
* {
    margin: 0;
    padding: 0;
    box-sizing: border-box;
}
```

The URLs for popular third party normalize and reset style sheets

https://necolas.github.io/normalize.css/

https://github.com/sindresorhus/modern-normalize

https://meyerweb.com/eric/tools/css/reset/

How to use a normalize or reset style sheet

1. Download the CSS file for the style sheet from the internet.
2. Save the CSS file in the correct location.
3. Code a <link> element for the style sheet in each HTML document. This element must be coded before the <link> elements for other style sheets.

Description

- A *normalize style sheet* makes minor adjustments to the default browser styles, and a *reset style sheet* sets the default browser styles to a blank slate.

- In the past, variations between default browser styles were common. In response, many developers used third-party normalize or reset style sheets to make sure their web pages looked the same on all browsers.

- Today, browser variations are so minimal that you can resolve most of them by using the universal selector to reset a few properties.

- If you need more than a reset selector, you can use a third-party normalize or reset style sheet.

Figure 5-6   How to use a reset selector

A web page that uses sizing and spacing

To show how to use the properties for sizing and spacing, figure 5-7 presents a web page that uses these properties. This web page is similar to the one presented at the end of the previous chapter. However, the CSS for this page centers the text in the browser window and improves the spacing between the elements.

The HTML for this web page is mostly the same as the HTML for the page at the end of the last chapter. As a result, you shouldn't have much trouble understanding how it works.

A web page that uses sizing and spacing

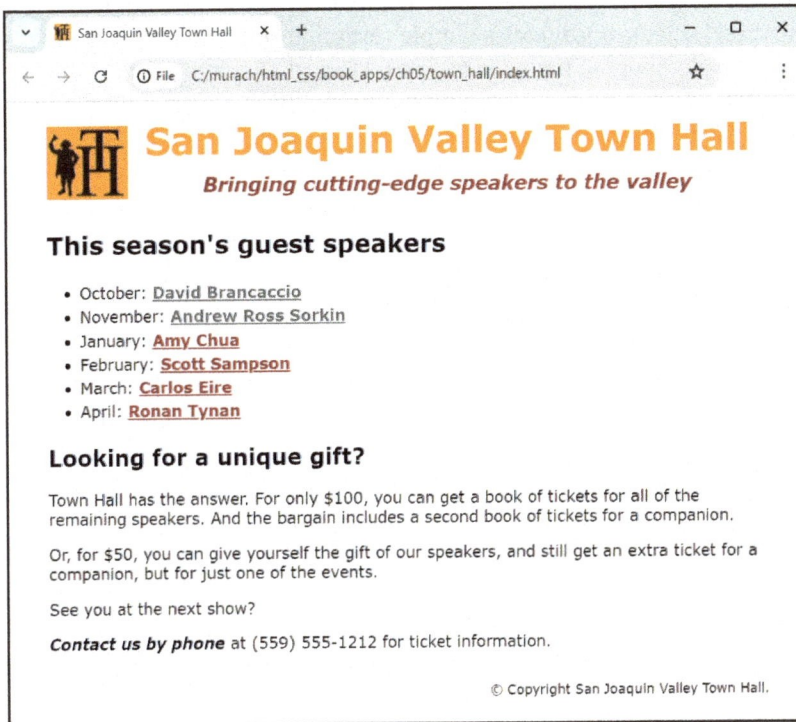

The HTML for the <body> element

```html
<body>
    <header>
        <img src="images/logo.gif" alt="Town Hall Logo" width="80">
        <h2>San Joaquin Valley Town Hall</h2>
        <h3>Bringing cutting-edge speakers to the valley</h3>
    </header>
    <main>
        <h1>This season's guest speakers</h1>
        <nav>
            <ul>
                <li>October: <a class="date_passed"
                  href="speakers/brancaccio.html">David Brancaccio</a></li>
                ...
            </ul>
        </nav>
        <h2>Looking for a unique gift?</h2>
        <p>Town Hall has the answer. For only $100, ...</p>
        <p>Or, for $50, you can give yourself the gift ...</p>
        <p>See you at the next show?</p>
        <p><em>Contact us by phone</em> at ...</p>
    </main>
    <footer>
        <p>&copy; Copyright San Joaquin Valley Town Hall.</p>
    </footer>
<body>
```

Figure 5-7   A web page that uses sizing and spacing (part 1)

Parts 2 and 3 of figure 5-7 presents the CSS for the web page. Here, the style rules that affect the size and spacing of the elements on the page are highlighted.

To start, a universal selector provides a simple reset that sets the margins and padding for all the elements in the page to zero. Then, the style rules that follow can provide the desired spacing. In addition, this reset sets the box-sizing property to border-box.

The default styles for the page begin by setting the width of the <body> element to 90%. Then, they set the top and bottom margins to 1em so there's space between the body and the top of the browser window. In addition, they set the left and right margins to auto. This causes the left and right margins to be calculated automatically so the page is centered in the browser window.

The style rule for the <h1>, <h2>, <h3>, <p>, and <ul> elements sets the top and bottom padding for these elements to .5em while keeping the right and left padding at zero. This provides spacing between headings, paragraphs, and unordered lists.

The style rule for the <ul> element sets the left margin to 2em. This restores a default indentation for unordered lists that's provided by most browsers but cleared by the reset selector.

The styles for the <header> element begin by setting the top and bottom margins for the header image to .5em while keeping the right and left padding at zero. This does a better job of aligning the image with the headings in the header.

The style rule for the <h2> element in the header sets the padding for this element to 0, while the style rule for the <h3> element increases the bottom padding to 1em. This makes the headings in the header closer together, and adds additional space between the header and the main content.

The CSS

```css
:root {
    --light: #f2972e;
    --dark: #931420;
}

/* reset */
* {
    margin: 0;
    padding: 0;
    box-sizing: border-box;
}

/* default styles */
body {
    font-family: Verdana, Arial, Helvetica, sans-serif;
    font-size: 100%;
    width: 90%;
    margin: 1em auto;
}
h1, h2, h3, p, ul { padding: .5em 0; }
h1 { font-size: 150%; }
h2 { font-size: 135%; }
h3 { font-size: 120%; }
li, p { font-size: 95%; }
a, em { font-weight: bold; }
a:link, a:visited { color: var(--dark); }
a:hover, a:focus { color: var(--light); }
ul {
    line-height: 1.5;
    margin-left: 2em;
}

/* the styles for the header */
header {
    text-align: center;
    & img {
        float: left;
        margin: .5em 0;
    }
    & h2 {
        font-size: 225%;
        color: var(--light);
        padding: 0;
    }
    & h3 {
        font-style: italic;
        color: var(--dark);
        padding-bottom: 1em;
    }
}
```

Figure 5-7   A web page that uses sizing and spacing (part 2)

Part 3 of figure 5-7 presents the style rules for the <main> and <footer> elements. The style rule for the <main> element adds .5em of padding to the bottom of this element. Then, the style rule for the <footer> element adds .5em of padding to the top of this element. This adds space between these two elements.

If you study the code in parts 2 and 3 of this figure, you can see that it avoids the problems of collapsing margins. It does that first with a reset selector that sets the margins for the elements to zero and then by using padding to provide the spacing before and after elements.

The CSS (continued)

```
/* the styles for the main content */
main {
    clear: left;
    padding-bottom: .5em;
    & a.date_passed {
        color: gray;
    }
}

/* the styles for the footer */
footer {
    padding-top: .5em;
    & p {
        font-size: 80%;
        text-align: right;
    }
}
```

Figure 5-7   A web page that uses sizing and spacing (part 3)

How to set borders and backgrounds

Now that you know how to size and space elements using the box model, you're ready to learn how to apply other formatting to boxes. That includes adding borders and setting background colors and images.

How to set borders

Figure 5-8 presents the properties for setting borders and illustrates how they work. To start, if you want the same border on all four sides of a box, you can use the shorthand border property. This property lets you set the width, style, and color for the borders.

The shorthand border property is illustrated in the first example after the syntax. Here, the first declaration creates a thin, solid, green border. However, different browsers may interpret the thin keyword, as well as the other keywords for width, differently. Because of that, most programmers set the width of a border to an absolute value as shown in the second and third declarations. Notice that the third declaration doesn't specify a color. In that case, the border is the same color as the element's text, which is often what you want.

To set the border for just one side of a box, you can use the shorthand property for a border side. This is illustrated by the second example. Here, the first declaration sets the top border, and the second declaration sets the right border.

It's common to use the shorthand properties to set the borders of a box. However, you can use the other properties in this figure to set the widths, styles, and colors for the sides of a border. This is shown by the third, fourth, fifth, and sixth examples. In some cases, you may want to use these properties to override another border setting for an element.

For instance, to set the width for each side of a border, you can use the border-width property as shown in the third example. Here, you can specify one, two, three, or four values. This works like the shorthand margin and padding properties you learned about earlier in this chapter. As a result, the values are applied to the top, right, bottom, and left sides of the border.

The border-style and border-color properties work similarly. You can use them to set the style and color for each side of a border by specifying one to four values. This is shown in the fourth and fifth examples. In the last example for the border-style property, the left and right borders are set to "none". As a result, the browser only displays the top and bottom borders.

The sixth example shows another way to set properties for a border on just one side of a box. For instance, the first declaration sets the bottom border width to 4 pixels, the second declaration sets the right border style to dashed, and the third declaration sets the left border color to gray.

Properties for setting borders

Property	Description
border	A shorthand property for setting the border width, style, and color for all four sides of the box.
border-*side*	A shorthand property for setting the border width, style, and color for the specified side of a box.
border-width	One to four values or keywords that specify the width for each side of a border. Possible keywords are thin, medium, and thick.
border-style	One to four keywords that specify the style for each side of a border. Possible values are dotted, dashed, solid, double, groove, ridge, inset, outset, and none. The default is none.
border-color	One to four color values or keywords that specify the color for each side of a border. The default is the color of the element.
border-*side*-width	A value or a keyword that specifies the width of the indicated side of a border.
border-*side*-style	A keyword that specifies the style for the indicated side of a border.
border-*side*-color	A color value or keyword that specifies the color of the indicated side of a border.

The syntax for the shorthand properties

```
border: [width] [style] [color];
border-side: [width] [style] [color];
```

Set the same border on all four sides

```
border: thin solid green;
border: 2px dashed #808080;
border: 1px inset;               /* uses the element's color property */
```

Set a border on one side

```
border-top: 2px solid black;
border-right: 4px double blue;
```

Set widths

```
border-width: 1px;                /* all four sides */
border-width: 1px 2px;            /* top and bottom 1px, right and left 2px */
border-width: 1px 2px 2px;        /* top 1px, right and left 2px, bottom 2px */
border-width: 1px 2px 2px 3px;  /* top 1px, right 2px, bottom 2px, left 3px */
```

Set styles

```
border-style: dashed;     /* dashed line all sides */
border-style: solid none; /* solid top and bottom, no border right and left */
```

Set colors

```
border-color: #808080;
border-color: black gray; /* black top and bottom, gray right and left */
```

Set a property for a specific side

```
border-bottom-width: 4px;
border-right-style: dashed;
border-left-color: gray;
```

Figure 5-8   How to set borders

How to add rounded corners and shadows

Figure 5-9 shows how to use the CSS features for adding rounded corners to borders and shadows to boxes. This provides a way to create these graphic effects without using images.

To round the corners of a border, you use the border-radius property. If you supply one value for this property, it applies to all four corners of the border. But if you supply four values as shown in the example, you can apply specific rounding to each corner. Here, the upper-left corner has a radius of 10 pixels, the upper-right corner has a radius of 20 pixels, the lower-right corner has a radius of 0 so it isn't rounded, and the lower-left corner has a radius of 20 pixels.

To add shadows to a box, you use the box-shadow property. With the first two values, you specify the offset for the shadow. With the third value, you specify the blur radius. With the fourth value, you specify how far the blur is spread. And with the fifth value, you specify a color for the shadow. To get the effects that you want, you usually need to experiment with these values, but this is easier and more flexible than using an image to get those effects.

When you specify the color for a border or its shadow, you can use the currentColor keyword to apply the same color as the color that's used for the element's text. This can make it easier to consistently apply colors to borders and shadows. In addition, since you only code the color in one place, this makes your code easier to maintain. Alternately, you could not specify a color at all. However, using the currentColor keyword makes your CSS easier to read and understand.

In this figure, for example, the CSS sets the color for the <section> element's text to blue. Then, the CSS sets the color for the border to the currentColor keyword. As a result, the browser uses blue for both the text and the border, even though the CSS only sets the text color to blue. By contrast, the CSS sets the color for the shadow to red. However, if you wanted to set it to blue, you could not code a color at all, or you could use the currentColor keyword.

Incidentally, there are some options for rounding corners that aren't presented in this figure. For instance, you can use a separate property like border-top-left-radius for each corner. You can also set the curvature of a corner by supplying two values for a single corner like this:

```
border-top-left-radius: 50px 20px;
```

So, if you want to go beyond what this figure offers, you can search the internet for more information.

The syntax for the border-radius and box-shadow properties

```
border-radius: radius;  /* applies to all four corners */
border-radius: topLeft topRight lowerRight lowerLeft;
box-shadow: horizontalOffset verticalOffset blurRadius spread color;
```

The HTML for a section

```
<section>
    <a href="ebooks_index.html">$10 Ebooks!</a>
</section>
```

The CSS for the section

```
section {
    color: blue;
    font-size: 160%;
    text-align: center;
    font-weight: bold;
    padding: 1em;
    width: 8em;
    border: 5px double currentColor;
    border-radius: 10px 20px 0 20px;
    box-shadow: 3px 3px 4px 4px red;
}
```

How the section looks in a browser

$10 Ebooks!

Description

- When you code the border-radius property, you can assign one rounding radius to all four corners or a different radius to each corner.
- When you code the box-shadow property, positive values offset the shadow to the right or down, and negative values offset the shadow to the left or up.
- The third value in the box-shadow property determines how much the shadow is blurred, and the fourth value determines how far the blur is spread.
- The fifth value in the box-shadow property specifies the color of the shadow. If this is omitted, it is the same color as the border.
- You can use the currentColor keyword to set the color of a border or a shadow to the same color that's used for the text of the element. This can make it easier to apply colors consistently.

Figure 5-9   How to add rounded corners and shadows

How to set background colors and images

Figure 5-10 presents the properties you can use to set the background for a box. When you set a background, it's displayed behind the content, padding, and border for the box, but it isn't displayed behind the margin.

Although this figure presents all the properties for using background colors and images, please remember the need for web accessibility. In brief, that means that the background colors and images should never make the text more difficult to read, especially for visually-impaired users. That's why most websites use high contrast between their foreground and background colors and only use background images for special purposes.

However, if you want to provide a background for an element, this figure shows how. To start, you can set a background color, a background image, or both. If you set both, the browser displays the background color behind the image. As a result, you can only see the background color if the image has areas that are transparent or the image doesn't repeat.

To set all five properties of a background, you can use the shorthand property. When you use this property, you don't have to specify the individual properties in a specific order, but it usually makes sense to use the order that's shown. If you omit a property, the browser uses its default value.

The first example shows how this works. The first declaration sets the background color to blue. The second declaration sets the background color and specifies the URL for an image. And the third declaration specifies all five background properties. You can also code each of these properties individually.

By default, a background image that's added to a box repeats horizontally and vertically to fill the box. If that isn't what you want, you can set the background-repeat property so the image only repeats horizontally, only repeats vertically, or doesn't repeat at all. This is illustrated by the first four declarations in the second example.

If an image doesn't repeat, you may need to set additional properties to determine where the image is positioned and whether it scrolls with the page. By default, an image is positioned in the top left corner of the box. However, you can change that by using the background-position property.

The next three declarations show how this works. The first declaration positions the image at the top left corner of the box, which is the default. The second declaration centers the image at the top of the box. And the third declaration positions the image starting 90% of the way from the left side to the right side of the box and 90% of the way from the top to the bottom of the box.

By default, a background image moves with the document as you scroll. If that's not what you want, you can set the background-attachment property to fixed as shown by the last declaration in the figure.

The properties for setting the background color and image

Property	Description
background	A shorthand property for setting background color, image, repeat, attachment, and position values.
background-color	A color value or keyword that specifies the color of an element's background. You can also specify the transparent keyword if you want elements behind the element to be visible. This is the default.
background-image	A relative or absolute URL that points to the image. You can also specify the keyword none if you don't want to display an image. This is the default.
background-repeat	A keyword that specifies if and how an image is repeated. Possible values are repeat, repeat-x, repeat-y, and no-repeat. The default is repeat, which causes the image to be repeated both horizontally and vertically to fill the background.
background-attachment	A keyword that specifies whether an image scrolls with the document or remains in a fixed position. Possible values are scroll and fixed. The default is scroll.
background-position	One or two relative or absolute values or keywords that specify the initial horizontal and vertical positions of an image. Keywords are left, center, and right; top, center, and bottom. If a vertical position isn't specified, center is the default. If no position is specified, the default is to place the image at the top-left corner of the element.

Accessibility guideline

* Don't use a background color or image that makes the text that's over it difficult to read.

The syntax for the shorthand property

```
background: [color] [image] [repeat] [attachment] [position];
```

How to use the shorthand property

```
background: blue;
background: blue url("../images/texture.gif");
background: #808080 url("../images/header.jpg") repeat-y scroll center top;
```

How to control image repetition, position, and scrolling

```
background-repeat: repeat;          /* repeats both directions */
background-repeat: repeat-x;        /* repeats horizontally */
background-repeat: repeat-y;        /* repeats vertically */
background-repeat: no-repeat;       /* doesn't repeat */

background-position: left top;      /* 0% from left, 0% from top */
background-position: center top;    /* centered horizontally, 0% from top */
background-position: 90% 90%;       /* 90% from left, 90% from top */

background-attachment: scroll;      /* image moves as you scroll */
background-attachment: fixed;       /* image does not move as you scroll */
```

Figure 5-10   How to set background colors and images

How to set background gradients

Figure 5-11 shows the basics of how to use *linear gradients* to provide a background that gradually morphs from one color to another. This feature lets you provide interesting backgrounds without using images. To show how linear gradients work, this figure presents three <div> elements that set their background image to a gradient.

For the first <div> element, the color changes from left to right. It begins with white on the far left (0%) and gradually morphs into red on the far right (100%).

For the second <div> element, the direction is 45 degrees and there are three color groups. It begins with red on the far left (0%), morphs into white in the middle (50%), and ends with blue on the right (100%).

The third <div> element works much like the second one. However, the CSS specifies both a starting and ending point for each color. As a result, the colors don't morph into each other, and the browser displays three solid stripes of red, white, and blue.

Typically, a gradient starts at 0% and ends at 100%. However, you can use values outside this range if that produces the effect you want as shown later in this chapter.

With a little experimentation, you can probably create many attractive linear gradients. In addition, you may want to experiment with other types of gradients such as radial and conical gradients. For more information about those, you can search the internet for the type of gradient you're interested in.

The syntax for using a linear gradient in the background-image property

```
background-image: linear-gradient(direction, color %, color %, ... );
```

The HTML for three divisions

```
<div id="gradient1"></div>
<div id="gradient2"></div>
<div id="gradient3"></div>
```

The CSS for the three divisions

```
#gradient1 {
    background-image: linear-gradient(to right, white 0%, red 100%);
}
#gradient2 {
    background-image: linear-gradient(45deg, red 0%, white 50%, blue 100%);
}
#gradient3 {
    background-image: linear-gradient(
        45deg, red 0%, red 33%, white 33%, white 66%, blue 66%, blue 100%);
}
```

The linear gradients in a browser

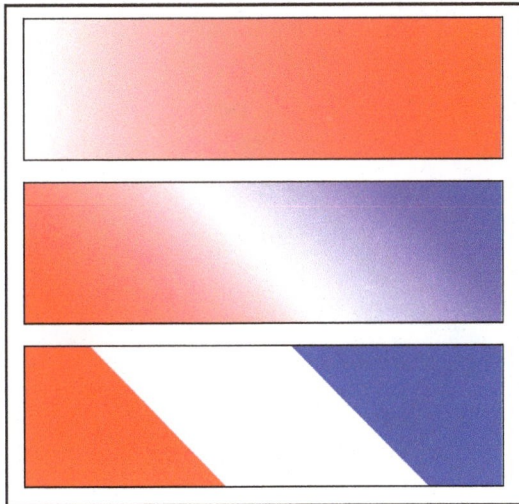

Description

- You can create *linear gradients* for backgrounds without using images.

- The first parameter of the linear-gradient() function specifies the direction of the gradient. For vertical or horizontal gradients, you can specify "to right", "to bottom", "to left", or "to top". For gradients that should be on an angle, you can specify a number of degrees.

- After the first parameter of the linear-gradient() function, you can code two or more parameters that consist of a color and a percent for each parameter. The first percent specifies where the first color should start, the last percent specifies where the last color should end, and the percents between specifies where other colors should be.

- Gradients typically start at 0% and end at 100%, but you can use values outside that range to get the effect you want.

Figure 5-11   How to set background gradients

A web page that uses borders and backgrounds

Figure 5-12 presents a web page that's similar to the one presented earlier in this chapter. In fact, the HTML for these two pages is identical. Because of that, the HTML for the page isn't presented here. However, the web page shown in this figure has some additional formatting.

First, this page displays a gradient behind the body, and it uses white as the background color for the body. That way, the gradient doesn't show behind the body content.

Second, this page displays some borders. To start, it displays a border around the body with rounded corners, and it rounds the upper left corner of the logo image to match the rounded border of the body. Then, it displays a border below the header and a border above the footer.

A web page that uses borders and backgrounds

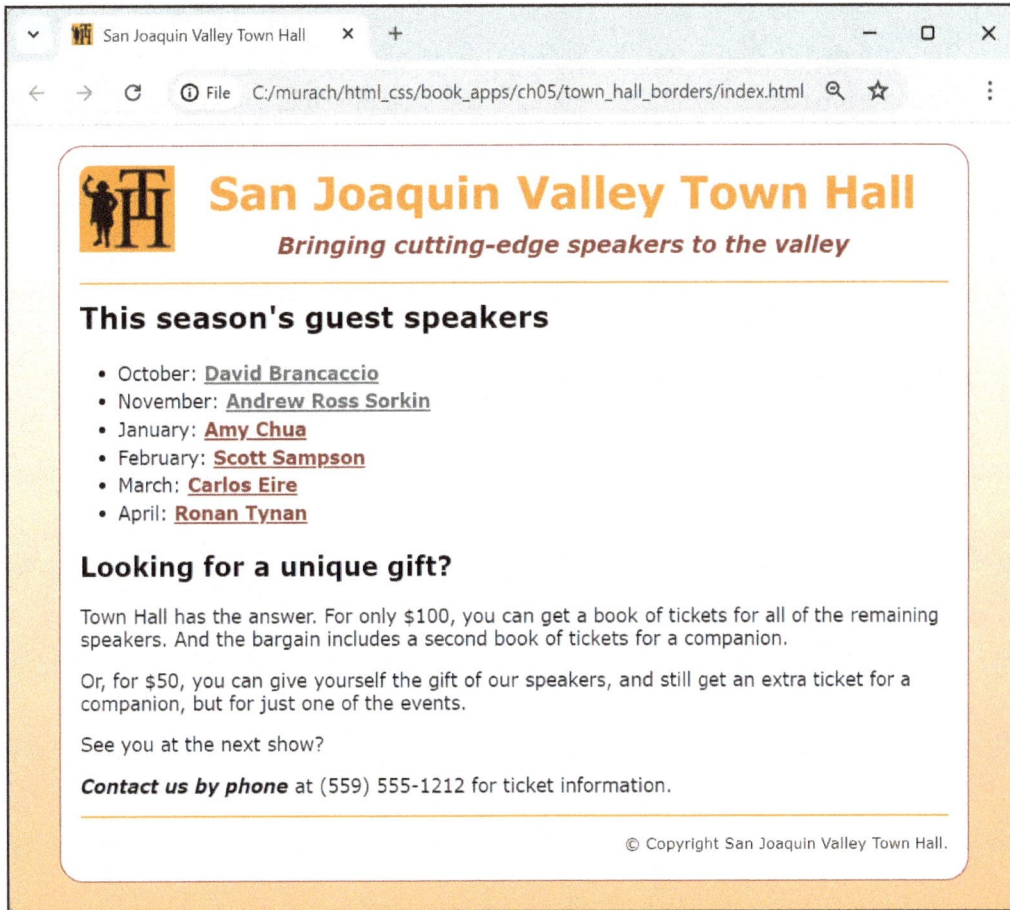

Figure 5-12   A web page that uses borders and backgrounds (part 1)

Part 2 of figure 5-12 presents the CSS for the web page. Here, all the style rules that affect the borders and background of the elements on the page are highlighted.

The style rule for the <html> element sets the background image to a gradient that ranges from white to the light color variable that's defined earlier. The ending percent for this gradient is 200%, rather than the typical 100%. As a result, the gradient color is lighter than it would otherwise be.

This style rule also sets the minimum height of the <html> element to 100%. This makes the box for this element stretch vertically to fill the entire screen. Without this, the gradient would repeat at the end of the box for the <html> element, and that isn't visually appealing for this web page.

The style rule for the <body> element sets the background color to white. That way, the contents of the body display on a white background. Since the top and bottom margins are set to 1em, the gradient shows above and below the body. And since the right and left margins are set to auto, the body is centered in the browser window and the gradient shows to the right and left of the body.

In addition, this style rule adds a border with rounded corners around the body. This border uses the dark color variable used by other elements in the page. Finally, to add some space between the border and the content, the style rule sets the padding of all four sides of the body to 1em.

The nested style rule for the <img> element in the header rounds the top left corner of the logo image. To do that, it sets the border-radius property with a 10-pixel radius for the upper-left corner and a zero radius for the remaining corners.

The style rules for the <header> and <footer> elements add a border below the header and above the footer. Both of these borders use the light color variable used by other elements in the page. Note that there is space between each border and the content above and below it.

For the header border, the space above is provided by the <h3> element in the header, which has bottom padding of 1em. The space below is provided by the default <h1> element, which has top padding of .5em.

For the footer border, the space above the border is provided by the <main> element, which has bottom padding of .5em, and the last <p> element, which also has bottom padding of .5em. The space below is provided by the <footer> element, which has top padding of .5em, and the <p> element, which also has top padding of .5em.

The CSS for the web page

```
/* variables and reset same as figure 5-7 */

/* default styles */
html {
    min-height: 100%;        /* so the box stretches across the entire screen */
    background-image: linear-gradient(to bottom,
                                      white 0%, var(--light) 200%);
}
body {
    font-family: Verdana, Arial, Helvetica, sans-serif;
    font-size: 100%;
    width: 90%;
    margin: 1em auto;
    background-color: white;
    border: 1px solid var(--dark);
    border-radius: 20px;
    padding: 1em;        /* adds space between border and content */
}

/* rest of default styles same as figure 5-7 */

header {
    text-align: center;
    border-bottom: 2px solid var(--light);
    & img {
        float: left;
        border-radius: 10px 0 0 0;      /* round upper-left corner */
    }
    & h2 {
        font-size: 225%;
        color: var(--light);
        padding: 0;
    }
    & h3 {
        font-style: italic;
        color: var(--dark);
        padding-bottom: 1em;
    }
}
main {
    clear: left;
    padding-bottom: .5em;
    & a.date_passed { color: gray; }
}
footer {
    padding-top: .5em;
    border-top: 2px solid var(--light);
    & p {
        font-size: 80%;
        text-align: right;
    }
}
```

Figure 5-12   A web page that uses borders and backgrounds (part 2)

Perspective

Now that you've completed this chapter, you should understand how the box model works for margins, padding, borders, and backgrounds. These are the skills you need to control the spacing, borders, and backgrounds for your web pages.

Terms

box model	collapsed margins
padding	normalize style sheet
margin	reset style sheet
fluid layout	reset selector
containing block	linear gradient
shorthand property	

Exercise 5-1   Enhance the Town Hall home page

In this exercise, you'll enhance the formatting of the Town Hall home page. When you're through, the page should look like this:

Add a reset selector and some default styles

1. Use your text editor to open these HTML and CSS files:

   ```
   exercises\ch05\town_hall\index.html
   exercises\ch05\town_hall\styles\main.css
   ```

2. Run the web page in Chrome to view it.

3. Add a reset selector that sets all margins and padding to zero and the box-sizing property to border-box.

4. Refresh the page to view your changes. Note that the reset has removed all the spacing.

5. Add a style rule that sets default top and bottom padding of .5em for headings, paragraphs, and unordered lists.

6. Add a style rule that sets a default left margin of 2em for unordered lists.

7. Refresh the page to view your changes.

Center the body and add some borders

From this point on, refresh the page after each change to make sure it looks the way you want it. If you have any problems, use the Developer Tools presented in the previous chapter to help you debug them.

8. Enhance the style rule for the body by setting the width to 80%, the margin to auto, and adding a 2-pixel solid border that uses the accent color variable.

9. Add a bottom border to the header that's the same as the border around the body.

10. Add top and bottom borders to the <h1> heading in the <main> element. Both borders should be the same as the border for the body and header.

11. Set the background color of the footer to the same color as the borders, and then set the font color for the nested paragraph to white so it's easier to read.

Set the padding for the header, main, and footer sections

12. For the header, delete the text-indent declarations for the <h2> and <h3> elements. Then, for the <img> element, set top and bottom margins of .75em and right and left margins of 1em. Then, set the bottom padding for the <h2> element to zero and the bottom padding for the <h3> element to 1em.

13. For the <main> element, add 1em of padding to the right, left, and bottom.

14. For the footer, set the top and bottom padding of the <p> element to 1em.

Add some finishing touches

15. Italicize the blockquote element to make it stand out. Then, use padding to indent it.

16. Add a linear gradient as the background for the header. The one that's shown here uses #f6bb73 at 0%, white at 30%, white at 65%, and #f6bb73 at 100% at a 30 degree angle. But experiment with this until you get it the way you want it.

17. Add a 4 pixel double border with rounded corners around each speaker image in the <main> element. The color should be the same as the other borders.

18. Refresh the page to make sure it looks like the one displayed at the start of this exercise.

19. Experiment on your own to see if you can improve the formatting.

Chapter 6

How to use flexbox for page layout

In this chapter, you'll learn how to use CSS to control the layout of a page. That means that you can control where each of the HTML elements appears on the page. When you finish this chapter, you should be able to implement effective 2- and 3-column page layouts that use a fluid design.

How to get started with flexbox

Flexible box layout, or *flexbox layout*, provides an efficient way to lay out, align, and distribute space among the elements of a web page, especially when the page has to adapt to different screen sizes. This is an important part of responsive web design.

Flexbox layout concepts

Figure 6-1 introduces some basic concepts for using flexbox layout. The diagrams in this figure show that a *flex container* can contain one or more *flex items*. These items are laid out along the *main axis*, which can have either horizontal or vertical orientation depending on the *flex direction*. In the first diagram, the flex container lays out two flex items in a horizontal row.

In addition to the main axis, a flex container has a *cross axis* that's perpendicular to the main axis. The flex items within a container can be aligned along either of these axes.

This first diagram also shows that the start of the main axis is called the *main start*, and the end of the main axis is called the *main end*. Similarly, the start of the cross axis is called the *cross start*, and the end of the cross axis is called the *cross end*. Last, the width of the container from the main start to the main end is called the *main size*, and the height of the container from the cross start to the cross end is called the *cross size*.

What flexbox does best is simplify how elements on a page are laid out side by side without using floats. It also makes it easy to change a layout from horizontal to vertical orientation when you're developing a responsive design. In this figure, for example, the second diagram lays out three flex items horizontally in a row, and the third diagram uses the same container to lay out the same three flex items vertically in a column.

When you use flexbox, you can still use traditional box model properties like width, height, margin, padding, and border. These properties can use fixed or relative units of measure.

The flexible box layout

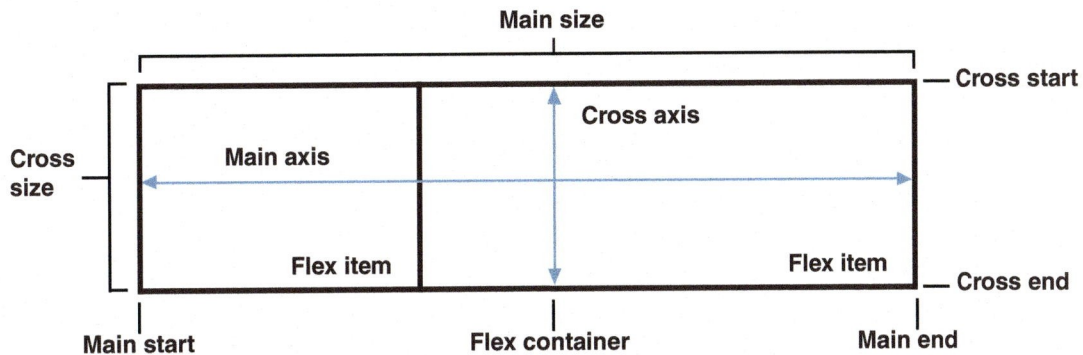

Flex items within a flexbox container displayed as a row

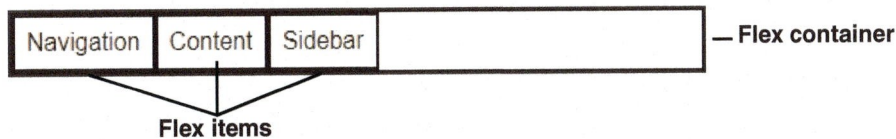

Flex items within a flexbox container displayed as a column

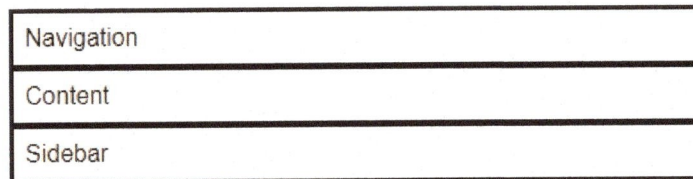

Description

- *Flexible box layout*, or *flexbox layout*, is a CSS module that can be used to develop page layouts for varying screen sizes.
- A *flexbox* is a container that contains one or more *flex items*.
- Flex items are laid out within a flex container along the *main axis* depending on the *flex direction*. Flex items can also be aligned along the *cross axis*.
- Width and height properties are used to set the dimensions of the flex container.
- If the width for a flex container isn't set, the container defaults to 100% of the parent element.

Figure 6-1   Flexbox layout concepts

How to create your first flexible box

Figure 6-2 shows how to create a simple 2-column page layout using a flexbox. To start, you set the display property for a block element to either flex or flex-inline. In most cases, you'll set this property to flex. Then, the block element becomes a flex container, and any block-level child elements within the container become flex items.

By default, a flex container sets the flex-direction property to row. This lays out its items horizontally from left to right. However, you can change the flex-direction property to change how a container lays out its items. For example, if you want to display the flex items in a column, you can set the flex-direction property to column. This changes the direction of the flex container's main and cross axes.

By default, the initial length of an item in a flex container is the same as the width of the item if the width is set. Otherwise, the initial length of an item is determined by the content of the item. However, you can use the flex-basis property to change the initial length of an item.

To illustrate how the flexbox properties work, this figure shows the HTML and CSS for a flexbox container and two items that display horizontally. Here, the HTML consists of a <main> element that contains a <section> element and an <aside> element.

In the CSS, the first declaration for the <main> element sets the display to flex. As a result, the <section> and <aside> elements become flex items. Then, because the flex-direction property is row by default, the section and aside display horizontally in a row.

The CSS for the <section> and <aside> elements sets the flex-basis property for these elements to 65% and 35%. As a result, the <section> element uses 65% of the available space within the container, while the <aside> element uses 35% of the available space. This causes a browser to display these elements side by side in a way that uses 100% of the flexbox container.

Properties for working with a flexbox

Property	Description
display	The type of box. Possible values include flex (rendered as a block) or flex-inline (rendered as inline content).
flex-direction	The direction of the flex items within the flex container. Possible values include row, column, row-reverse, and column-reverse. The default is row.
flex-basis	The initial length of the flex item. Possible values include auto and units of measure such as pixels, ems, or percents. The default is auto, which means that the length is the same as the width of the item if the width is set or is determined by the content if the width isn't set.

The HTML for the <main> element

```
<main>
    <section>
        <p>Welcome to San Joaquin Valley Town Hall. We have some fascinating
            speakers for you this season!</p>
    </section>
    <aside>
        <p>The luncheon starts 15 minutes after the lecture ends.</p>
    </aside>
</main>
```

The CSS for the <main> element

```
main {
    display: flex;           /* flex-direction is row by default */
    padding-top: 1em;
}
section {
    flex-basis: 65%;
}
aside {
    flex-basis: 35%;
    background-color: lightblue;
    border-radius: 1em;
    padding: 1em;
}
```

The <main> element in a browser

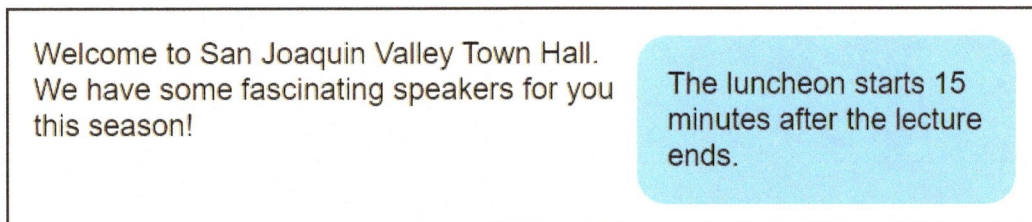

Welcome to San Joaquin Valley Town Hall. We have some fascinating speakers for you this season!

The luncheon starts 15 minutes after the lecture ends.

Figure 6-2   How to create your first flexible box

How to create page layouts

To create a page layout with columns, you can use flexbox layout to position the elements that make up the columns of the page.

How to create a 1-column layout

Figure 6-3 shows how to create a 1-column page layout. Here, the HTML consists of <body>, <header>, <main>, and <footer> elements. In addition, the <main> element contains <section> and <aside> elements.

The CSS for the <main> element sets the display property to flex, and it sets the flex-direction property to column. This causes a browser to display the single column shown in this figure. In addition, the CSS in this figure sets the background colors for the flexbox items to make it easy to see their layout.

For a 1-column layout, you don't need to use a flexbox. That's because the block elements defined in an HTML document flow from the top of the page to the bottom of the page by default. In other words, the browser already lays out the block elements in a single column.

However, it's a common design pattern to create a 1-column layout for small screen sizes and a two- or three-column layout for larger screen sizes. In that case, using flexbox to create the 1-column layout makes it easier to switch to other layouts as the screen size changes. You'll learn more about that in chapter 8.

A 1-column layout

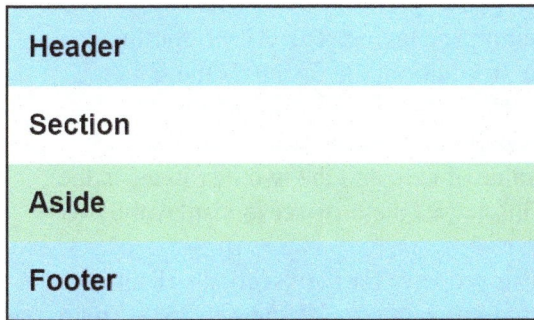

The HTML for the \<body\> element

```
<body>
    <header><h2>Header</h2></header>
    <main>
        <section><h2>Section</h2></section>
        <aside><h2>Aside</h2></aside>
    </main>
    <footer><h2>Footer</h2></footer>
</body>
```

The CSS

```
body {
    font-family: Arial, sans-serif;
    margin: 1em;
}
main {
    display: flex;
    flex-direction: column;
}
section {
    background-color: azure;
}
aside {
    background-color: aquamarine;
}
header, footer {
    background-color: lightskyblue;
}
```

Description

- Many websites use a 1-column layout for devices with small screens.
- By default, browsers lay out elements using one column, so you don't need to use a flexbox to modify the layout. However, setting up a flexbox often makes it easier to switch between different layouts.

Figure 6-3   How to create a 1-column layout

How to create a 2-column layout

Figure 6-4 shows how to create a 2-column page layout. The HTML for this layout is the same as the previous figure. In addition, the CSS uses the same background colors for the flexbox items. However, the CSS for the <main> element doesn't set the flex-direction property to column. As a result, the flex-direction property uses its default value of row, and the two flex items in the <main> element display horizontally. This causes the browser to display the two columns shown in this figure.

In addition, the CSS sets the flex-basis property for the <section> element to 65%, and it sets the flex-basis property for the <aside> element to 35%. This causes the aside to take up less space than the section, which is a common page layout. Finally, the CSS sets the min-height property of the <section> element to 10em. This adds some height to the layout example so it's proportioned more like a real-world web page.

Occasionally, you may need to convert a web page that uses fixed widths to use relative widths instead. This often happens when a graphic designer creates the design for a web page using software that renders the design as an image that uses pixels. To convert the fixed widths to the percents you need, you can use the following formula:

```
target ÷ context x 100 = percent
```

Here, *target* is the width of the element in pixels that you want to convert, and *context* is the width of its containing element in pixels.

For example, suppose you need to convert a web page that has a <main> element with a width of 960 pixels, a <section> element with a width of 600 pixels, and an <aside> element with a width of 360 pixels. To get the flex-basis value for the <section> element, you perform the following calculation:

```
570 ÷ 960 x 100
```

This results in 62.5%. Then, to get the flex-basis value for the <aside> element, you perform the following calculation:

```
330 ÷ 960 x 100.
```

This results 37.5%. Then, if the percentages don't need to be exact, you can make the CSS easier to write by adjusting these percents to whole numbers like 62% and 38%.

A 2-column layout

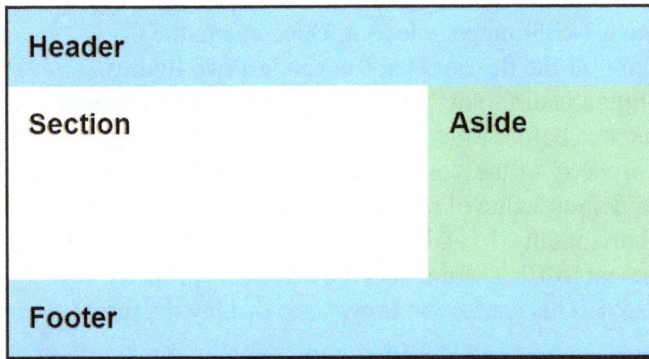

Header	
Section	Aside
Footer	

The HTML for the \<body\> element

```
<body>
    <header><h2>Header</h2></header>
    <main>
        <section><h2>Section</h2></section>
        <aside><h2>Aside</h2></aside>
    </main>
    <footer><h2>Footer</h2></footer>
</body>
```

The CSS

```
body {
    font-family: Arial, sans-serif;
    margin: 1em;
}
main {
    display: flex;
}
section {
    flex-basis: 65%;
    min-height: 10em;       /* just to add some height */
    background-color: azure;
}
aside {
    flex-basis: 35%;
    background-color: aquamarine;
}
header, footer {
    background-color: lightskyblue;
}
```

Description

- Many websites use a 2-column layout for devices with large screens.
- You can use flexbox to create a 2-column layout.

Figure 6-4   How to create a 2-column layout

How to create a 3-column layout

Figure 6-5 shows how to create a 3-column page layout. Once again, the CSS uses the same background colors for the flexbox items as the last two figures. However, the HTML for this figure is different. Specifically, the <main> element contains two <aside> elements, one before the <section> element and one after.

The CSS sets the display property of the <main> element to flex. Then, the flex-direction property uses its default value of row to display the three flex items in the <main> element horizontally. In addition, the CSS sets the flex-basis property for the <section> element to 50%, and it sets the flex-basis property for both <aside> elements to 25%. This causes the browser to display the three columns shown in this figure.

A 3-column layout

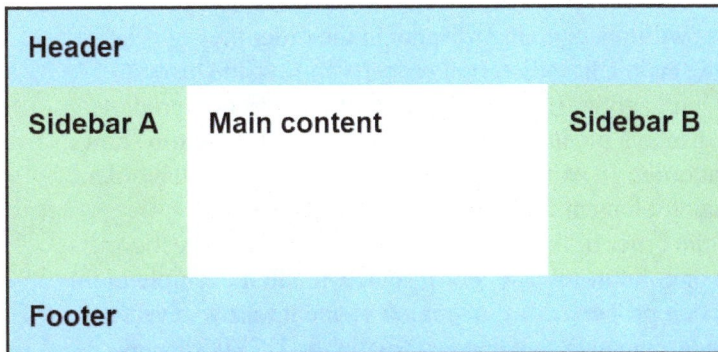

The HTML for the <body> element

```
<body>
    <header><h2>Header</h2></header>
    <main>
        <aside><h2>Sidebar A</h2></aside>
        <section><h2>Main content</h2></section>
        <aside><h2>Sidebar B</h2></aside>
    </main>
    <footer><h2>Footer</h2></footer>
</body>
```

The CSS

```
body {
    font-family: Arial, sans-serif;
    margin: 1em;
}
main {
    display: flex;
}
section {
    flex-basis: 50%;
    min-height: 10em;        /* just to add some height */
    background-color: azure;
}
aside {
    flex-basis: 25%;         /* applies to both <aside> elements */
    background-color: aquamarine;
}
header, footer {
    background-color: lightskyblue;
}
```

Description

- Many web pages use a 3-column layout, and you can use flexbox to create one.

Figure 6-5   How to create a 3-column layout

How to change the order of flex items

By default, the flex items within a container display in the order they appear in the HTML. If the CSS sets the flex-direction property to row, the browser displays the items from left to right. If the CSS sets the flex-direction property to column, the browser displays the items from top to bottom. Figure 6-6 shows how you can change that order. In particular, it shows how to change the order of the flex items in the <main> element from the previous figure.

One way to change the order of the flex items is to set the flex-direction property to row-reverse or column-reverse. For instance, the first example in this figure sets the flex-direction property for the <main> element to row-reverse. Because of that, the <aside> element that appears first in the HTML displays last, and the <aside> element that appears last in the HTML displays first.

Another way to change the order of flex items is with the order property that's shown in this figure. This gives you more options for setting the order of the flex items, including the ability to set an order that doesn't exist in the HTML.

For instance, the second example sets the flex-direction property for the <main> element to column. Because of that, the flex items display vertically. Then, the second example sets the order property for the <section> element to 1 and the order property for the <aside> element to 2. Because of that, the browser displays the <section> element first and the two <aside> elements display second.

In the second example, the <aside> elements still display in the order that they appear in the HTML. If you wanted to change that, you could select individual <aside> elements by id, class, or pseudo-class and give them different order values. For instance, if you want to display the first <aside> element last, you could add the following style rule:

```
aside:first-child {
    order: 3;
}
```

This would cause the browser to display Sidebar A last, after Sidebar B.

The <main> element with its flex items in reverse order

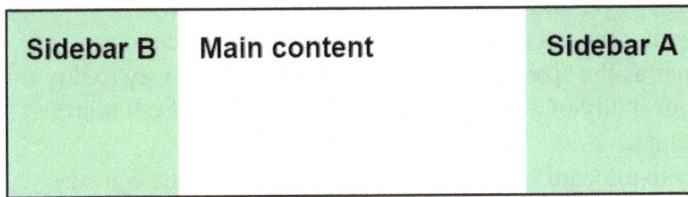

Sidebar B	Main content	Sidebar A

The CSS

```
main {
    display: flex;
    flex-direction: row-reverse;
}
```

A property for changing the order of flex items

Property	Description
order	Sets the order of a flex item relative to other flex items within a flex container.

The <main> element with flex items manually reordered

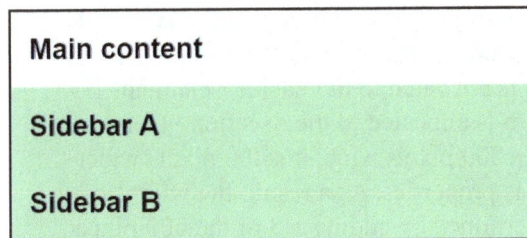

Main content
Sidebar A
Sidebar B

The CSS

```
main {
    display: flex;
    flex-direction: column;
}
section {
    order: 1;
    background-color: azure;
}
aside {
    order: 2;
    background-color: azure;
}
```

Description

- By default, the items within a flex container display in the order in which they appear in the HTML. But flexbox layout makes it easy to change the order.

Figure 6-6   How to change the order of flex items

How to allocate space to flex items

When you use flexbox layout, you can determine how space is allocated to the flex items within a container as the size of the container changes. One way to do this is to set the flex-basis property of a flex item to a percentage of the container as shown earlier in this chapter.

However, another way to allocate space to flex items is to use the flex-grow and flex-shrink properties presented in figure 6-7. To start, the first example in this figure shows some of the HTML that's used by the second and third examples. This HTML consists of a <main> element that contains <nav>, <section>, and <aside> elements.

The second example uses the flex-grow property to specify the proportion that you want flex items to grow relative to each other. Here, the flex-basis properties of the <nav>, <aside>, and <section> elements are set to fixed widths of 100 pixels, 100 pixels, and 200 pixels, respectively. As a result, when the container is 400 pixels wide, the browser allocates 100 pixels to the <nav> and <aside> elements and 200 pixels to the <section> element.

As the size of the container grows, the browser allocates extra space to each flex item based on the ratio of its flex-grow value to the total of the flex-grow values for all of the flex items. Here, the flex-grow properties of the <nav> and <aside> elements are set to 1 and the <section> element is set to 3, for a total of 5 flex-grow units. So, 1/5 of extra space is allocated to the <aside> element, 1/5 is allocated to the <nav> element, and 3/5 is allocated to the <section> element.

In this case, if the container becomes 800 pixels wide, it's 400 pixels wider than the widths specified by the flex-basis properties. As a result, the widths of the <nav> and <aside> elements are determined by adding 1/5 of the 400 pixels to the flex-basis of 100 pixels for a total of 180 pixels. Similarly, the width of the <section> is determined by adding 3/5 of 400 pixels to the flex-basis of 200 pixels for a total of 440 pixels. This causes the width of the content to increase more than the sizes of the navigation and sidebar.

The third example shows how to use the flex-shrink property to determine the size of flex items when the width of a container is less than the widths of the flex items. Here, the CSS sets the flex-basis property of the <nav>, <section>, and <aside> elements to 250 pixels, for a total of 750 pixels. Then, it sets the flex-shrink property of these elements to 0, 1, and 1, respectively. This means there are two flex-shrink units: 1 for the <section> and 1 for the <aside>. So, as the container gets narrower, 1/2 of the negative space is subtracted from the <section> and <aside> elements. However, since the flex-shrink property of the <nav> element is set to 0, that element doesn't shrink.

As a result, if the container is 750 pixels wide, the width of the elements is the same as their flex-basis properties. However, if the container shrinks to 500 pixels wide, which is 250 pixels less than the width specified by the flex-basis properties, the browser subtracts 1/2 of that space (125 pixels) from the width of the <section> and <aside >elements. This causes the size of the content and sidebar to shrink while the navigation remains the same size.

The flex-grow and flex-shrink properties

Property	Description
flex-grow	A number that indicates how much a flex item will grow relative to other flex items in the container. The default is 0, which means the item won't grow.
flex-shrink	A number that indicates how much a flex item will shrink relative to other flex items in the flex container. A value of 0 means that the item won't shrink. The default is 1.

The HTML for a simple page layout

```
<main>
    <nav>Navigation</nav>
    <section>Content</section>
    <aside>Sidebar</aside>
</main>
```

Use the flex-basis and flex-grow properties

```
nav, aside {
    flex-basis: 100px;
    flex-grow: 1;
}
section {
    flex-basis: 200px;
    flex-grow: 3;
}
```

The layout at 400 pixels wide

Navigation	Content	Sidebar

The layout at 800 pixels wide

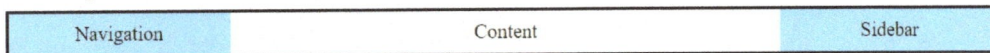

Navigation	Content	Sidebar

Use the flex-basis and flex-shrink properties

```
nav, section, aside { flex-basis: 250px; }
nav { flex-shrink: 0; }
section, aside { flex-shrink: 1; }
```

The layout at 750 pixels wide

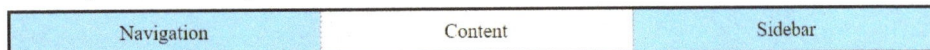

Navigation	Content	Sidebar

The layout at 500 pixels wide

Navigation	Content	Sidebar

Description

- The flex-grow property is used when there's more space available within a container than what's specified by the flex-basis properties for all flex items.
- The flex-shrink property works like the flex-grow property, but it's used when the flex items occupy more space than what's available within a container.

Figure 6-7   How to allocate space to flex items

A web page that uses a 2-column layout

To show a more realistic example of using flexbox layout, figure 6-8 presents the HTML and CSS for a web page that uses a 2-column layout. To start, part 1 shows the page in a browser. Except for the aside, most of the content is the same as the web page presented in the previous chapter. However, this page uses a 2-column layout to display the main content for the page in the first column and the aside in the second column.

A web page that uses a 2-column layout

San Joaquin Valley Town Hall

Bringing cutting-edge speakers to the valley

This season's guest speakers

- October: David Brancaccio
- November: Andrew Ross Sorkin
- January: Amy Chua
- February: Scott Sampson
- March: Carlos Eire
- April: Ronan Tynan

Looking for a unique gift?

Town Hall has the answer. For only $100, you can get a book of tickets for all of the remaining speakers. And the bargain includes a second book of tickets for a companion.

Or, for $50, you can give yourself the gift of our speakers, and still get an extra ticket for a companion, but for just one of the events.

See you at the next show?

Contact us by phone at (559) 555-1212 for ticket information.

Lecture notes

Event change for November

SJV Town Hall is pleased to announce the addition of award-winning author Andrew Ross Sorkin. The appearance of previously scheduled speaker, Greg Mortenson, has been postponed.

Day, time, and location

All one-hour lectures are on the second Wednesday of the month beginning at 10:30 a.m. at William Saroyan Theatre, 700 M Street, Fresno, CA.

© Copyright San Joaquin Valley Town Hall.

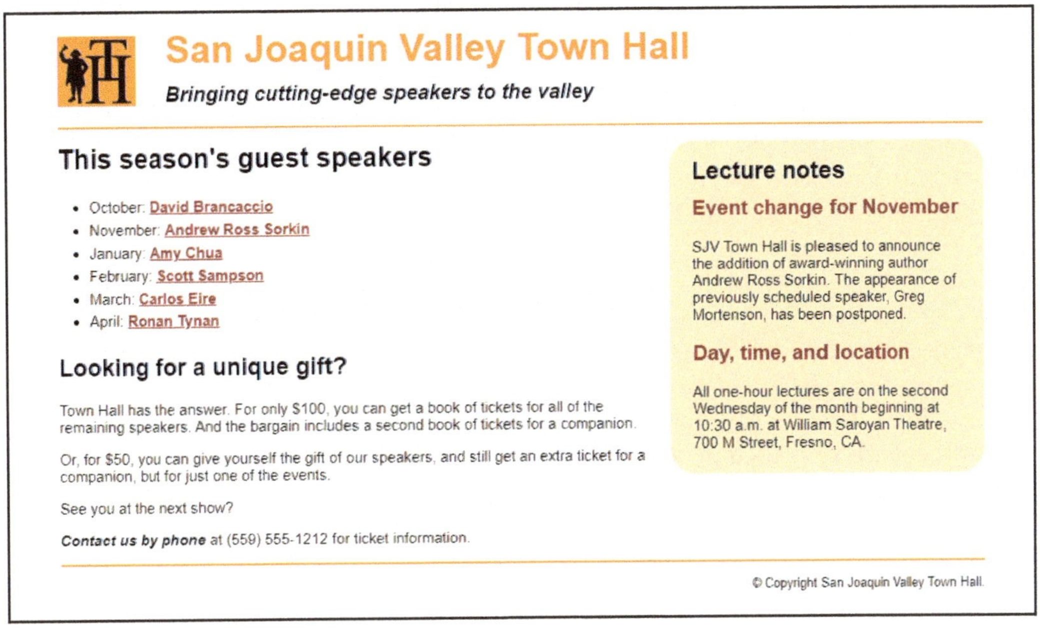

Description

- This web page uses a page layout that includes a header, two columns, and a footer.
- The columns for this page are coded as a <section> element and an <aside> element within a <main> element.
- The columns are created with flexbox layout.
- A bottom border is applied to the header, and a top border is applied to the footer.

Figure 6-8   A web page that uses a 2-column layout (part 1)

Part 2 of figure 6-8 presents the HTML for this web page. Here, the only new content is the <aside> element. In addition, the content that was previously in the <main> element is now in a <section> element within the <main> element.

Before looking at the CSS, take a minute to note the use of the HTML semantic elements for this page. To start, the <header> and <footer> elements identify the header and footer for the page, and the <main> element identifies the rest of the content for the page. Within the <main> element, the <aside> and <section> elements provide the content for the two columns. In addition, the <nav> element within the <section> element identifies some navigation links.

Parts 3 and 4 present the CSS for this web page. Since most of the CSS has been presented earlier in this book, this figure highlights some of the code that applies to the flexbox layout.

First, there's a new variable for the background color for the aside. Since this color is only used in one place, you could code the hex value directly in the CSS for the aside. However, it's a good practice to use variables to define all colors together so they're easy to change. Plus, you never know when you might want to use a color in another location. To that end, it might be good to give the variable a less specific name than "aside". However, for the purposes of this example, this name is OK.

Second, the display property of the <main> element is set to flex. This displays the <section> and <aside> elements horizontally as two columns.

Third, the <section> and <aside> elements have a flex-basis of 65% and 35%, respectively. As a result, the content of the section, which is the main content of the page, is wider than the content for the aside.

Fourth, the <aside> element sets the height property to fit-content. As a result, this flexbox item is only as tall as its content. This is necessary because, by default, flexbox items stretch to fill the entire height of the flexbox container.

The HTML

```
<head>
    <title>San Joaquin Valley Town Hall</title>
    <meta name="viewport" content="width=device-width, initial-scale=1.0">
    <meta charset="UTF-8">
    <link rel="icon" href="images/favicon.ico">
    <link rel="stylesheet" href="styles/main.css">
</head>
<body>
    <header>
        <img src="images/logo.jpg" alt="Town Hall Logo" width="80">
        <h2>San Joaquin Valley Town Hall</h2>
        <h3>Bringing cutting-edge speakers to the valley</h3>
    </header>
    <main>
        <section>
            <h1>This season's guest speakers</h1>
            <nav>
                <ul>
                    <li>October: <a href="speakers/brancaccio.html">
                        David Brancaccio</a></li>

                        .
                        .
                    <li>April: <a href="speakers/tynan.html">
                        Ronan Tynan</a></li>
                </ul>
            </nav>

            <h2>Looking for a unique gift?</h2>
            <p>Town Hall has the answer...</p>
            <p>Or, for $50, you can give...</p>
            <p>See you at the next show?</p>
            <p><em>Contact us by phone</em> at ...</p>
        </section>
        <aside>
            <h2>Lecture notes</h2>

            <h3>Event change for November</h3>
            <p>SJV Town Hall is pleased to announce...</p>

            <h3>Lecture day, time, and location</h3>
            <p>All one-hour lectures are on...</p>
        </aside>
    </main>
    <footer>
        <p>&copy; Copyright San Joaquin Valley Town Hall.</p>
    </footer>
</body>
</html>
```

Figure 6-8   A web page that uses a 2-column layout (part 2)

The CSS

```css
/* variables for theme colors */
:root {
    --light: #f2972e;
    --dark: #931420;
    --aside: #ffebc6;
}

/* reset */
* {
    margin: 0;
    padding: 0;
    box-sizing: border-box;
}

/* default styles */
body {
    font-family: Arial, Helvetica, sans-serif;
    font-size: 100%;
    width: 90%;
    margin: 1em auto;
}
h1, h2, h3, p, ul { padding: .5em 0; }
h1 { font-size: 150%; }
h2 { font-size: 135%; }
h3 { font-size: 120%; }
li, p { font-size: 95%; }
a, em { font-weight: bold; }
a:link, a:visited { color: var(--dark); }
a:hover, a:focus { color: var(--light); }
ul {
    line-height: 1.5;
    margin-left: 2em;
}

header {
    border-bottom: 2px solid var(--light);
    & img {
        float: left;
        margin: .5em 2em .5em 0;
    }
    & h2 {
        font-size: 225%;
        color: var(--light);
        padding: 0;
    }
    & h3 {
        font-style: italic;
        padding-bottom: 1em;
    }
}

main {
    display: flex;
    padding-bottom: .5em;
}
```

Figure 6-8   A web page that uses a 2-column layout (part 3)

The CSS (continued)

```
section {
    flex-basis: 65%;
}

aside {
    flex-basis: 35%;
    background-color: var(--aside);
    height: fit-content;
    padding: 1em 1.5em;
    margin: 1em 0 1em 1.5em;
    border-radius: 25px;
    & h2 { padding: 0 }
    & h3 { color: var(--dark); }
}

footer {
    border-top: 2px solid var(--light);
    padding-top: .5em;
    & p {
        font-size: 80%;
        text-align: right;
    }
}
```

Figure 6-8   A web page that uses a 2-column layout (part 4)

How to implement a fluid design

Now that you know how to use flexbox to lay out a web page, you're ready to learn how to implement a *fluid design* that adjusts the web page to the size of the screen. To do that, you can use relative font sizes, relative line heights, and scalable images. This is another important part of responsive web design.

How to control the viewport

The *viewport* is the part of a web page that's displayed on the screen. When you develop a responsive website, you need to be sure that you configure the viewport appropriately so it works for both large and small screens. To do that, you can use the <meta> element that's presented in figure 6-9.

On a device with a large screen such as a desktop computer, the viewport is the content that's displayed by the browser. However, the user can change the size of the viewport by changing the size of the browser window. In this figure, for example, two desktop browser windows simulate a small viewport.

On a device with a small screen such as a mobile phone, the viewport is usually the size of the screen. However, it can also be larger or smaller than the screen. If the width of the viewport is larger than the screen, the user can scroll to view the entire viewport.

In this figure, the first web page displays in a way where the entire width of the page is visible. Since this page doesn't include a <meta> tag that provides viewport metadata, the desktop browser automatically scales the images and font sizes to fit the browser width so that the user doesn't have to scroll left and right to view the content. Some mobile browsers also scale a page like this. However, other mobile browsers don't scale the page at all, which causes the viewport to extend beyond the width of the screen. Then, the user needs to scroll left and right to see the entire viewport, which isn't what you usually want.

In most cases, you want to make sure that the web page isn't scaled automatically. To do that, you add a <meta> element like the one in this figure to the page. When you use this <meta> element, the second page in this figure displays. Here, the text displays at its original size, so it rolls over as necessary to fit in the screen.

In the <meta> element, the name attribute is set to viewport to indicate that the metadata applies to the viewport. Then, the content attribute specifies two properties for the viewport. Here, the width property sets the width of the viewport to the width of the device, and the initial-scale property sets the initial zoom factor, or *scale*, for the viewport to 1. This represents the default width for the viewport and is what prevents the browser from scaling the page automatically.

If you set the user-scalable property to "no", the user can't zoom in or out of the display. Similarly, if you set the minimum-scale and maximum-scale properties, you can limit how much the user can zoom in or out. However, you generally shouldn't use these properties because restricting a user's ability to zoom in can negatively affect the visually impaired. For this reason, most browsers allow users to override settings that restrict zooming.

A web page on a small viewport

Without viewport metadata

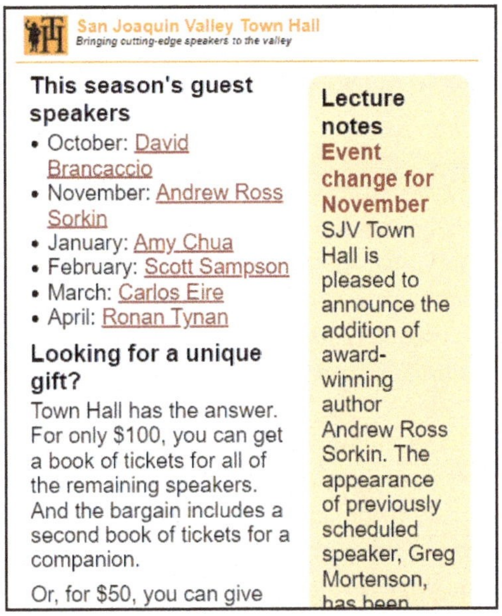

With viewport metadata

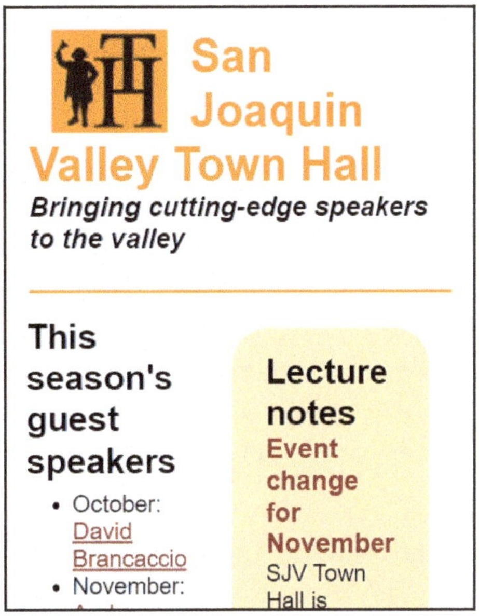

Content properties for viewport metadata

Property	Description
width	The logical width of the viewport specified in pixels. You can also use the device-width keyword to indicate that the viewport should be the width of the screen in CSS pixels at a scale of 100%.
height	The logical height of the viewport specified in pixels. You can also use the device-height keyword to indicate that the viewport should be the height of the screen in CSS pixels at a scale of 100%.
initial-scale	A number that indicates the initial zoom factor used to display the page.
minimum-scale	A number that indicates the minimum zoom factor for the page.
maximum-scale	A number that indicates the maximum zoom factor for the page.
user-scalable	Indicates whether the user can zoom in and out of the viewport. Possible values are yes and no.

A <meta> element that sets viewport properties

```
<meta name="viewport" content="width=device-width, initial-scale=1.0">
```

Description

- The *viewport* is the part of a web page that's displayed on the screen. It can be larger or smaller than the screen.
- You use a <meta> element to control the viewport. You add this element within the <head> element of a page.

Figure 6-9   How to control with the viewport

How to use units of measure based on the viewport

Besides the units of measure that you've learned about so far, figure 6-10 presents some units that were developed especially for responsive design. These units are based on the size of the viewport.

The units of measure that are based on the viewport are relative to the height or width of the viewport. For example, one vw is equal to 1/100 of the width of the viewport. As a result, if the viewport is 600 pixels wide, one vw is equal to 6 pixels. Similarly, if the viewport is 800 pixels tall, one vh is equal to 8 pixels.

The vmin and vmax units are similar, but vmin corresponds to the smallest dimension of the viewport and vmax corresponds to the largest dimension of the viewport. So if the viewport is 600 pixels wide by 800 pixels tall, one vmin is equal to 6 pixels and one vmax is equal to 8 pixels.

The code examples in this figure show how these units of measure work. Here, the CSS sets the width of the body to 90vw. As a result, the width of the body is always 90/100 of the viewport width. In addition, since the left and right margins of the body are set to auto, it is centered in the viewport.

The two screens in this figure show how the page displays when the viewport is 580 and 412 pixels wide. As you can see, as the width of the viewport changes, the width of the body also changes.

Relative units of measure based on the viewport

Symbol	Name	Description
vh	viewport height	One vh is equal to 1/100 of the height of the viewport.
vw	viewport width	One vw is equal to 1/100 of the width of the viewport.
vmin	viewport minimum	One vmin is equal to 1/100 of the height or width of the viewport, depending on which is smaller.
vmax	viewport maximum	One vmax is equal to 1/100 of the height or width of the viewport, depending on which is larger.

The HTML for a simple page

```
<body>
    <header>
        <h1>San Joaquin Valley Town Hall</h1>
    </header>
    <main>
        <p>Welcome to San Joaquin Valley Town Hall. We have some fascinating
        speakers for you this season!</p>
    </main>
</body>
```

Set the width of the body based on the width of the viewport

```
body {
    width: 90vw;
    margin: 1em auto;
    padding: 1em;
    border: 1px dotted black;
}
```

The page when the viewport is 580 pixels wide

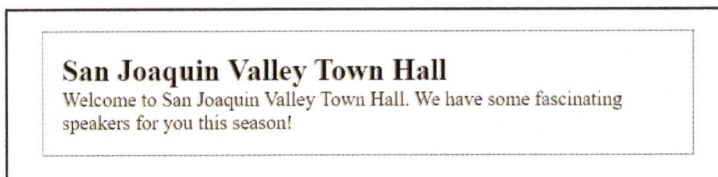

San Joaquin Valley Town Hall
Welcome to San Joaquin Valley Town Hall. We have some fascinating
speakers for you this season!

The page when the viewport is 412 pixels wide

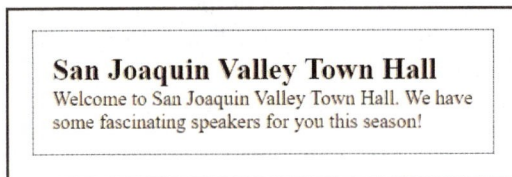

San Joaquin Valley Town Hall
Welcome to San Joaquin Valley Town Hall. We have
some fascinating speakers for you this season!

Figure 6-10   How to use units of measure based on the viewport

How to size fonts relative to the viewport

Earlier in this book, we recommended that you use relative measurements for your font sizes. That way, the font size for an element is relative to the size of the font used by the parent element. As a result, if you want to change the font size that's used by a parent element and all of its child elements, you just need to change the font size for the parent element.

Now, figure 6-11 shows how to make that font size responsive by using the CSS calc() function. This function executes the expression it receives and returns the result. The expression within this function can use the addition (+), subtraction (-), multiplication (*), and division (/) operators and can work with absolute units of measure like pixels and points, relative units of measures like ems and percents, and literal values like integers or decimal numbers.

The first example in this figure shows how to use the calc() function to set the width property of the <body> element. Here, the code sets the width property to the result of 100% of the containing block minus 40 pixels. Since the margin property is set to auto, this centers the body on the screen with 20 pixels of space on either side. This creates an effect that's similar to setting the width to a percent such as 90%. However, when you use this technique, there are always 20 pixels of space on each side of the body, regardless of the width of the screen.

The second example shows how to use the calc() function to set a responsive font size. Here, the code sets the font-size property of the <body> element to 100% of the default font size for the browser plus 1 viewport width. The screens below the example show that this causes the browser to use a larger font size when the viewport is wider and a smaller font size when the viewport is smaller.

You can achieve a similar effect by simply coding a viewport width value, such as

```
font-size: 4vw;
```

However, this technique can cause the font size to become too small when displayed on viewports that have small screen widths. By contrast, if you use the calc() function shown in the second example, the first value of the expression functions as a minimum value. For instance, in this example, the font size will never be smaller than the default font size for the browser, no matter how narrow the viewport.

Earlier in this book, you learned how to use the line-height property to set the amount of space between lines. By default, most browsers set this property to 1.2. However, a value of 1.5 is often recommended to improve readability for long chunks of text by adding more space between lines. This also improves accessibility for the visually impaired.

The last example in this figure sets the line-height property of the <body> element to 1.5. Then, the other elements on the web page inherit this line height. If you find that you need more or less space between lines for some elements such as headings, you can override this value for those elements. Note that this example uses a *unitless measure*. In other words, it's not set as 1.5em or 150%. As a result, the line height scales as the viewport changes.

One of the CSS functions

Function	Description
calc(expression)	Returns the result of the expression it receives. The expression can use the **+**, **-**, *****, and **/** operators and can work with absolute units of measure, relative units of measure, and literal values.

Set the width property

```
body {
    width: calc(100% - 40px);
    margin: auto;
}
```

Set a responsive font size based on the viewport width

```
body {
    ...
    font-size: calc(100% + 1vw);
}
```

The font size when the viewport is 580 pixels wide

Welcome to San Joaquin Valley Town Hall. We have some fascinating speakers for you this season!

The font size when the viewport is 412 pixels wide

Welcome to San Joaquin Valley Town Hall. We have some fascinating speakers for you this season!

Set the line height

```
body {
    ...
    line-height: 1.5;  /* unitless measure */
}
```

Description

- You can use the calc() function and relative units of measure to calculate font size based on the width of the screen. That way, the font size scales as the viewport changes.
- When using calc() to set the font size, a larger second value creates a larger font and a more dramatic change when the viewport changes. If you want the change to be more subtle, use a smaller value. It's common to add a fraction of a viewport width, such as 0.2vw.
- A larger line height value adds more space between lines, which is better for accessibility. Most browsers use a default value of 1.2.
- If you use a *unitless measure* for the line height, it scales as the viewport changes.

Figure 6-11   How to size fonts relative to the viewport

How to scale images

As the width of an element changes, you typically want the width of any images within that element to change too. In other words, you want the image to scale with the element. To do that, you can use *scalable images* as shown in figure 6-12.

To start, the first example shows the HTML that displays an image on a web page. To make this image scalable, the CSS in the second example sets its max-width property to 100%. As a result, the image always scales to be as wide as the element that contains it.

Some developers also like to set the height property to auto when they scale images. That way, the aspect ratio of the height to the width remains the same regardless of the scale of the image. However, most browsers set the height property to auto by default. As a result, you can usually omit the height property.

In general, when you use a scalable image, you shouldn't display it any larger than its native size since that can lead to poor image quality. To limit an image to its native size, you can add a max-width property as shown in the third example in this figure. Here, the first declaration sets the width property to 100%, and the second declaration sets the max-width property to 600px. As a result, the image scales to be as wide as the containing block if the containing block is 600px or less. However, the image never scales to be wider than 600 pixels.

When you use small images like logos, you may also want to include the min-width property. That way, the image won't get too small.

This figure also presents three CSS functions that you can use to set the minimum and maximum widths of an image. The min() and max() functions accept a list of values and return the smallest or largest value in the list. By contrast, the clamp() function accepts a minimum value, a default value, and a maximum value. It returns the default value if it's between the minimum and maximum values. Otherwise, it returns the minimum or maximum value.

The fourth example in this figure uses the min() function to set the width property. To do that, it passes values of 100% and 600px to the function. As a result, when the width of the containing block is greater than 600px, the function returns 600px because it's the lower value. This example creates the same effect as the third example. As a result, you can use either technique to specify a maximum value for an image. The advantage of the fourth example is that the code is shorter. However, the third example is easier to read and understand.

The fifth example uses the clamp() function to set the width property. To do that, it passes values of 300px, 100%, and 600px to the function. As a result, if the width of the containing block is greater than 300px and less than 600px, the function returns 100%. But when the width of the containing block is greater than 600px, the function returns the maximum value of 600px. And when the width of the containing block is less than 300px, the function returns the minimum value of 300px. This example creates the same effect as the third example if the third example also had a min-width property set to 300px.

A scalable image at different screen sizes

The HTML for the image

```
<img src="surf.jpg" alt="rocky surf">
```

The CSS for the image

```
img { width: 100%; }
```

CSS that limits the width of the image

```
img {
    width: 100%;
    max-width: 600px;
}
```

Three CSS functions for setting minimum and maximum sizes

Function	Description
min(list-of-values)	Returns the smallest value in the list.
max(list-of-values)	Returns the largest value in the list.
clamp(min, val, max)	Returns val unless it's below the min or above the max. In those cases, this function returns the min or max value.

CSS that limits the maximum width of the image

```
img { width: min(100%, 600px); }
```

CSS that limits the minimum and maximum widths of the image

```
img { width: clamp(300px, 100%, 600px); }
```

Description

- To create a *scalable image*, set the max-width property of the image to a percent of its containing block.
- To limit the size of an image to its native width, set the max-width property to the native width in pixels. This maintains the image quality.

Figure 6-12   How to scale images

A fluid web page

Figure 6-13 presents a web page on the Town Hall website that uses a fluid design. This page displays some information about an upcoming speaker. To do that, it uses flexbox to create a 2-column layout, with the flex-basis properties for the flex items for the columns set to percents. In addition, it sets the widths of the other structural elements to percents, it sets the font size and line height based on the size of the viewport, and it uses scalable images. As a result, if you open this page in a browser and change the width of the browser window, the page adjusts to the size of the window as shown by the two screen captures in part 1.

A fluid web page at different screen sizes

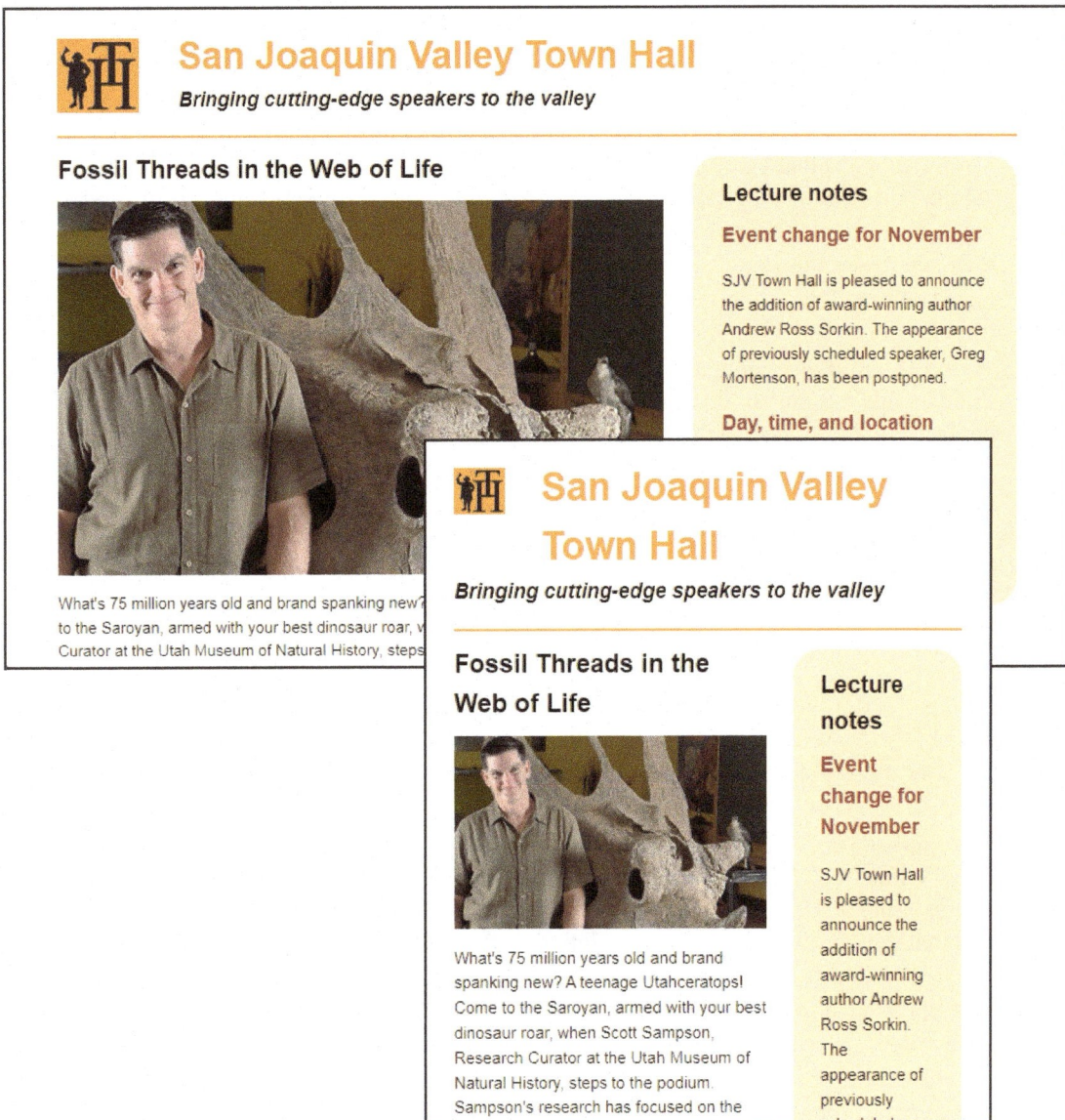

Description

- This page uses a fluid layout so the widths of the elements change as the width of the viewport changes.
- This page scales the images and sets the font sizes and line height based on the width of the viewport.

Figure 6-13   A fluid web page (part 1)

Part 2 of figure 6-13 shows some of the HTML and CSS for this web page. To help you focus on the most important parts, some lines of code are highlighted.

In the HTML, the head section contains a <meta> element that specifies viewport metadata. In addition, the HTML omits the widths of the images because this web page uses CSS to scale these images.

In the CSS, the style rule for the <body> element uses the calc() function to set the font size. The first value passed to this function is 90%. Because of this value, the font size will never be less than 90% of the default font size for the browser. Then, this code adds a second value that uses a relative measurement based on the viewport to the first value. Because of this, the font-size changes as the viewport width changes. However, since the second value is a fraction of the viewport width, the change in font size is subtle.

The line height for the <body> element is set to 1.5. This overrides the default value provided by the browser, which is usually 1.2. As a result, this page displays more space between lines, which improves readability and makes the page more accessible for the visually impaired.

The width of the <body> element is set with the min() function. To start, this code passes a value of 90% to the min() function to set its width to 90% of the containing element. Then, it passes a value of 1024 pixels to set its maximum width. Because of that, the page won't expand beyond this width even if the browser window can accommodate a wider page. This is useful because most fluid pages don't look good when displayed beyond a given width, which is possible on large screens.

The style rule for the <header> element sets the line height to 1.2. That's because the larger line height of 1.5 that's set for the <body> element adds too much space between the headings in the <header> element. As a result, this code sets the line height to 1.2 for these two headings so they align better with the image in the header. This shows there's no one-size-fits-all solution to line height and that you can always override a default setting to get the look you want.

The nested style rule for the logo image in the <header> element uses the clamp() function to set the image width. This code starts by passing 40 pixels to set a minimum width for the image, and it ends by passing a value of 80 pixels to set a maximum width that's equal to the actual width of the image. Between these values, the code passes a relative value of 10%. As a result, when the image width is greater than 40 pixels and less than 80 pixels, the image scales as the viewport width changes. Otherwise, the image width is set with the specified minimum or maximum value.

The nested style rule for the image in the <section> element works similarly. However, because it doesn't set a minimum width, it uses the min() function instead of the clamp() function. This code passes a first value of 100% to set the width the same as the containing element and a second value of 600 pixels to set a maximum width that's equal to the actual width of the image.

The <head> and <body> elements for the speaker page

```
<head>
    <title>San Joaquin Valley Town Hall | Scott Sampson</title>
    <meta charset="UTF-8">
    <meta name="viewport" content="width=device-width, initial-scale=1.0">
    <link rel="stylesheet" href="../styles/main.css">
</head>
<body>
    <header>
        <img src="../images/logo.gif" alt="Town Hall Logo">
        <h2>San Joaquin Valley Town Hall</h2>
        <h3>Bringing cutting-edge speakers to the valley</h3>
    </header>
    <main>
        <section>
            <img src="../images/sampson.jpg" alt="Scott Sampson"/>
            <h1>Fossil Threads in the Web of Life</h1>
            <p>What's 75 million years old ...</p>
        </section>
        <aside>...</aside>
    </main>
    <footer>...</footer>
</body>
```

The updated CSS

```
body {
    font-family: Arial, Helvetica, sans-serif;
    font-size: calc(90% + 0.3vw);
    line-height: 1.5;
    width: min(90%, 1024px);
    margin: 1em auto;
}
h1, h2, h3, p, ul { padding: .25em 0; }          /* padding decreased */
ul { margin-left: 2em; }              /* line height no longer set for ul */
header {
    border-bottom: 2px solid var(--light);
    line-height: 1.2;
    & img {
        width: clamp(40px, 10%, 80px);
        float: left;
        margin: .5em 2em .5em 0;
    }
    & h2 {
        font-size: 225%;
        color: var(--light);
        padding: 0;
    }
    & h3 {
        font-style: italic;
        padding-bottom: 1em;
    }
}
section {
    flex-basis: 65%;
    & img { width: min(100%, 600px); }
}
```

Figure 6-13   A fluid web page (part 2)

Perspective

Now that you've completed this chapter, you should be able to use flexbox layout to develop fluid web pages with layouts that use multiple columns. This is the first step toward developing web pages that provide for responsive web design.

Terms

flexible box layout	cross end
flexbox layout	main size
flex container	cross size
flex items	target
main axis	context
flex direction	fluid design
cross axis	viewport
main start	scale
main end	scalable image
cross start	

Exercise 6-1   Enhance the Town Hall home page

In this exercise, you'll enhance the formatting of the Town Hall home page. You'll also format the Featured Speaker part of the page that's added to the HTML. When you're through, the page should look like this:

Open the HTML and CSS files and review the HTML

1. Use your text editor to open these HTML and CSS files:

```
exercises\ch06\town_hall\index.html
exercises\ch06\town_hall\styles\main.css
```

2. In the HTML file, note the viewport metadata in the head section.

3. Note that the <main> element now contains an <aside> element and a <section> element.

4. Note that the <aside> element contains the "Guest speakers" heading, links, and images, and that the heading is now an <h2> element.

5. Note that the <section> element contains the "Our mission" heading, paragraph, and blockquote, as well as the "Our Ticket Packages" heading and unordered list.

6. Note that the <section> element also contains a "Featured Speaker" heading that's an <h1> element, followed by an <article> element with headings, an image, and a paragraph that describes the speaker. In addition, it has a link for more information.

7. Note that none of the <img> elements have a width attribute.

8. Run the web page and view it.

Enhance the CSS file to create a 2-column layout

From this point on, refresh the page after each change to make sure it looks the way you want it. If you have any problems, use the Developer Tools to help you fix them.

9. Enhance the style rule for the <main> element by setting the display property to flex. The <aside> and <section> elements should display side by side. Make sure the <aside> element displays before the <section> element.

10. Add style rules for the <aside> and <section> elements. Set the flex-basis property for the aside to 25% and for the section to 75%.

11. Adjust the margins and padding as needed so the columns lay out the way you want.

Add the CSS for the featured speaker

12. In the HTML file, note that the <article> element is assigned to the featured class. In the CSS file, add a style rule for this class.

13. In the style rule for the <main> element, find the nested style rule for the <h1> element. Copy the border-top and border-bottom declarations to the style rule for the featured class. Then, delete the nested style rule for the <h1> element.

14. In the style rule for the <section> element, add a nested style rule for the <img> element that floats the image to the right.

15. Adjust the margins and padding as needed so it looks the way you want.

Make the web page fluid

16. Enhance the style rule for the <body> element as follows:

    - Set the width to 80% with a maximum width of 1024 pixels.
    - Set the font-size to 90% plus .2 viewport widths.
    - Set the line height to 1.5. To accommodate this, change the default top and bottom padding for headings, paragraphs, and unordered lists from .5em to .25em and set the line height in the <header> element to 1.2.

17. Make the images scalable, with the following minimum and maximum widths:

    Logo image: min 40 pixels, max 80 pixels

    Featured speaker image: min 100 pixels, max 250 pixels

    Guest speaker images: min 40 pixels, max 75 pixels

    For each image, you can use 100% as the width of the image when it's between the minimum and maximum widths. However, that value may keep the image from actually scaling. If that's the case, experiment with values that are less than 100% until the image scales the way you want it to.

18. Make any final adjustments to the margins or padding and test the page.

Exercise 6-2   Enhance a speaker page

In this exercise, you'll enhance the page for one speaker. This formatting for this page is similar to the home page.

San Joaquin Valley Town Hall
Celebrating our 75th Year

Guest speakers

October
David Brancaccio

November
Andrew Ross Sorkin

January
Amy Chua

February
Scott Sampson

Fossil Threads in the Web of Life

February
Scott Sampson

What's 75 million years old and brand spanking new? A teenage Utahceratops! Come to the Saroyan, armed with your best dinosaur roar, when Scott Sampson, Research Curator at the Utah Museum of Natural History, steps to the podium. Sampson's research has focused on the ecology and evolution of late Cretaceous dinosaurs and he has conducted fieldwork in a number of countries in Africa.

Scott Sampson is a Canadian-born paleontologist who received his Ph.D. in zoology from the University of Toronto. His doctoral work focused on two new species of ceratopsids (horned dinosaurs) from the Late Cretaceous of Montana, as well as the growth and function of certopsid horns and frills.

Following graduation in 1993, Sampson spent a year working at the American Museum of Natural History in New York City, followed by five years as assistant professor of anatomy at the New York College of Osteopathic Medicine on Long Island. He arrived at the University of Utah accepting a dual position as assistant professor in the Department of Geology and Geophysics and curator of vertebrate paleontology at the Utah Museum of Natural History. His research interests largely revolve around the phylogenetics, functional morphology, and evolution of Late Cretaceous dinosaurs.

In addition to his museum and laboratory-based studies, Sampson has conducted paleontological work in Zimbabwe, South Africa, and Madagascar, as well as the United States and Canada. He was also the on-air host for the Discovery Channel's Dinosaur Planet and recently completed a book, Dinosaur Odyssey: Fossil Threads in the Web of Life, which is one of the most comprehensive surveys of dinosaurs and their worlds to date.

© San Joaquin Valley Town Hall, Fresno, CA 93755

Open and review the HTML file for the speaker

1. Use your text editor to open this HTML file:

   `exercises\ch06\town_hall\speakers\sampson.html`

2. Review the HTML file and see that it's structured the same way as the HTML file for the home page and uses the same CSS file.

3. Run the web page and view it. It should only need a few adjustments.

Enhance the CSS file

4. In the HTML file, note that the <section> element is assigned to the speaker class. In the CSS file, add a style rule for the speaker class.

5. In the style rule for the speaker class, do the following:

   - Add a nested style rule for the <h1> element that sets the color to the accent color variable.

   - Add a nested style rule for the <h2> element that sets the color to black.

   - Add a nested style rule for the <p> element that sets the top and bottom padding to .5em. This overrides the default padding for the <p> element that you adjusted in the previous exercise.

6. Make any final adjustments to the margins or padding and test the page.

7. Test all the links to make sure they work correctly.

8. Make sure all images display correctly.

Chapter 7

How to use flexbox with navigation menus

In the last chapter, you learned how to use flexbox for page layout. In this chapter, you'll learn how to use flexbox to create navigation menus. In the process, you'll learn even more about flexbox including how to wrap and align flex items. And you'll learn how to use positioning and symbols to create a 2-tier navigation menu.

Basic skills for navigation menus

This chapter begins by presenting some basic skills for creating navigation menus for your website. In particular, it shows how to use flexbox to lay out the items in a navigation menu.

How to display a navigation menu

Figure 7-1 starts with the HTML for a simple *navigation menu*, or *navigation bar*, which is a menu that provides links to the most important sections of a website. The <a> elements for the links are typically coded within an unordered list, because a list is a logical container for links. Then, the unordered list is coded within a <nav> element, because the <nav> element conveys that the purpose of the HTML is navigation. This makes your HTML more readable, for both humans and screen readers. As a result, it makes your web page more accessible.

The first CSS example displays a *horizontal navigation menu*. To start, the style rule for the <nav> element sets the background color to black. As a result, the navigation menu displays on a black background.

The style rule for the <ul> element sets the display property to flex. By default, the flex-direction property is set to row. As a result, this displays the <li> elements, which are the flex items, horizontally. In addition, the style rule for the <ul> element sets the list-style-type property to none. This removes the bullets from the unordered list items.

The style rule for the <li> element uses the padding property to add space around each list item. Then, the style rule for the <a> element sets the text-decoration property to none. This removes the underline from the links. In addition, this style rule sets the color of the links to white so they're visible on the black background of the parent <nav> element.

When a browser displays this navigation menu, it lays out the menu items horizontally with white text on a black background, no underlines, and appropriate spacing around each link.

The second CSS example displays a *vertical navigation menu*. The style rule for the <nav> element sets the background color to black. In addition, it sets the width to 40%. Otherwise, the <nav> element would be as wide as the element that contains it. Sometimes that's what you want, but if it isn't, you can set the width property as needed.

The style rule for the <ul> element sets the display property to flex and the flex-direction property to column. This causes the <li> elements to display vertically. This style rule also removes the bullets from the unordered list items. Finally, the style rules for the <li> and <a> elements are the same as the first example.

When a browser displays this navigation menu, it lays out the menu items vertically with white text on a black background, no underlines, and appropriate spacing around each item.

HTML for a navigation menu

```
<nav>
    <ul>
        <li><a href="index.html">Home</a></li>
        <li><a href="speakers.html">Speakers</a></li>
        <li><a href="aboutus.html">About Us</a></li>
    </ul>
</nav>
```

CSS that displays the menu horizontally

```
nav {
    background-color: black;
}
ul {
    display: flex;                  /* flex-direction is row by default */
    list-style-type: none;
}
li {
    padding: .5em 1em;
}
a {
    text-decoration: none;
    color: white;
}
```

How it looks in a browser

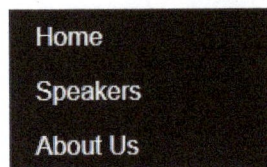

CSS that displays the menu vertically

```
nav {
    background-color: black;
    width: 40%;
}
ul {
    display: flex;
    flex-direction: column;
    list-style-type: none;
}
/* the styles for the <li> and <a> are the same as above */
```

How it looks in a browser

Description

- A *navigation menu* allows the user to navigate the website. It typically contains links that are coded within an unordered list that's coded within a <nav> element.
- You can use flexbox layout to display a navigation menu horizontally or vertically.

Figure 7-1   How to display a navigation menu

How to style the links in a navigation menu

Figure 7-2 presents some good practices for styling the links in a navigation menu. First, it shows how to make the entire list item that contains the link clickable, not just the text. Second, it shows how to change the appearance of a menu item when a user hovers the mouse over it or tabs to it. Third, it shows how to highlight the menu item for the current page so users know where they are in the website.

To make the entire menu item clickable, you add padding to the <a> element, not the <li> element. This makes the text for the link and the padded area clickable. However, by default, an <a> element is an inline element, and inline elements don't allow top and bottom padding. So, to apply padding to all four sides of an <a> element, you must first set the display property to block or inline-block as shown in this figure. Here, the style rule for the <a> element sets the display property to block and applies padding to all four sides of the link.

To change the appearance of a menu item when the user hovers the mouse over it or tabs to it, you can use the style rule for the :hover and :focus pseudo-classes of the <a> element. In this figure, the CSS turns the underlining for the link back on. As a result, when a browser displays this navigation menu, it displays an underline under the link when the mouse hovers over that link or the user tabs to it. In addition, when the mouse hovers anywhere over the menu item, the cursor becomes a hand rather than a pointer. This shows that the entire item is clickable.

To highlight the menu item for the current page, you can assign it to a CSS class. Then, you can write a style rule for that class to highlight that menu item. In this figure, the HTML assigns the link to the Speakers page to a class named current. Since this menu item is for the current page, there's no need to link to another page. As a result, this HTML omits the href attribute from the <a> element.

The CSS adds a style rule for all <a> elements of the current class. This style rule highlights the menu item by changing its color and background color, which provides a visual indication that the current page is in the Speakers section of the website. In addition, this style rule turns off underlining for the current class. As a result, if the user hovers the mouse over this menu item, the cursor won't become a hand pointer. This shows that the link for the current item isn't clickable.

The style rule for the <li> element applies a small margin to all four sides of the link. As a result, some of the black background of the navigation menu shows around the highlighted link.

HTML for a navigation menu that has a current link

```
<nav>
    <ul>
        <li><a href="index.html">Home</a></li>
        <li><a class="current">Speakers</a></li>
        <li><a href="about.html">About Us</a></li>
    </ul>
</nav>
```

CSS that formats the links

```
ul {
    display: flex;
    list-style-type: none;
}
a {
    text-decoration: none;
    color: white;
    display: block;
    padding: .5em 1em;
    border-radius: 10px;
}
a:hover, a:focus {
    text-decoration: underline;
}
a.current {
    color: black;
    background-color: lightgray;
    text-decoration: none;
}
li {
    margin: .25em;
}
```

How it looks in a browser

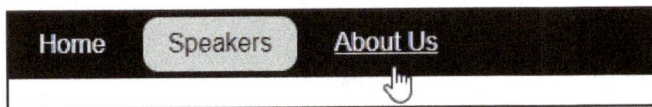

Description

- To make the entire box for a link clickable, you can set the link's display property to block and use padding to provide space around the link.

- It's a good practice to highlight the link for the current page so users know where they are in the website, and it's common to omit the href attribute for the current link.

Figure 7-2   How to style the links in a navigation menu

How to wrap menu items

If a flex container for a horizontal menu isn't wide enough for the flex items it contains, those items don't wrap to multiple lines by default. Instead, the sizes of the flex items are reduced as much as possible and the content within each flex item wraps. This can cause the browser to cut off some of the menu items on small screens as shown by the first example in figure 7-3.

To avoid this issue, you can set the flex-wrap property of the flex container to wrap. Then, when the flex container isn't wide enough for all of the flex items, the flex items wrap to a second row within the container, as shown by the second example.

You can also use the flex-wrap property with vertical menus. For instance, the third example creates a vertical menu by setting the flex-direction property to column. Then, it sets flex-wrap property to wrap. As a result, if the flex container isn't tall enough to display all of the flex items, it wraps them to a second column.

The flex-flow property in the table in this figure is a shorthand property that you can use to set both the flex-direction and flex-wrap properties in a single declaration. For instance, you could shorten the CSS in the third example by replacing the two highlighted declarations with this declaration:

```
flex-flow: column wrap;
```

This makes the code shorter, but it may also make the code more difficult to read for some developers.

Properties that wrap flex items

Property	Description
flex-wrap	Determines how flex items wrap if there isn't enough room for them in the flex container. Possible values are nowrap, wrap, and wrap-reverse. The default is nowrap.
flex-flow	The shorthand property for setting the flex-direction and flex-wrap properties.

A horizontal menu with no wrapping

```
ul {
    display: flex;
}
```

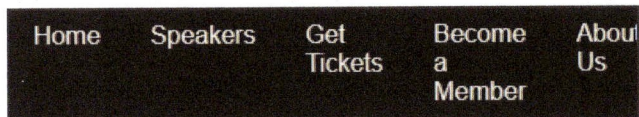

A horizontal menu with wrapping

```
ul {
    display: flex;
    flex-wrap: wrap;
}
```

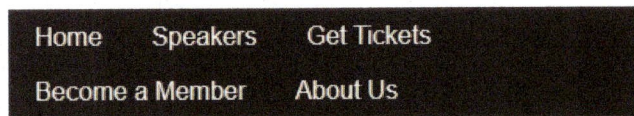

A vertical menu with wrapping

```
ul {
    display: flex;
    flex-direction: column;
    flex-wrap: wrap;
}
```

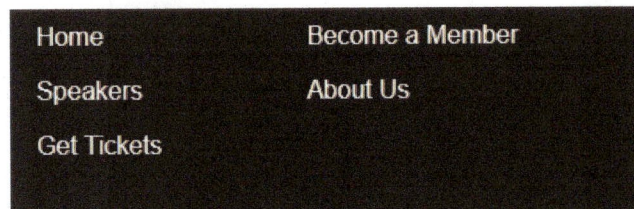

Description

- You can wrap menu items when the parent element of a navigation menu doesn't have enough width or height for all the menu items.

Figure 7-3   How to wrap menu items

How to align menu items along the main axis

Figure 7-4 shows how you can align flex items along the main axis of a flex container. By default, the browser aligns these items at the start of the container, but you can use the justify-content property to change that alignment.

The CSS in this figure shows how to align the items in a horizontal navigation menu. To do that, you can set the justify-content property to one of the values shown in this figure.

If you want an equal amount of space between the menu items, you can specify the space-between value. This displays the items with an equal amount of space between them with the first and last items aligning with the left and right sides of the container.

If you want to add some space before the first item and after the last item, you can set the justify-content property to space-around. This works like space-between, but it adds space before the first item and after the last item that's equal to half the space between the items. Or, you can set the justify-content property to space-evenly. This works like space-around except the space before the first item and after the last item is equal to the space between the items.

If you want to align all of the items with the right side of the container, you can set the justify-content property to flex-end. In this case, the flex items are aligned at the end of the container. Or, you can center all of the items in the middle of the container by setting the justify-content property to center.

The examples in this figure show how to work with the justify-content property when the flex direction is set to row. However, you can also use the justify-content property when the flex direction is set to column. In that case, the main axis extends from the top of the container to the bottom of the container. As a result, the justify-content property aligns items vertically rather than horizontally.

A property that aligns flex items along the main axis

Property	Description
justify-content	Aligns flex items within a flex container along the main axis.

Common values for the justify-content property

Value	Description
flex-start	Aligns items at the beginning of the flex container. This is the default.
flex-end	Aligns items at the end of the flex container.
center	Aligns items in the center of the flex container.
space-between	Allocates space evenly between flex items.
space-around	Allocates space evenly between flex items with half-size spaces before the first item and after the last item.
space-evenly	Allocates space evenly between flex items with full-size spaces before the first item and after the last item.

Align menu items along the main axis

```
nav { background-color: lightgray; }
ul {
    display: flex;
    justify-content: space-between;
    list-style-type: none;
}
li { background-color: black; }
a {
    text-decoration: none;
    color: white;
    display: block;
    padding: .5em;
}
```

Space-between

Space-around

Space-evenly

Flex-end

Description

- You can align menu items along the main axis when the parent element of a navigation menu has more space along the main axis than it needs for all the menu items.

Figure 7-4   How to align menu items along the main axis

How to align menu items along the cross axis

In addition to aligning the flex items along the main axis, you can align them along the cross axis. To do that, you use the align-items and align-self properties shown in figure 7-5.

Setting the align-items property to flex-start, center, and flex-end works like it does for the justify-content property. However, the flex items are aligned at the start, center, and end of the cross axis instead of the main axis.

The CSS in this figure shows how to align a horizontal navigation menu along the cross axis so the flex items are centered vertically in the container. To do that, it sets the align-items property to center. As a result, when a browser displays this navigation menu, it centers the menu vertically in the container.

This CSS also shows how you can override the alignment of individual flex items by setting the align-self property of that item. In this case, the CSS sets the align-self property of the :hover pseudo-class of the <li> element to stretch. As a result, when the mouse hovers over a flex item, it extends from the start of the cross axis to the end of the cross axis.

You can also align the content of each flex item at the bottom of the flex items. To do that, you code baseline for the align-items property. This is useful when the contents of the flex items vary in height.

Like the justify-content property that you learned about in the last figure, you can also use the align-items and align-self properties when the flex direction of a container is set to column. Then, these properties align flex items horizontally within the container.

Two properties that align flex items along the cross-axis

Property	Description
align-items	Aligns flex items within a flex container along the cross axis.
align-self	Overrides the container's cross-axis alignment for an individual flex item.

Common values for the align-items and align-self properties

Value	Description
stretch	The flex items extend from the start of the flex container to the end of the flex container. This is the default.
flex-start	The flex items are aligned at the start of the flex container.
flex-end	The flex items are aligned at the end of the flex container.
center	The flex items are centered in the flex container.
baseline	The content within each flex item is aligned at the bottom.
auto	For align-self items only: The item is aligned as specified by the align-items property of the flex container. This is the default.

Align menu items along the cross axis

```css
nav {
    background-color: lightgray;
}
ul {
    display: flex;
    align-items: center;
    list-style-type: none;
    min-height: 5em;
}
li {
    background-color: black;
}
li:hover {                      /* override center alignment on hover */
    align-self: stretch;
}
a {
    text-decoration: none;
    color: white;
}
```

How it looks in a browser

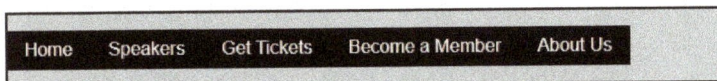

How it looks when the user hovers over a menu item

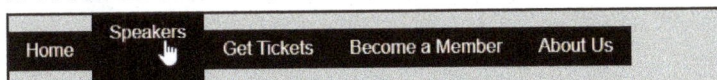

Description

- You can align menu items along the cross axis when the parent element of a navigation menu has more space along the cross axis than it needs for all the menu items.

Figure 7-5   How to align menu items along the cross axis

A web page with a navigation menu

Now that you've learned some basic skills for working with navigation menus, you're ready to see how to use these skills to create a web page that includes a navigation menu.

Part 1 of figure 7-6 shows the Town Hall web page from the last chapter, updated to include a horizontal navigation bar with five menu items. On larger screens, the menu items lay out side by side in a single row. But on smaller screens, the menu items wrap and display side by side on multiple lines.

The navigation menu also uses a rounded border to highlight the menu item for the current page. For example, the Home menu item is highlighted on the first screen, and the Speakers menu item is highlighted on the second screen.

In addition, this navigation menu adds underlining to a menu item when the mouse hovers over it or the user tabs to it. For example, the mouse is hovering over the Speakers menu item in the first screen.

The navigation menu at different screen sizes

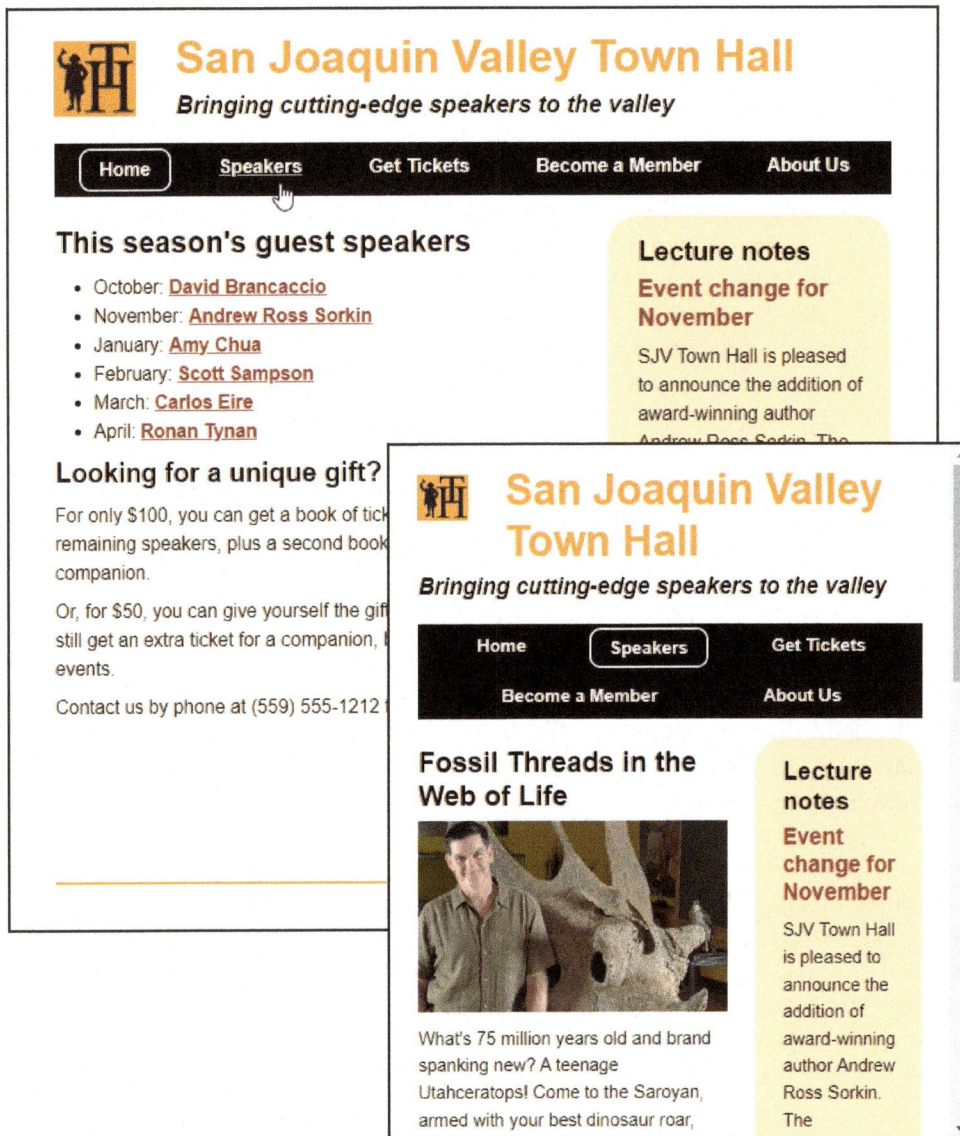

Description

- The web page displays the menu items horizontally along the top of the page and allocates space evenly between the menu items.
- The navigation menu wraps on screens that aren't wide enough to display all of its items.
- The web page identifies the current section by displaying its navigation menu item within a rounded border.
- When the user hovers the mouse over or tabs to a navigation menu item, the web page underlines the item.

Figure 7-6   A web page with a navigation menu (part 1)

Part 2 of figure 7-6 shows the most important HTML and CSS for the navigation menu. It starts by showing the structure of the web page. Here, the <nav> element that contains the navigation menu is in the body of the page between the header and the main content.

The HTML for the navigation menu assigns the <nav> element to the navbar class. In addition, it assigns the <ul> element to the menu class. This makes it easy to use CSS to style this menu without conflicting with other <nav> or <ul> elements on the page. Finally, it assigns the <a> element for the first menu item to the current class instead of providing an href attribute for it. This indicates that the menu item is for the current section of the website and should be highlighted.

The CSS for the navigation menu starts by adding two colors that the navigation menu uses to the three colors that are already used by the web page. Then, it declares the style rule for the navbar class as well as the style rule for the menu class.

The style rule for the navbar class starts by setting the background color of the <nav> element. Then, it provides several nested style rules.

The nested style rule for the <ul> element removes the bullets from the items of the list. After that, this style rule sets the margin and padding properties to 0. This overrides the defaults set for <ul> elements.

The nested style rule for the <a> element sets the display property to block. Then, it sets the padding and margins for all four sides of the link. This makes the entire padded area of the link clickable. After that, this style rule removes the underline from the link and sets its color to a light color to make it visible on the dark background of the navigation bar.

The nested style rule for the :hover and :focus pseudo-classes of the <a> element identify the link that the user is hovered over or has tabbed to by underlining this link. Then, the nested style rule for the current class of the <a> element highlights the current menu item by adding a 2px rounded border. In addition, this style rule sets the text-decoration property to none. This style rule overrides the style rule for the pseudo-classes. As a result, the current item isn't underlined when the mouse hovers over it.

The style rule for the menu class starts by setting the display property to flex to create a horizontal navigation menu. Then, it sets the flex-wrap property to wrap, so the menu items wrap to multiple lines if the menu isn't wide enough for them. Finally, it sets the justify-content property to space-evenly. As a result, the browser allocates space evenly between the menu items with full-size spaces before the first menu item and after the last menu item.

The structure of the web page

```
<body>
    <header>...</header>
    <nav>...</nav>
    <main>...</main>
    <footer>...</footer>
</body>
```

The HTML for the navigation menu

```
<nav class="navbar">
    <ul class="menu">
        <li><a class="current">Home</a></li>
        <li><a href="speakers.html">Speakers</a></li>
        ...
    </ul>
</nav>
```

The CSS

```
/* variables for theme and menu colors */
:root {
    --dark: #931420;
    --medium: #f2972e;
    --light: #ffebc6;
    --dark_menu: black;
    --light_menu: white;
}

/*** menu styles ***/
.navbar {
    background-color: var(--dark_menu);
    & ul {
        list-style-type: none;
        margin: 0;
        padding: 0;
    }
    & a {
        display: block;
        padding: .25em 1em;
        margin: .25em;
        text-decoration: none;
        color: var(--light_menu);
    }
    & a:hover, a:focus { text-decoration: underline; }
    & a.current {
        border: 2px solid;
        border-radius: 10px;
        text-decoration: none;
    }
}
.menu {
    display: flex;
    flex-wrap: wrap;
    justify-content: space-evenly;
}
```

Figure 7-6   A web page with a navigation menu (part 2)

How to create a 2-tier navigation menu

So far, this chapter has shown how to create a 1-tier navigation menu. For many websites, that's all you need. However, for some websites, you may need to create a *2-tier navigation menu*, where each item in the navigation bar can contain a *submenu* with additional items that can drop down from the navigation bar. This can also be called a *multi-tier navigation menu*.

How to position an element

To create a submenu that drops down, you have to position some of the elements of the navigation menu on the page. To do that, you use the property shown in the first table in figure 7-7.

The position property determines how the browser positions an element. This property can be set to the five values shown in the second table. The default value is static, which causes elements to be placed in the normal flow where block elements flow from top to bottom and inline elements flow from left to right.

The absolute value removes the element from the normal flow and positions it based on the top, bottom, right, and left properties that you specify. When you use *absolute positioning*, the element is positioned relative to the closest prior element that is positioned. If no prior element is positioned, the element is positioned relative to the browser window.

The fixed value works like the absolute value in that it removes the element from the normal flow and the position of the element is specified by using the top, bottom, left, and right properties. However, instead of being positioned relative to a containing block, an element that uses *fixed positioning* is positioned relative to the browser window. That means that the element doesn't move when you scroll.

In contrast, the relative value causes an element to be positioned relative to its normal position in the flow. So, when you use the top, bottom, left, and right properties with *relative positioning*, they specify the element's offset from its normal position.

The sticky value works like a combination of relative and fixed positioning. An element that uses *sticky positioning* acts like it has relative positioning and scrolls until it hits an offset set by the top, bottom, left, or right properties. At that point, the element acts like it has fixed positioning and doesn't move when you continue to scroll.

The example in this figure shows how a footer can be positioned so it's always visible at the bottom of the screen. To do that, the position property is set to fixed. Then, the bottom property positions the footer relative to the browser window so it's always zero pixels from the bottom.

When you position elements, you may need to set the margins or padding for other elements to make room for them. For instance, if you position a footer so it's fixed to the bottom of the browser window, you may need to adjust the main content so it isn't hidden by the footer.

Property for positioning elements

Property	Description
position	A keyword that determines how an element is positioned.

Possible values for the position property

Value	Description
static	The element is placed in the normal flow where block elements flow from top to bottom and inline elements flow from left to right. This is the default.
absolute	The element is removed from the flow and positioned relative to the closest ancestor that is positioned. As a result, static elements flow beneath the absolute element. The position is determined by the top, bottom, left, and right properties.
fixed	The element is removed from the flow and positioned relative to the browser window. As a result, the element doesn't move even when you scroll. The position is determined by the top, bottom, left, and right properties.
relative	The element is positioned relative to its position in the normal flow. The position is determined by the top, bottom, left, and right properties.
sticky	The element has relative positioning up to a specified offset, then has fixed positioning. As a result, it moves when you scroll until it hits the specified offset. Then, it doesn't move, even when you scroll. The offset is determined by the top, bottom, left, and right properties.

A footer positioned so it's always visible at the bottom of the screen

```
footer {
    position: fixed;
    bottom: 0;
}
```

Description

- You may need to make room for positioned elements by setting the margins or padding for other elements.

Figure 7-7   How to position an element

An example that uses positioning

To show how positioning works, figure 7-8 presents a web page that uses fixed, absolute, and sticky positioning. This web page should give you an idea of how positioning can be used. For instance, sticky positioning is often used to keep a navigation menu at the top of the browser after the header scrolls off, fixed positioning is often used to keep a footer visible at the bottom of the browser, and absolute positioning is often used with navigation submenus.

The <body> element for the web page contains a <header> element, a <nav> element, a <main> element, and a <footer> element. The <main> element contains a <section> element and an <aside> element.

The CSS for the web page sets the positions of the <nav>, <aside>, and <footer> elements. It also adds a bottom margin to the <main> element to make room for the footer.

The style rule for the <nav> element sets the position property to sticky and the top property to 0. As a result, the <nav> element has relative positioning until the top of the element scrolls to the top of the browser window. At that point, it has fixed positioning and stops scrolling. The three screens in this figure show how this works. In the first screen, the <nav> element is positioned like any other element. But in the second and third screens, it's fixed to the top of the browser window, on top of the other content of the web page, which scrolls underneath it.

The style rule for the <aside> element sets the position property to absolute. This takes the <aside> element out of the normal flow and positions it relative to the previous positioned element, the <nav> element. In addition, the top, left, and right properties set its position in the browser. Its left side is 5% from the left side of the browser window to align it with the left side of the <nav> element. Its right side is 65% from the right side of the browser window so its width is the remaining 30% of the browser window. And its top is 50% of the distance between the <nav> element and the bottom of the browser window.

The three screens show how this works. In these screens, the <aside> element is positioned outside the flow of the elements, on top of the other content of the web page. As the page scrolls up in the second screen, the <aside> element maintains its position relative to the <nav> element. However, it maintains its position as if the <nav> element wasn't sticky. That's why the <aside> element scrolls all the way out of view on the last screen.

The style rule for the <footer> element sets the position property to fixed and the bottom property to 0. As a result, the <footer> element always displays at the bottom of the browser window, on top of the other content of the web page, and doesn't scroll as the rest of the content scrolls. The three screens show how this works. In all three screens, the footer is fixed to the bottom of the browser window. In the first two screens, it displays on top of the other content of the web page, but on the last screen, it doesn't. That's because of the margin-bottom setting for the <main> element.

The style rule for the <footer> element also sets the width of the footer to 90%, to match the width set for the body (not shown). This is necessary because an element that uses fixed positioning doesn't inherit the width of the parent.

The HTML for the <body> element

```
<header><h1>Positioning</h1></header>
<nav>This nav scrolls to the top of the screen, then it sticks.</nav>
<main>
    <section>Lorem ipsum dolor sit amet...</section>
    <aside>This absolutely positioned sidebar isn't in the flow of
        the page.
    </aside>
</main>
<footer>This footer is fixed to the bottom of the screen</footer>
```

The CSS for the positioning

```
nav {
    position: sticky;
    top: 0
}
main { margin-bottom: 2em; }   /* to make room for the fixed footer */
aside {
    position: absolute;          /* relative to positioned <nav> element */
    top: 50%;
    left: 5%;
    right: 65%;
}
footer {
    position: fixed;
    bottom: 0;
    width: 90%;
}
```

How it looks in a browser at various scroll positions

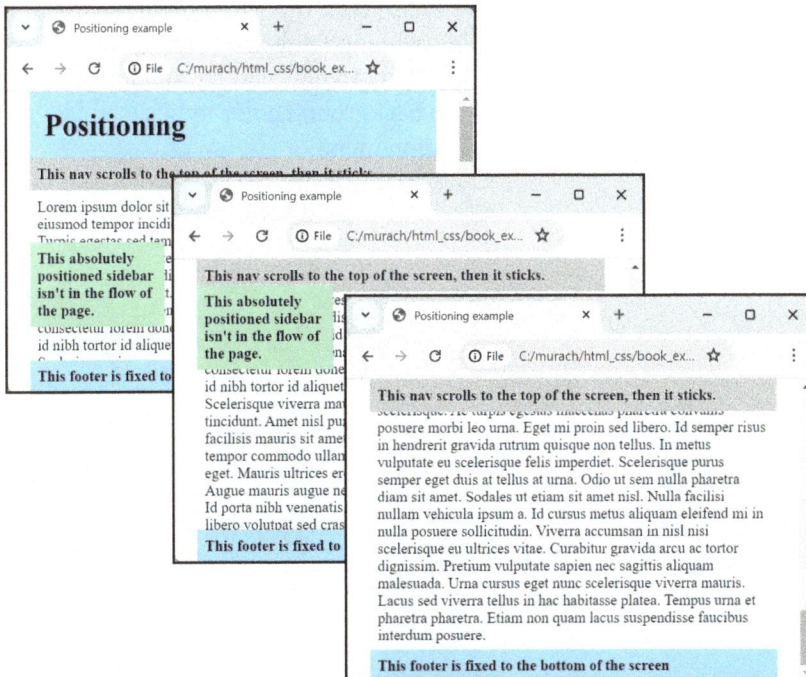

Figure 7-8   An example that uses positioning

How to create a submenu

Figure 7-9 presents a main navigation menu that has a submenu. This navigation menu uses an unordered list to define its items. In addition, one of these items contains a nested unordered list that defines a submenu.

The HTML for this menu shows that the <ul> element assigned to the menu class has an <li> element that contains a nested <ul> element that's assigned to the submenu class. This nested <ul> element defines the submenu shown in this figure.

The CSS shown in this figure begins by defining a style rule for the menu class. This style rule uses flexbox layout to display a horizontal navigation menu with items that have appropriate spacing.

The second style rule selects any <li> element in the menu class that has a submenu. To do that, this style rule uses a *functional pseudo-class selector* to check whether each <li> element in the menu class has a descendent element assigned to the submenu class. If so, the code for this style rule sets the position property of the <li> element to relative. As a result, the <li> element is a positioned element, not a static element.

Then, the nested style rule displays the submenu when the user hovers the mouse over the <li> element. To do that, this style rule uses the child selector (>) to select the submenu class. As a result, the browser only displays the submenu for the item that the mouse is hovering over.

The style rule for the submenu class sets the position property to absolute. As a result, the submenu uses absolute positioning and is positioned relative to the item that contains it. This works because the <li> element is a positioned element, not a static element.

In addition, the style rule for the submenu class sets a few more properties. It sets the left property to 0, so the submenu is positioned directly below the menu item that contains it. It sets the display property to none, so the submenu is hidden when the web page loads. And it sets the background color to black and the text color to white to match the main navigation menu.

A submenu

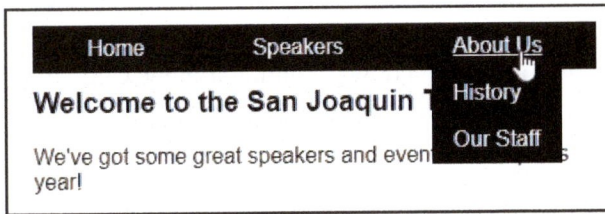

The HTML for the submenu

```
<nav>
    <ul class="menu">
        <li><a href="index.html">Home</a></li>
        <li><a href="speakers.html">Speakers</a></li>
        <li><a href="aboutus.html">About Us</a>
            <ul class="submenu">
                <li><a href="history.html">History</a></li>
                <li><a href="staff.html">Our Staff</a></li>
            </ul>
        </li>
    </ul>
</nav>
```

A functional pseudo-class

Pseudo-class	Description
:has(selector-list)	Selects elements that have elements that match the list of selectors.

The CSS for the submenu

```
.menu {
    display: flex;
    justify-content: space-around;
}
.menu li:has(.submenu) {
    position: relative;
    &:hover > .submenu {
        display: block;
    }
}
.submenu {
    position: absolute;   /* relative to the item that contains the submenu*/
    left: 0;
    display: none;
    background-color: black;
    color: white;
}
```

Description

- To create a submenu, you code it within a menu item, absolutely position it, and hide it initially. Then, you display the submenu when the mouse hovers over its menu item.

- A *functional pseudo-class* begins with a colon (:) like a pseudo-class but ends with a pair of parentheses that can contain a list of parameters, like a function.

Figure 7-9   How to create a submenu

Useful symbols for working with menus

When you work with navigation menus, it can be helpful to include symbols like the ones shown in figure 7-10. For instance, a down-pointing triangle can indicate that a menu item has a submenu, and a hamburger icon can be used to toggle the display of a navigation menu.

There are two ways to include symbols in your navigation menus. First, you can include the HTML code directly in the HTML for the menu as shown in the first example in this figure. Here, the text for a menu link includes a <small> element, which is used to define smaller text. Within the <small> element, this example includes the HTML code for the down triangle. As a result, when a browser displays this example, it displays a small down triangle after the text for the menu item.

Second, you can use the CSS entity value to add a symbol. To do that, you can set the content property of the ::before or ::after pseudo-element as shown in the second example. In this example, the ::after pseudo-element displays the down triangle after the content of the <a> element that's in the submenu-link class. Since this <a> element has content of "About Us", the browser displays the down triangle after this text.

In addition, the CSS in the second example sets the font-size for the symbol to small. This sizes the symbol so it's the same size as the symbol in the first example that's coded within the <small> element.

Now that you know both ways to add a symbol to a navigation menu, you might wonder which technique is a best practice. In general, if the web page doesn't need to change the symbol, it's easier to set it in HTML. However, sometimes a web page needs to change the symbol based on user action, such as hovering over a link or clicking a menu button. In those cases, it's better to set it by using CSS.

Although this figure presents six useful symbols, HTML and CSS provide for many more symbols such as the copyright and registered trademark symbols. As a result, if you need a symbol that isn't shown in this figure, you can usually find the code or entity for it by searching the internet for "HTML entity" or "CSS entity".

Some useful symbols for working with menus

Symbol	Description	HTML code	CSS entity
◄	Left triangle	`&#9668;`	`\25c0`
►	Right triangle	`&#9658;`	`\25b6`
▲	Up triangle	`&#9650;`	`\25b2`
▼	Down triangle	`&#9660;`	`\25bc`
≡	Hamburger icon	`&#9776;`	`\2630`
✖	Close	`&#10006;`	`\2716`

Use HTML to include a down triangle

The HTML

```
<a href="speakers.html">Speakers<small>&#9660;</small></a>
```

How it looks in the browser

Speakers▼

Use CSS to include a down triangle

The HTML

```
<a href="aboutus.html" class="submenu-link">About Us</a>
```

The CSS

```
.submenu-link::after {
    content: "\25bc";
    font-size: small;
}
```

How it looks in the browser

About Us▼

Description

- You can include symbols in your website using HTML or CSS.
- To include a symbol with CSS, you typically set the content property of the ::before or ::after pseudo-element.
- For a complete list of symbols, you can search the internet for "HTML entity" or "CSS entity".

Figure 7-10   Useful symbols for working with menus

A web page with a 2-tier navigation menu

At this point, this chapter has presented the skills you need to create a 2-tier navigation menu. Now, it shows how these skills work together by presenting a web page that creates a 2-tier navigation menu.

Part 1 of figure 7-11 shows the same Town Hall page from earlier in this chapter after it has been updated to include a submenu for the Speakers menu item. This submenu provides links to pages for each speaker.

The Speakers menu item has a down triangle symbol to indicate that it has a submenu. When the user hovers over the Speakers menu item, the browser displays the submenu directly below the item over the top of the main content of the page. However, the browser hides the submenu when the mouse moves away from the Speaker menu item or its submenu.

A navigation menu with a submenu

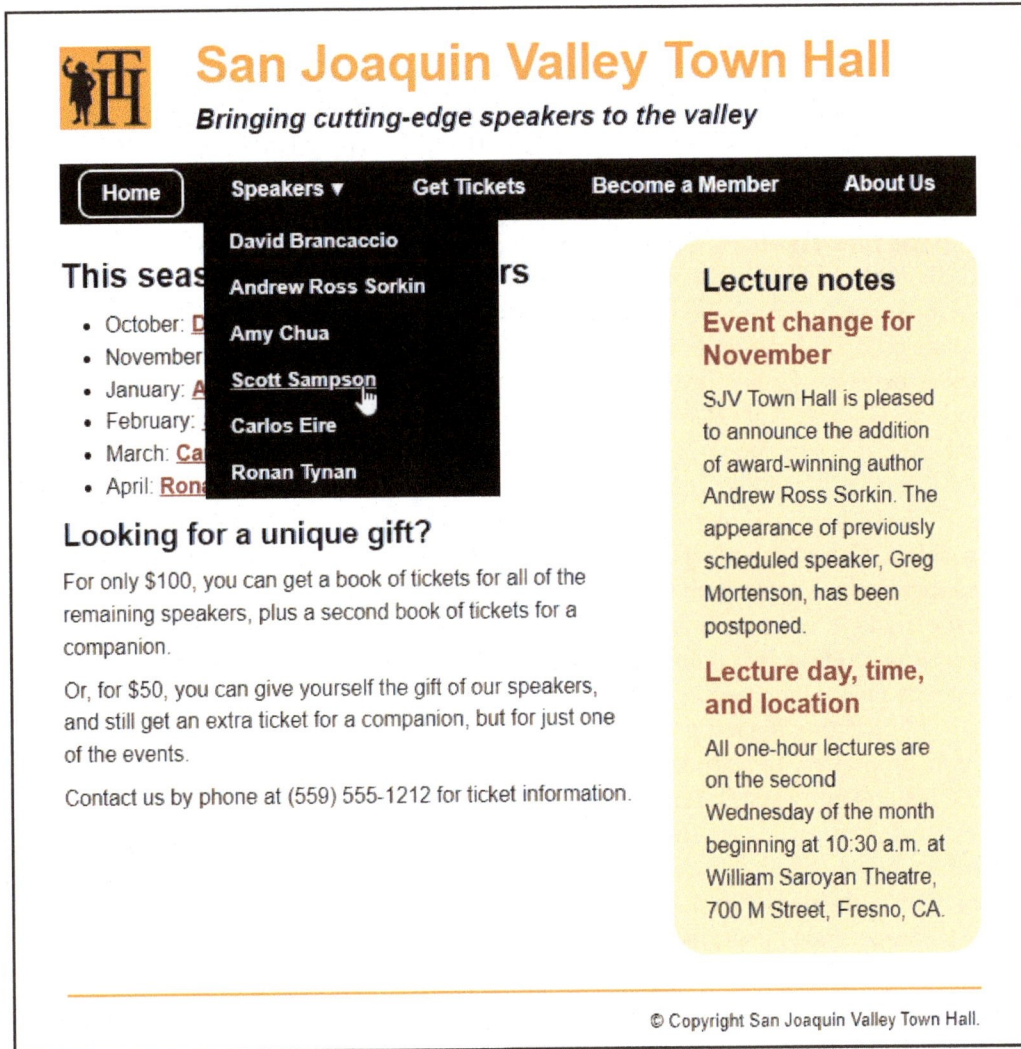

Description

- This web page uses a down triangle to identify a menu item that has a submenu.
- This web page hides the submenu unless the user hovers over its menu item.
- This web page positions the submenu below its menu item and displays it on top of the content for the page.

Figure 7-11   A web page with a 2-tier navigation menu (part 1)

Part 2 of figure 7-11 begins by presenting the HTML for the navigation menu. This navigation menu begins by defining a <ul> element and assigning it to the menu class.

Within this <ul> element, the <li> element for the Speakers item defines an <a> element that includes a <small> element that includes the HTML code for the down triangle. This displays the down triangle after the text of the Speakers link to indicate that the menu item has a submenu. In addition, this <li> element contains a nested <ul> element that's assigned to the submenu class.

After the HTML, this figure shows the CSS for the submenu. To start, it declares the style rule for each <li> element that has an element that's assigned to the submenu class. This style rule starts by setting the position property to relative. Then, it has a nested style rule for the <small> element that adds some space between the link text and the down triangle symbol. Next, it has a second nested style rule that selects the submenu that's a child of the item that the mouse is currently hovering over. This displays the submenu that's contained in the item that the mouse is hovering over.

The style rule for the submenu class starts by setting the position property to absolute. This absolutely positions the submenu relative to the prior positioned element, which is the item that contains the submenu. Then, it sets the left property to 0 so the submenu lines up on the left side of the menu item that contains it. These two properties cause the browser to display the submenu directly below the Speaker menu item.

After setting the left property, the style rule for the submenu class sets a few more properties. To start, it sets the text and background colors so they match the main menu. Then, it sets the width of the submenu to 175% of the containing element. This makes the submenu wider than the Speaker menu item so it can accommodate the length of the speaker names. Next, it sets the display property to none. This hides the submenu initially. However, when the user hovers over the Speaker menu item, the browser displays the submenu.

The HTML for the navigation menu

```
<nav class="navbar">
    <ul class="menu">
        <li><a class="current">Home</a></li>
        <li><a href="#">Speakers<small>&#9660;</small></a>
            <ul class="submenu">
                <li><a href="#">David Brancaccio</a></li>
                       .

                       .
                <li><a href="#">Ronan Tynan</a></li>
            </ul>
        </li>
        <li><a href="tickets.html">Get Tickets</a></li>
        <li><a href="members.html">Become a Member</a></li>
        <li><a href="about.html">About Us</a></li>
    </ul>
</nav>
```

The CSS for the submenu

```
.menu li:has(.submenu) {
    position: relative;
    & small {
        margin-left: .25em;    /* add space for arrow */
    }
    &:hover > .submenu {
        display: block;
    }
}
.submenu {
    position: absolute;   /* relative to the item that contains it */
    left: 0;
    color: var(--light_menu);
    background-color: var(--dark_menu);
    width: 175%;
    display: none;
}
```

Figure 7-11   A web page with a 2-tier navigation menu (part 2)

Perspective

Now that you've completed this chapter, you should be able to use flexbox layout to create navigation menus, including 2-tier navigation menus. This is another important step toward developing web pages that provide for responsive web design.

Terms

navigation menu	multi-tier navigation menu
navigation bar	absolute positioning
horizontal navigation menu	fixed positioning
vertical navigation menu	relative positioning
2-tier navigation menu	sticky positioning
submenu	functional pseudo-class

Exercise 7-1   Add a navigation menu

In this exercise, you'll add a navigation menu to the Town Hall home page. When you're through, your page should look like this:

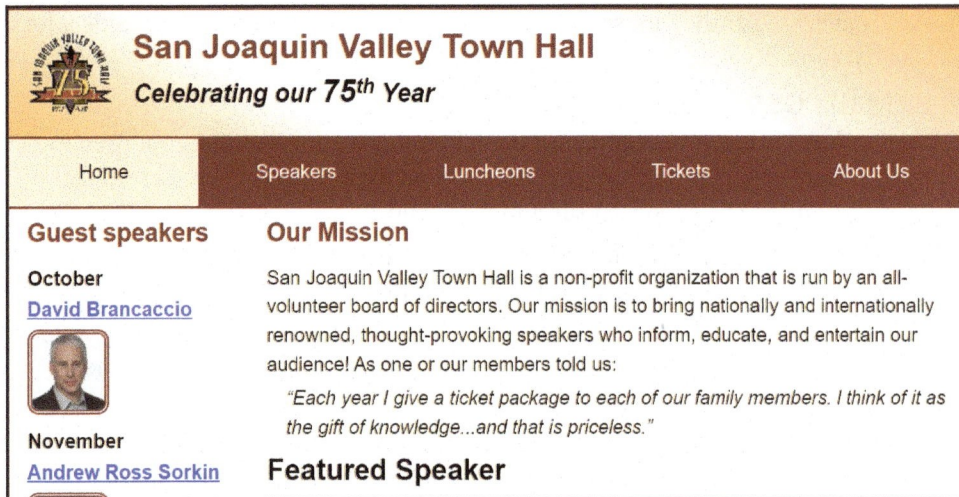

Open the HTML and CSS files for this page

1. Use your text editor to open these HTML and CSS files:

    ```
    exercises\ch07\town_hall\index.html
    exercises\ch07\town_hall\styles\main.css
    ```

Add the HTML for the navigation menu

2. Add a <nav> element between the <header> and <main> elements. Assign the <nav> element to the navbar class.

3. Add a <ul> element within the <nav> element. Assign the <ul> element to the menu class.

4. Add the first <li> element for this unordered list. Then, add an <a> element with the text set to "Home" within the <li> element. Assign the <a> element to the current class and don't include an href attribute.

5. Add a second <li> element that contains an <a> element. Set the text of the <a> element to "Speakers" and the href attribute to the hash character (#) so it doesn't link anywhere.

6. Copy the second <li> element and paste it three times to create the last three list items. Then, change the text for the links in each list item so they match the screen for this exercise.

7. Run the web page and view it to see how the navigation menu looks in the browser before any CSS is applied to it.

Add the CSS for the menu class

8. Add a style rule for the menu class that uses flexbox to display the menu items horizontally and wrap the menu items on small screens.

9. To space the menu items so they take up the entire navigation bar, add a nested style rule for the <li> element and set the flex-basis property to 20%.

10. Refresh the web page to see how it looks. The navigation menu should lay out horizontally with space between the links. However, it still needs a lot of formatting.

Add the CSS for the navbar class

11. Add variables for a light, mid, and dark menu colors to the :root pseudo-class. For the mid color, you can choose any light color. The screen that's shown for this exercise uses antique white.

12. Add a style rule for the navbar class. Set the background color for the navbar class to the accent color. Then, add the following nested style rules:

    • For the <a> element, display links as block elements, set the padding for all four sides to 1em, remove the underline from the links, and set the text color to the light menu color.

    • For the :hover pseudo-class, :focus pseudo-class, and current class of the <a> element, set the background color to the mid menu color and the text color to the dark menu color.

    • For the <ul> element, set margins and padding to 0 and remove the bullets from the list items.

13. Refresh the web page and view it. Notice that there's no space between the menu items or between the menu items and the bottom of the navigation bar. As a result, the background color of the navigation menu isn't shown.

Add the finishing touches

14. To add some space around the menu items, update the nested <a> element style rule in the navbar class. Set a top margin of 0, left and right margins of 1 pixel, and a bottom margin of 2 pixels.

15. To center the text in the menu items, update the nested <li> element style rule in the menu class. Set the text-align property to center.

16. Refresh the web page to view it.

Exercise 7-2   Add submenus

In this exercise, you'll add submenus to the Town Hall home page from the last exercise. When you're through, the page should look like this:

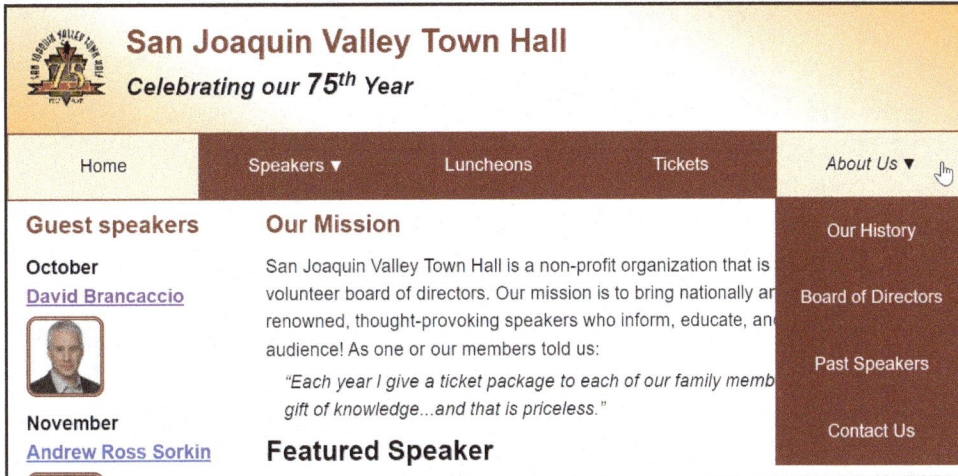

Add the HTML for the submenu

1. Update the <a> element within the <li> element for the Speakers item so the text includes a <small> element that contains the HTML code for the down triangle symbol.
2. Add a <ul> element within the <li> element. Assign the <ul> element to the submenu class.
3. Add four <li> elements with <a> elements for each speaker. To do that, you can copy and paste the <a> elements from the <aside> element.
4. Refresh the web page and view it to see how the navigation menu looks in the browser before any CSS is applied to the submenu.

Add the CSS for the item that contains the submenu

5. Add a style rule for the menu item that contains the submenu class using the :has() pseudo-class and set the position property to relative.
6. Add a nested style rule for the <small> element that adds some space between the link text and the down triangle symbol.
7. Add another nested style rule that selects the .submenu class that's a child of the :hover pseudo-class.
8. Set the display property so the element assigned to the .submenu class displays when the user hovers over the item that contains the submenu class.

Add the CSS for the submenu

9. Add a style rule for the submenu class and set the position property to absolute.
10. Set the left property to 0 and the right property to 0.
11. Set the background color to the accent color.

12. Set the text color to the light menu color.

13. Set the display property to hidden so that it only displays when the user hovers over the menu item.

14. Refresh the web page and test the submenu by hovering your mouse over the menu item. If everything is working, the submenu should display when you do so.

Add another submenu

15. Follow steps 1 and 2 to modify the HTML for the <li> element that contains the About Us link so it contains a submenu.

16. Add <li> elements with <a> elements that match the drop-down in the screen for this exercise. You can set the href attribute of those links to the hash character (#).

17. Refresh the page. This second submenu should work correctly because it uses the same CSS as the other submenu that you already coded and tested.

18. To make the left and right borders of the submenu items match the main menu items, add the following nested style rule to the submenu class:

```
& > li a { margin: 0 2px 2px; }
```

19. To remove the extra space from the left side of the first menu item and the right side of the last menu item, including submenus, add the following style rules:

```
.menu > li:first-child a { margin-left: 0; }
.menu > li:last-child a { margin-right: 0; }
```

Chapter 8

How use media queries for responsive web design

So far, this book has presented some of the skills for creating a responsive web design such as using scalable images with a fluid design. Now, this chapter shows how to use media queries to modify the styling of your web pages depending on the size of the screen. This allows you to create web pages that look just the way you want on devices with various screen sizes from mobile phones to tablets to desktop computers.

Introduction to responsive web design

A web page that uses *responsive web design* (*RWD*) adapts to the size of the screen while maintaining the overall look-and-feel of the site. The next two figures introduce you to the components of a responsive design and present some of the ways that you can test a responsive design.

The components of a responsive web design

Figure 8-1 describes the components of a responsive design. To start, a responsive design uses *fluid design* as shown in chapter 6. To make a fluid design work, the web pages also need to use *scalable images* that adjust to the size of their containing elements. That way, the web page adjusts to the size of the screen. If, for example, you drag the right border of the browser window on a desktop computer, the page adapts to the various widths.

Although using a fluid design allows the browser to adjust the elements on a page to different screen sizes, it doesn't give you complete control. To get that, you need to use *media queries*.

The three screens in this figure show how media queries can be used to adjust a web page for screen sizes of various devices such as desktop computers, tablets, and mobile phones. For example, the largest screen has a header that displays the logo with three navigation menu items across the top of the screen, and it displays an image between the header and the form for booking a flight. That makes sense because this screen has enough room to display the menu items and the image while still looking good and being easy to use.

On the other hand, the medium-sized screen displays a hamburger menu symbol instead of the three navigation menu items, and it displays less image between the header and the form. That makes sense because this screen isn't wide enough to display the menu items across the top of the page in a way that looks good or tall enough to display as much image while still displaying the form.

Finally, the smallest screen only displays a thin slice of image between the header and the form. That makes sense because this screen doesn't have enough height to display the image while still displaying the form. Also, this screen displays the controls on the form in a single column since it isn't wide enough to use multiple columns to display these controls.

A website that uses responsive web design

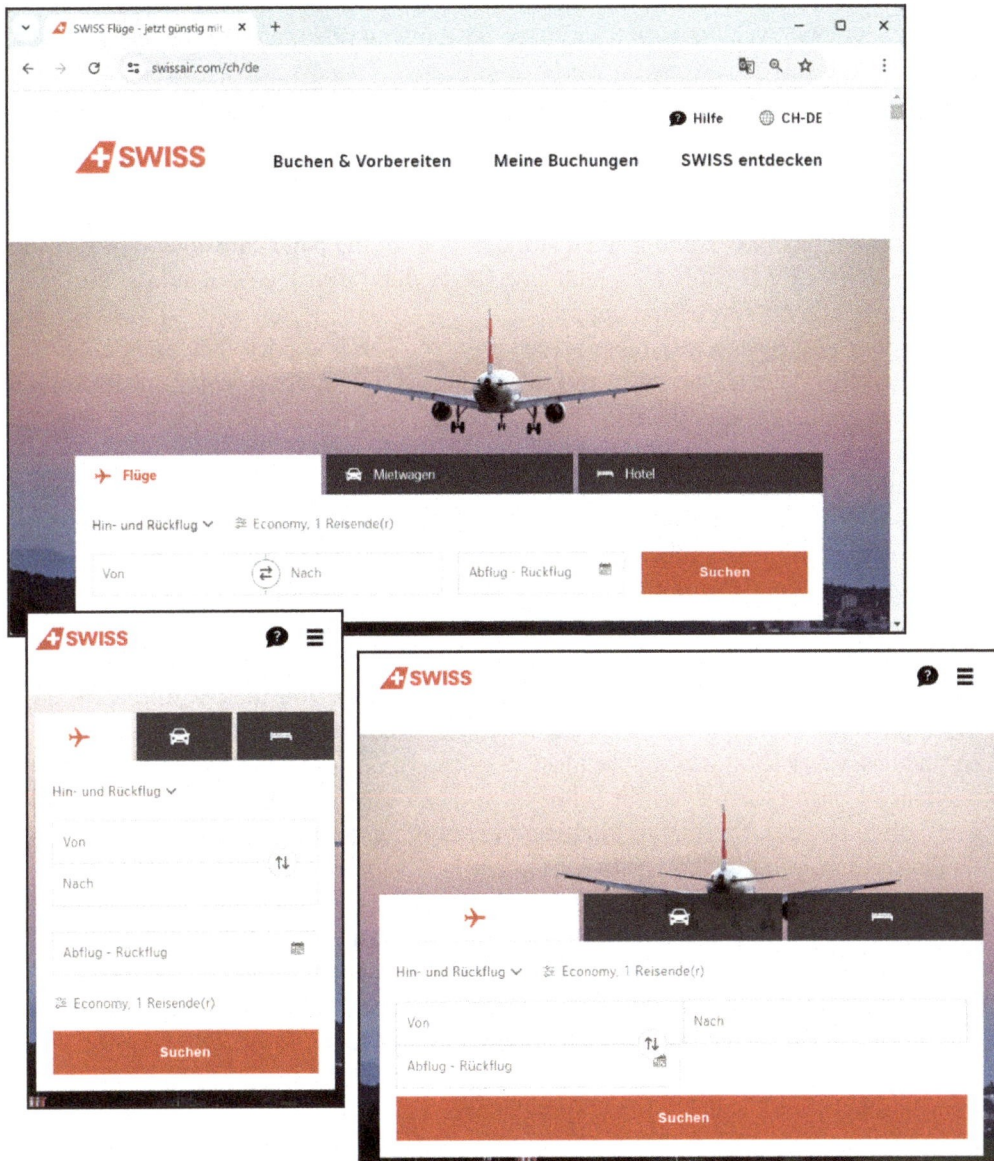

Components of responsive web design

- *Fluid design* so the web pages adjust to the screen size of the device.
- *Scalable images* that adjust to the size of the elements that contain them.
- *Media queries* that adjust the page layouts to the screen size of the device.

Description

- *Responsive web design* (RWD) means that web pages respond to the devices that access them, from desktop computers to tablets in landscape and portrait modes to mobile phones in landscape and portrait modes.

Figure 8-1   The components of responsive web design

How to test a responsive design

As you develop a website with responsive design, you need to test it on devices of various sizes to be sure that it works as expected. To get started, you can use the Developer Tools for a browser as shown in figure 8-2. This figure shows how to use Chrome, but most browsers work similarly.

To access the Developer Tools for a browser, you start by displaying the page you want to test. Then, you press F12 to open the Developer Tools. In this example, the tools are displayed at the bottom of the page. However, you can change that by clicking the menu icon (three dots) near the right side of the toolbar and select a Dock Side option.

To display the page in another screen size, you can you click the "Toggle device toolbar" icon as shown by the callout in this figure. Then, you can select a device from the drop-down list at the top of the window. For example, you can select one of the iPhone or iPad options to simulate displaying the page at the specified iPhone or iPad.

If you select Responsive from the drop-down list instead of a specific device, you can drag the edges of the screen to create a custom size. In this figure, for example, the Responsive option displays the page at a custom screen size.

Chrome also provides a variety of other options. For example, after you select a specific device, you can click the Rotate icon in the toolbar to change the orientation between portrait and landscape. Similarly, you can use the Zoom drop-down list to change the zoom percent. With a little experimentation, you should be able to learn how to use these options.

Once a page is displayed for a device, you can use the mouse to test it. For mobile devices, this means clicking and dragging with a mouse instead of tapping and swiping with your finger. Otherwise, the page works the same as if you were running it on the specified device.

After you test a responsive design with the Developer Tools, you typically want to test it on the actual devices and browsers. To do that, you must deploy the website to a server. Unfortunately, there are so many different devices that you won't be able to test your site on all of them. However, if you can get your pages working correctly on a several devices that cover the range of screen sizes for your target audience, it's likely that your pages will work correctly on most devices.

A web page displayed by Chrome's Developer Tools

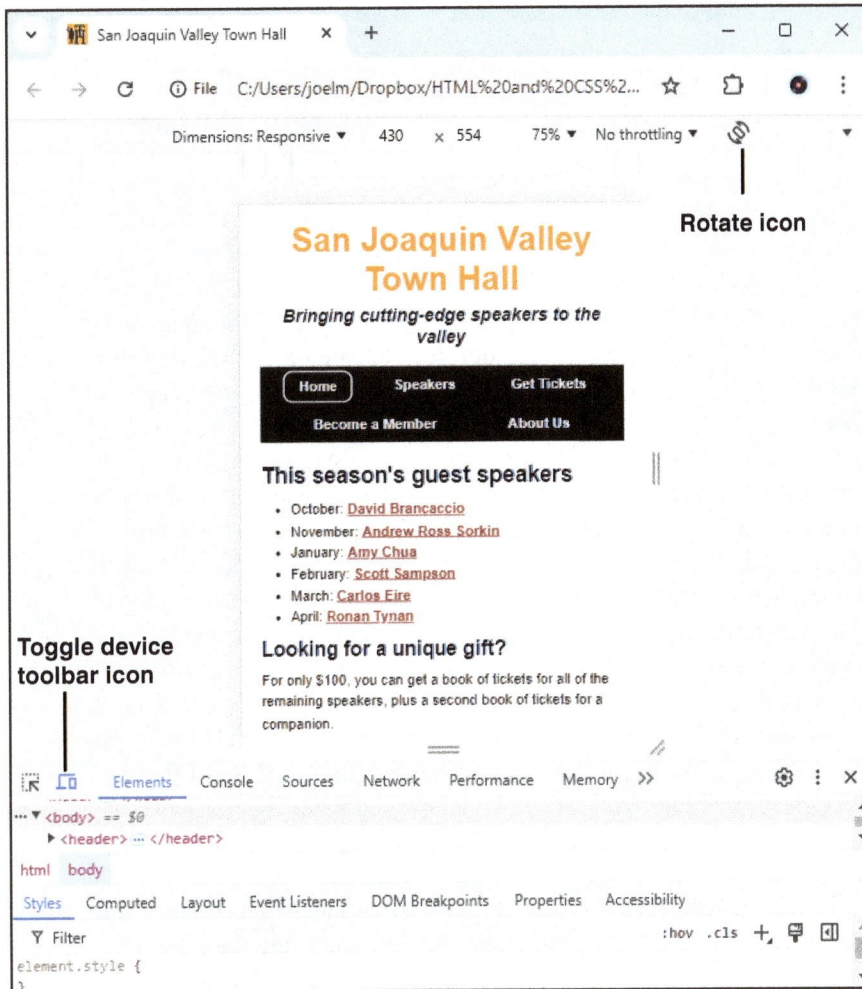

Two ways to test a responsive design

- Use the Developer Tools provided by most modern browsers.
- Deploy your website to a server and then test it on multiple devices.

How to test a responsive design with Chrome

1. Display the page in Chrome.
2. Press F12 to display the Developer Tools.
3. Click the "Toggle device toolbar" icon near the left side of the toolbar.
4. Select the device you want to emulate from the Dimensions drop-down list in the toolbar above the page.
5. If you select Responsive from the Dimensions drop-down list, create a custom size by dragging the edges of the screen or by entering a size in the available boxes.
6. Use the mouse to test the page.

Figure 8-2   How to test a responsive design

How to work with media queries

Media queries are a CSS feature that lets you write conditional expressions within your CSS code. These expressions can be used to query various characteristics of a device such as the screen size or orientation. This provides a way for you to change the appearance of a web page as the screen size changes.

How to code media queries

Figure 8-3 shows how to code media queries for a responsive design. The top of this figure presents the basic syntax of a media query. It starts with an *@media selector*, followed by a *media type*, which defines the category of the device. Common media types are all, print, and screen. For a responsive web page, you set the media type to screen.

The media type can be preceded by the optional only keyword. This keyword prevents older browsers that don't support media queries from applying the given styles. However, the only keyword isn't required for modern browsers. Because of that, some developers omit this keyword for new media queries. For the sake of completeness, the examples in this chapter include the only keyword.

After the media type, you code one or more conditional expressions. Each expression specifies a *media feature*. A media feature describes a specific characteristic of a device such as the width or height of its viewport. The table in this figure lists some of the common media features for the screen media type.

The first example in this figure shows two media queries. Both of these media queries check that the width of the viewport is greater than or equal to 768 pixels. The first media query uses the min-width feature to do this, while the second uses the width feature and the >= logical operator.

The second example shows two media queries that identify a screen width that's within a range. In the first media query, the first conditional expression checks that the width of the viewport has a minimum width of 600 pixels, and the second conditional expression checks that the width has a maximum width of 768 pixels. In other words, this media query checks that the width is greater than or equal to 600 pixels and less than or equal to 768 pixels. Then, the second media query uses the width feature and the <= operator to do the same thing.

Within a media query, you code the CSS that adjusts the format of the web page so it's appropriate for the specified condition. For example, you can change the page layout for the specified screen size.

The specific condition which requires a change in the appearance of a page is called a *breakpoint*. Some common breakpoints based on the typical width of various devices are presented in the second table in this figure. However, the breakpoints for a web page typically depend on how the content looks at various screen sizes, not on the predetermined width of a device. As a result, you typically need to use the technique presented in the next figure to determine the breakpoints for your web pages.

The basic syntax of a media query

```
@media [only] media-type [and (media-feature-1)] [and (media-feature-2)]... {
    /* style rules go here */
}
```

Common media features for the screen media type

Media feature	Description
width	The width of the viewport.
min-width	The minimum width of the viewport.
max-width	The maximum width of the viewport.
height	The height of the viewport.
min-height	The minimum height of the viewport.
max-height	The maximum height of the viewport.
orientation	Landscape or portrait.

Two ways to specify a viewport width of 768 pixels or greater

```
@media only screen and (min-width: 768px) { ... }
@media only screen and (width >= 768px) { ... }
```

Two ways to specify a viewport width between 600 and 768 pixels

```
@media only screen and (min-width: 600px) and (max-width: 768px) { ... }
@media only screen and (600px <= width <= 768px) { ... }
```

Common breakpoints for viewport width

Pixels	Description
below 600px	Extra-small screens, such as small phones.
600px	Small screens, such as large phones or tablets in portrait mode.
768px	Medium screens, such as tablets in landscape mode.
992px	Large screens, such as laptop and desktop computers.
1200px	Extra-large screens, such as large laptop and desktop computers.

Description

- You can use a CSS *@media selector* to define a *media query*. This selector specifies the *media type* for the query.

- For the screen media type, a media query can include one or more conditional expressions where each conditional expression can check a *media feature* such as the minimum or maximum width of the viewport. If all of the conditions in a media query are true, the browser applies the styles within the media query to the page.

- You can use <, <=, >, and >= operators to check the width and height of the viewport.

- The condition that causes a media query to change the appearance of a page can be referred to as a *breakpoint*.

- It's a best practice to use a *mobile-first design* where the default CSS is for the smallest screen size, and the media queries adjust that CSS for progressively larger screens.

Figure 8-3   How to code media queries

How to determine the breakpoints for media queries

With responsive web design, it's considered a best practice to use a *mobile-first design*. That means you start by coding your CSS for the smallest screen size that your site supports, which is typically for a mobile phone. Then, you add media queries to change the CSS for progressively larger screens.

However, some developers prefer to use a *desktop-first design*. That means you start by coding your CSS for the largest screen size that your site supports, which is typically a desktop computer. Then, you add media queries to change the CSS for progressively smaller screens.

Figure 8-4 shows how you can use a browser's Developer Tools to determine the breakpoints for a web page based on its content. This works whether you have a mobile-first design or a desktop-first design. Either way, you start by opening the page in a browser and displaying the Developer Tools.

For a mobile-first design, you start with the browser window at a narrow width. Then, you widen the browser window until the design looks bad, or breaks. At that point, record the width that's shown in the upper-right corner of the page. Next, continue to do that until you've recorded all of the breakpoints.

For a desktop-first design, start with the browser window at maximum width. Then, narrow the browser window until the design looks bad, or breaks. At that point, record the width that's shown in the upper-right corner of the page, and continue until you've recorded all of the breakpoints.

The example in this figure is a desktop-first design that's been narrowed until the navigation menu and the headings in the header wrap. So, this is a breakpoint, and the width is 688 pixels.

If the Developer Tools are docked at the bottom of the browser window in Chrome, you can only narrow the window to 500 pixels. If your web page needs to support screens that are narrower than that, you can dock the Developer Tools at the right side of the window. Then, you can make the window narrower.

Once you've determined the breakpoints for your web page, you can create media queries for the various screen sizes. For instance, for the page shown here, you can create a media query for a minimum width of 688 pixels and move the CSS for large screens to that media query. Then, you can adjust the default CSS, so it looks good for smaller screens. For example, on small screens, you can hide the logo image so there's more room for the headings and center the headings so they look better when they wrap.

When you determine breakpoints, you don't have to be exact. You just want to make sure that the breakpoint occurs before whatever makes the web page look bad. For instance, in the example here, you might want to set the breakpoint at a minimum of 700 pixels or more to make sure the logo and the headings look good on large screens.

Also, when you're adding breakpoints to a mobile-first design, there may be a bit of guesswork involved. For instance, the technique presented here might not show you exactly where you want to change from a 1-column to a 2-column layout. So, you can pick a breakpoint that seems plausible and code a media query for it that converts the page to a 2-column layout. Then, you can narrow and widen the browser window to see if the breakpoint you chose works, or if there's a better one.

A web page in Chrome as the headings and navigation menu wraps

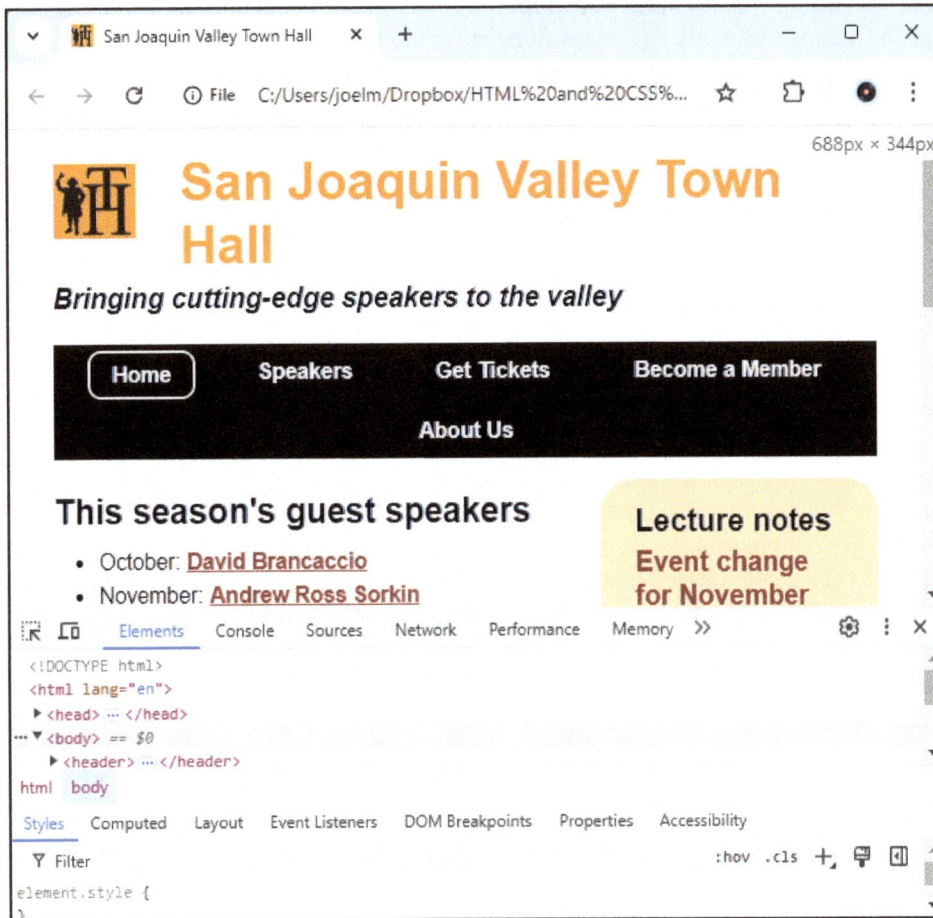

How to determine the breakpoints for a web page

1. Open the page in Chrome and press F12 to open the Developer Tools.
2. Drag the right edge of the browser window to the right or left, and watch for a design break.
3. When a design break occurs, note the width that's shown in the upper right corner of the window.
4. Continue to change the size of the window to determine other breakpoints.

Description

- Once you create a page with fluid design, you can use a browser's Developer Tools to determine where the *breakpoints* for your media queries should be.
- Breakpoints don't have to be exact, and they may be unique for each web page.
- When the Developer Tools are docked at the bottom of the browser window in Chrome, the window can only be narrowed to 500 pixels. To display the page at narrower widths, you can dock the Developer Tools at the right side of the window by clicking the meatball menu (three dots) in the toolbar and selecting a docking side.

Figure 8-4   How to determine the breakpoints for media queries

You can also have as many or as few breakpoints as you want. For example, maybe you want separate breakpoints for when the menu items and the headings wrap. Or, maybe that feels like overkill because the breakpoints are so close together. In that case, you can set a single breakpoint before either the menu items or the headings wrap.

A responsive web page

Figure 8-5 presents a version of the Town Hall home page that uses responsive web design. This web page uses the fluid design presented in chapter 6 and the navigation menu presented in chapter 7. However, it also uses media queries to make it look better on various screen sizes.

Part 1 shows three screens that simulate how this web page appears when displayed on a browser for a desktop computer, a tablet, and a mobile phone. If you compare the appearance of this page on the three screens, you'll notice some differences. For starters, the fluid design makes the font size and logo image smaller for smaller screens and larger for larger screens.

Beyond that, the mobile phone's small screen doesn't display a logo and the headings in the header are centered, not left aligned. In addition, it uses a single column to display its main content. Then, for the tablet's medium screen, the header adds the logo and left aligns the headings. Next, the desktop computer's large screen uses two columns, while the tablet and mobile phone screens use one column.

A web page in desktop, tablet, and mobile layouts

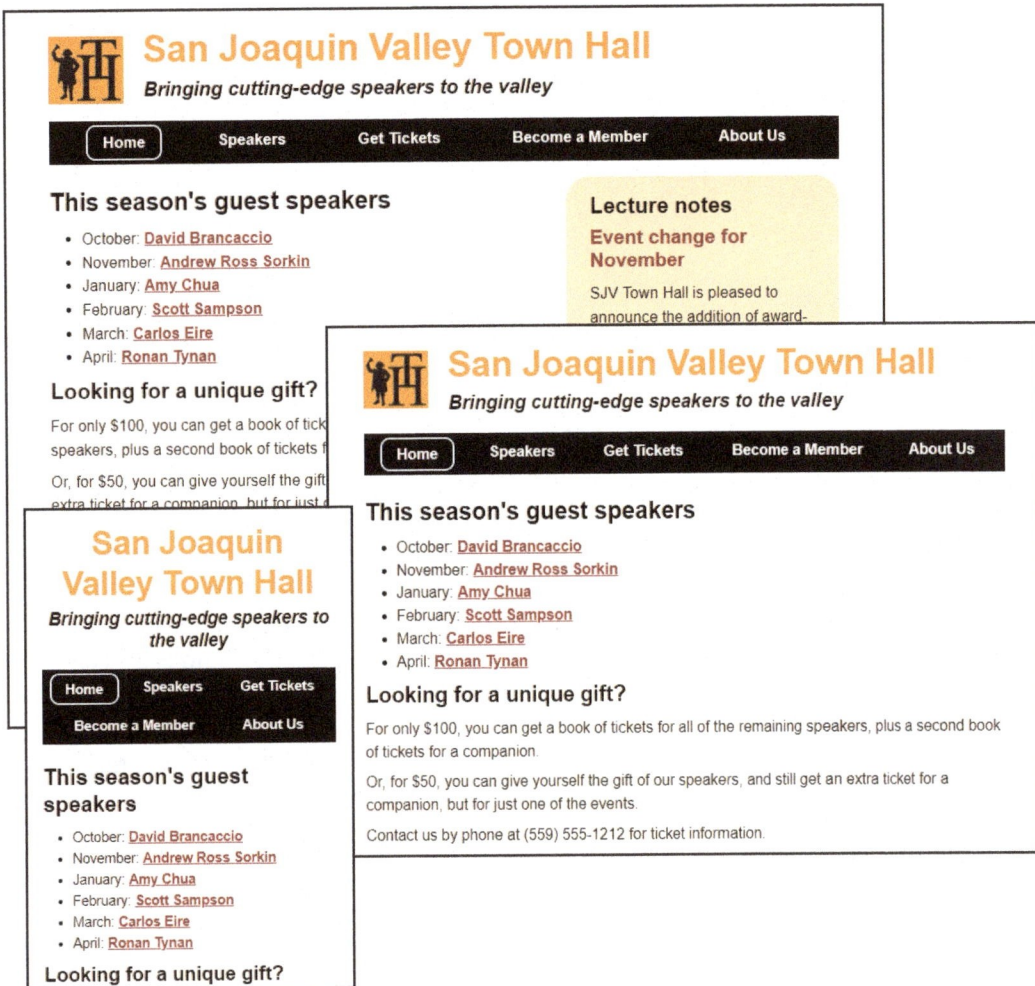

Description

- This web page uses a fluid layout and a navigation menu.
- This web page wraps the navigation menu.
- The web page uses media queries to change its appearance depending on the screen size.

  Small screen: The header section doesn't display a logo and centers its two headings, and the main section uses a one-column layout.

  Medium screen: The header section adds the logo and left aligns the two headings.

  Large screen: The main section uses a two-column layout.

Figure 8-5   A responsive web page (part 1)

Part 2 of figure 8-5 presents the CSS for this web page. However, it doesn't present the CSS for the color variables, the reset, and the default styles because that CSS was already presented in chapter 6. In addition, it doesn't present the CSS for the .navbar and .menu classes because that CSS was already presented in chapter 7.

The CSS in this figure uses mobile-first design. In other words, the CSS in part 2 is for the smallest screens. Then, the styles in the media queries presented in part 3 override some of these styles for larger screens.

To get the CSS presented in chapter 6 to look good on small screens, it was only necessary to make a few changes. Those changes are indicated by the comments in the figure.

The style rule for the <header> element sets the text-align property to center so the headings in the header are centered in the screen. In addition, it sets the display property of the nested image to none to hide the logo.

The style rule for the <main> element sets the flex-direction property to column. This lays out the main content in a single column. Another way to create a 1-column layout is to omit the display and flex-direction properties here. However, setting these properties here makes it easier to override them in media queries later.

The style rules for the <section> and <aside> elements don't include a flex-basis property. That's because the default flex-basis value of auto is all you need for a 1-column layout. In addition, the style rule for the <aside> element has no left margin.

The style rule for the nested <p> element in the <footer> element sets the text-align property to center. This centers the footer text in the screen.

The mobile-first CSS for the smallest viewports

```css
/* variables, reset, and default styles same as chapter 6 */

header {
    line-height: 1.2;
    text-align: center;        /* center headings */
    & img {
        display: none;         /* hide logo image */
    }
    & h2 {
        font-size: 225%;
        color: var(--light);
        padding: 0;
    }
    & h3 {
        font-style: italic;
        padding-bottom: 1em;
    }
}

main {
    display: flex;
    flex-direction: column;    /* 1-column layout */
    padding-bottom: .5em;
}

section {                      /* no flex-basis property */
    padding: 1em 0;
    & img { width: min(100%, 600px); }
}

aside {                        /* no flex-basis property */
    background-color: var(--aside);
    height: fit-content;
    padding: 1em 1.5em;
    margin: 1em 0;             /* no left margin */
    border-radius: 25px;
    & h2 { padding: 0; }
    & h3 { color: var(--dark); }
}

footer {
    border-top: 2px solid var(--light);
    padding-top: .5em;
    & p {
        font-size: 80%;
        text-align: center;    /* center text */
    }
}

/* CSS for .navbar and .menu classes same as chapter 7 */
```

Figure 8-5   A responsive web page (part 2)

Part 3 of figure 8-5 presents the media queries for larger screens. The breakpoint for the first media query is a min-width value of 745 pixels. This means that the styles in this media query are applied when the viewport is 745 pixels wide or greater. The breakpoint for the second media query is a min-width value of 768 pixels, which means these styles are applied when the viewport is 768 pixels wide or greater.

Since these breakpoints are so close together, you could also combine them into a single media query. If you wanted to do that, you'd move the styles for the larger viewport into the media query for the smaller viewport. However, it's also fine to have multiple media queries at close breakpoints.

The CSS in the first media query starts with a style rule for the <header> element. First, this style rule sets the text-align property to left. This overrides the text-align property set in the mobile-first CSS to left align the headings in the header.

The nested style rule for the <img> element displays the image for the logo. This overrides the mobile-first setting that hides the image. After that, it floats the image to the left, adds some margins around the image, and uses the clamp() function to make the image scalable but with a minimum width of 40 pixels and a maximum width of 80 pixels.

The style rule for the <p> element nested in the <footer> element sets the text-align property to right. This overrides the text-align property set in the mobile-first CSS to right align the text in the footer.

The CSS in the second media query starts with a style rule for the <main> element that sets the flex-direction property to row. This overrides the flex-direction property set in the mobile-first CSS, and changes the layout from vertical to horizontal.

The style rules for the <section> and <aside> elements set the flex-basis property of these two elements. This sets the desired width of the columns in a 2-column layout. In addition, the style rule for the <aside> element sets a left margin so there's some space between the columns. This overrides the left margin of 0 set in the mobile-first CSS.

The media queries for larger screens

```
@media only screen and (min-width: 745px) {
    header {
        text-align: left;          /* left-align headings */
        & img {
            display: block;        /* show and format logo image */
            float: left;
            margin: .5em 1.25em .5em 0;
            width: clamp(40px, 10%, 80px);
        }
    }

    footer p {
        text-align: right;         /* right-align footer text */
    }
}

@media only screen and (min-width: 768px) {
    main {
        flex-direction: row;       /* 2-column layout */
    }
    section {
        flex-basis: 65%;           /* set flex-basis */
    }
    aside {
        flex-basis: 35%;           /* set flex-basis */
        margin-left: .5em;         /* add left margin */
    }
}
```

Figure 8-5   A responsive web page (part 3)

How to create a responsive navigation menu

So far, this book has shown how to create responsive navigation menus by using flexbox and wrapping the menu items on small screens. Now, the following figures show how to provide a navigation bar that has a menu button that displays a drop-down navigation menu for small screens.

How to create a menu button

Figure 8-6 presents the HTML for a navigation menu with a menu button. This is similar to the HTML for navigation menus presented in chapter 7. However, this HTML includes an <input> element and a <label> element.

The <input> element has a type attribute of checkbox, which means it's a checkbox control. In addition, it has an id value of checkbox-toggle.

The <label> element has a for attribute with the same value as the id of the checkbox. Because of this, the browser binds the label to the checkbox. This is important because it makes clicking the label the same as clicking on the element, even if that element is hidden. As a result, clicking on this label checks and unchecks the checkbox.

The <label> element is also assigned to the menu-btn class. This class allows the CSS to format the label to look like a hamburger symbol or a close symbol depending on whether the checkbox is checked or unchecked.

The CSS in this figure starts with the style rule for the menu class. This style rule sets the display property to none. As a result, the browser hides the navigation menu when the page loads.

The style rule for the menu-btn class sets the display to block. Since a <label> element is an inline element, this is necessary so the subsequent padding property can add padding to all four sides. Then, this style rule sets the color property to white.

The style rule for the ::after pseudo-element of the menu-btn class sets the content property to the hamburger symbol. This symbol is commonly used for menu buttons.

The style rule for the #checkbox-toggle selector sets the display property to none. This hides the checkbox. However, since the checkbox is bound to the label, clicking on that label checks and unchecks the checkbox, even though the checkbox is hidden.

The last two style rules apply when the user clicks the label to check the checkbox. Both style rules use the :checked pseudo-class of the checkbox-toggle id selector. The first rule selects the next sibling that's assigned to the menu class and sets its display property to block to display the menu items. The second rule selects the ::after pseudo-element of the next sibling that's assigned to the menu-btn class and sets its content property to the close (X) symbol to display that symbol.

The first screen shows the navigation bar when the checkbox isn't checked. In that case, the browser hides the checkbox, hides the menu items, displays the hamburger symbol in the label, and adds padding around the label.

The second screen in figure 8-6 shows the navigation bar when the user clicks the label to check the checkbox. In that case, the browser displays the menu items

The HTML for a navigation menu with a menu button

```
<nav class="navbar">
    <input type="checkbox" id="checkbox-toggle" />
    <label for="checkbox-toggle" class="menu-btn"></label>

    <ul class="menu"><!-- navigation links --></ul>
</nav>
```

Some of the CSS for the navigation menu

```
.menu {
    display: none;                          /* menu is initially hidden */
}


.menu-btn {                                 /* format menu button */
    display: block;
    padding: 1em;
    color: white;
}
.menu-btn::after {
    content: "\2630";                       /* hamburger symbol */
}


#checkbox-toggle { display: none; }         /* checkbox is always hidden */

/* when checkbox is checked, display the menu and the close symbol */
#checkbox-toggle:checked ~ .menu {
    display: block;
}
#checkbox-toggle:checked ~ .menu-btn::after {
    content: "\2716";                       /* close (X) symbol */
}
```

The menu when the checkbox is not checked

The menu when the checkbox is checked

Description

- You can use <input> and <label> elements to create a button that displays and closes a menu. To do that, use the <input> element to add a check box. Then, set the for attribute of the label to the id of the checkbox. That way, clicking the label checks and unchecks the checkbox, even if the checkbox is hidden.

Figure 8-6   How to create a menu button for a responsive navigation menu

and the label displays the close symbol. When the user clicks the close symbol, the hidden checkbox is unchecked, and these last two style rules are no longer applied. As a result, the browser displays a navigation bar that looks like the first screen.

A responsive navigation menu

Figure 8-7 presents the Town Hall home page with a responsive navigation menu. On large screens, this page displays a navigation menu like the one presented in chapter 7. But on small screens, it displays a navigation bar with a menu button like the one in the last figure. Then, when the user clicks on the menu button, the navigation menu displays as a drop-down menu and the hamburger symbol changes to a close symbol. In addition, this navigation menu uses positioning so the menu displays on top of the content of the page.

The HTML for the navigation menu consists of a <nav> element assigned to the navbar class that contains a <ul> element assigned to the menu class. In addition, the <nav> element contains a checkbox that's assigned to the menu-btn class as well as a label that's bound to the checkbox. As a result, when users click the label, the checkbox is checked and unchecked.

A responsive navigation menu on a large screen

The same navigation menu on a smaller screen

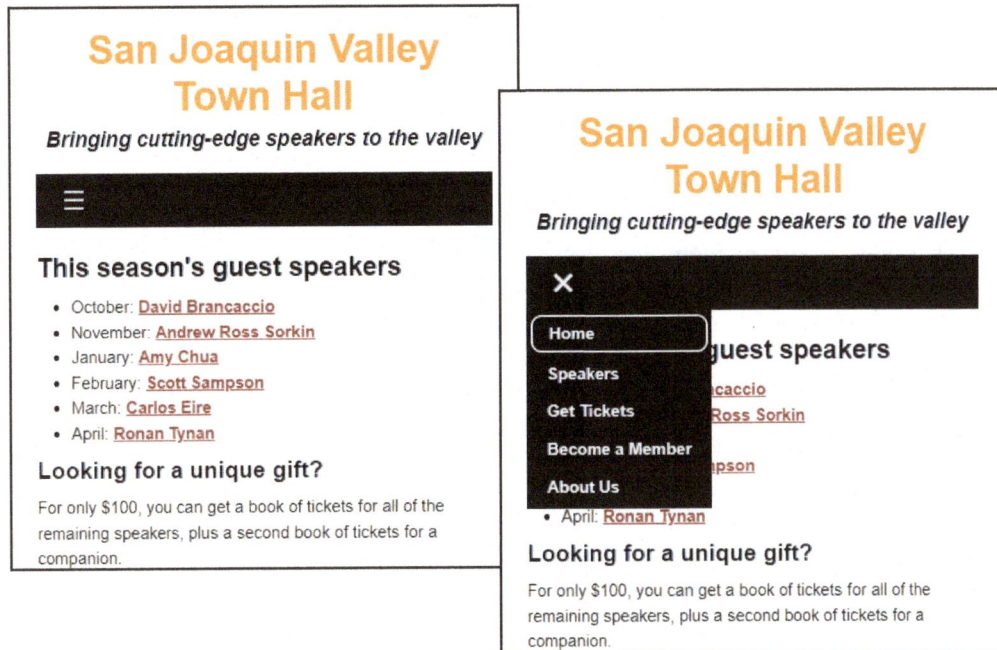

The HTML for the navigation menu

```
<nav class="navbar">
    <input type="checkbox" id="checkbox-toggle" />
    <label for="checkbox-toggle" class="menu-btn"></label>

    <ul class="menu">
        <li><a class="current">Home</a></li>
        <li><a href="#">Speakers</a></li>
        <li><a href="#">Get Tickets</a></li>
        <li><a href="#">Become a Member</a></li>
        <li><a href="#">About Us</a></li>
    </ul>
</nav>
```

Figure 8-7   A responsive navigation menu (part 1)

Part 2 of figure 8-7 starts by presenting the mobile-first CSS for the smallest viewports. Most of this CSS is the same as the CSS presented in chapter 7. However, there are a few differences.

The style rule for the navbar class uses relative positioning. This allows the CSS to position the drop-down menu relative to the navigation bar.

The style rule for the menu class starts by specifying absolute positioning. This positions the drop-down menu relative to the navigation bar and also makes it display on top of the page content. Then, this style rule sets the left property to 0. As a result, the drop-down menu aligns with the left edge of the navigation bar. Finally, it sets the display property to none so the menu is initially hidden.

The style rule for the checkbox-toggle id selector sets the display property of the checkbox to none. Since no other CSS changes this property, the checkbox is always hidden. But, since it's bound to the label, users can check and uncheck the checkbox by clicking on its label.

The style rule for the menu-btn class sets the display property to block so it can add padding to all four sides of the <label> element assigned to this class. It sets the color to the light menu value and also sets the font-size in pixels to control the size of the hamburger and close symbols. Then, the ::after pseudo-element of the menu-btn class sets the content property to the hamburger symbol.

The next two style rules apply when the checkbox is checked. The first rule displays the menu, and the second rule displays the close symbol.

Part 2 finishes by presenting the media query for the navigation menu for larger screens. The first two style rules in this media query display the navigation menu horizontally and hide the button for the drop-down menu.

More specifically, the style rule for the menu class overrides the absolute positioning of the menu for small screens and uses relative positioning instead. As a result, the navigation menu becomes part of the normal flow of elements. Then, this style rule sets the display property to flex to override the mobile-first value of none. This displays the menu items horizontally. Finally, it sets the justify-content property to control the horizontal spacing of the menu items.

The style rule for the menu-btn class sets the display property to none. This hides the label that's bound to the hidden checkbox.

The last style rule handles a situation that can occur when you check the checkbox on a small screen and then change the width of the browser window from small to large. Without this style rule, the menu items display vertically on a large screen. That's because the mobile-first style rule sets the display property of the menu class to block, which displays items vertically by default. To fix that, this style rule overrides that mobile-first CSS and sets the display property of the menu class to flex.

The mobile-first CSS for the smallest viewports

```css
.navbar {
    position: relative;
    background-color: var(--dark_menu);
    & a {
        display: block;
        padding: .25em 1em;
        margin: .25em;
        text-decoration: none;
        color: var(--light_menu);
    }
    & a:hover, a:focus {
        text-decoration: underline;
    }
    & a.current {
        border: 2px solid;
        border-radius: 10px;
        text-decoration: none;
    }
    & ul {
        list-style-type: none;
        margin: 0;
        padding: 0;
    }
}
.menu {
    position: absolute;    /* in relation to .navbar */
    left: 0;
    display: none;         /* menu initially hidden */
    background-color: var(--dark_menu);
}
#checkbox-toggle { display: none; }        /* checkbox always hidden */
.menu-btn {
    display: block;                        /* show menu button */
    padding: 0.25em 1em;
    color: var(--light_menu);
    font-size: 24px;
}
.menu-btn::after { content: "\2630"; }     /* hamburger symbol */
#checkbox-toggle:checked ~ .menu {
    display: block;
}
#checkbox-toggle:checked ~ .menu-btn::after {
    content: "\2716";                      /* close (X) symbol */
}
```

The media query for larger screens

```css
@media only screen and (min-width: 745px) {
    .menu {
        position: relative;                /* change position */
        display: flex;                     /* show horizontal menu */
        justify-content: space-evenly;     /* adjust menu spacing */
    }
    .menu-btn { display: none; }           /* hide menu button */

    /* make sure menu displays horizontally if menu checkbox is checked */
    #checkbox-toggle:checked ~ .menu { display: flex; }
}
```

Figure 8-7   A responsive navigation menu (part 2)

A 2-tier responsive navigation menu

Figure 8-8 presents the Town Hall home page with a responsive navigation menu that has a submenu. On large screens, such as desktops and large tablets, the page displays a 2-tier navigation menu like the one you saw in chapter 7. On small screens, it displays the menu button and drop-down menu as shown in the last figure, but the drop-down menu also includes a submenu item.

In other words, on small screens, the page displays a 3-tier navigation menu. In this case, the first tier is the navigation bar and menu button, the second tier is the menu displayed as a drop-down menu, and the third tier is the submenu.

The menu and submenu have a couple differences in how they display on large and small screens. First, the submenu displays below the menu item on large screens, but it displays to the right of the menu item on small screens. Second, the triangle symbol for a submenu points down on large screens, but it points to the right on small screens.

The HTML for the navigation menu consists of a <nav> element that contains a checkbox, a label that's bound to the checkbox, and an unordered list of menu items.

In addition, the text for the Speakers link includes a <small> element, and the <li> element contains a nested <ul> element assigned to the submenu class. This is similar to the HTML for a submenu presented in chapter 7. However, the <small> element doesn't contain an HTML code for a triangle symbol. This allows the CSS to change the symbol for the triangle so it points down or to the right depending on screen size.

A 2-tier responsive navigation menu at large and small screen sizes

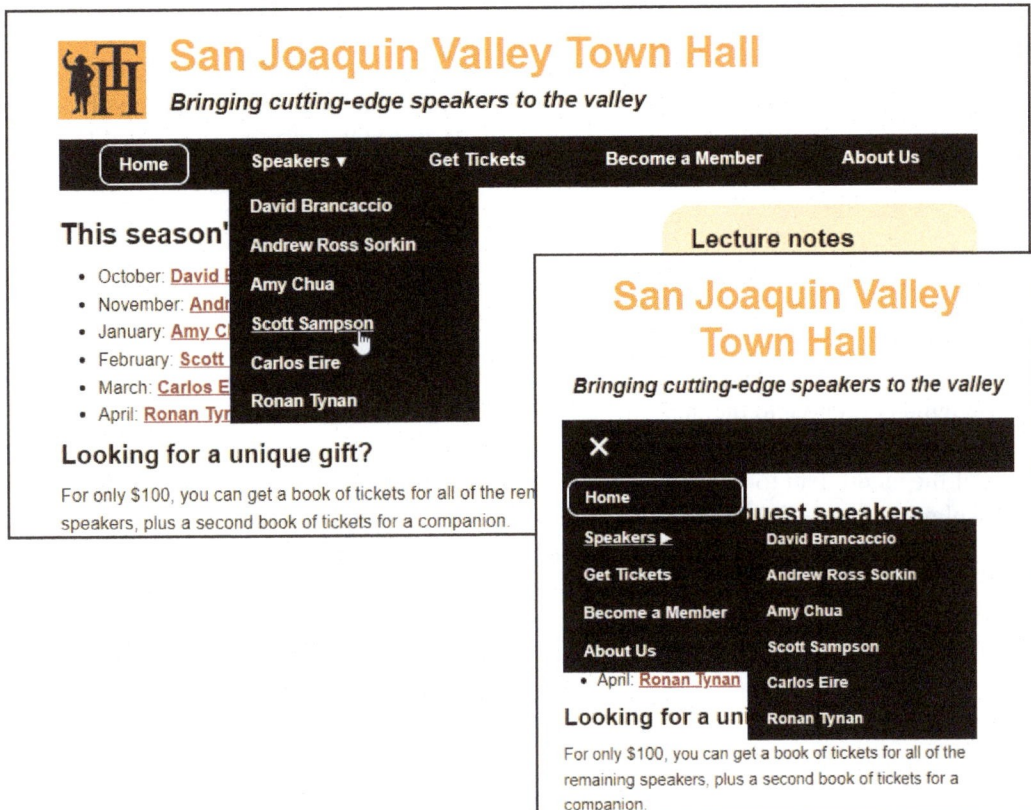

The HTML for the navigation menu

```html
<nav class="navbar">
    <input type="checkbox" id="checkbox-toggle" />
    <label for="checkbox-toggle" class="menu-btn"></label>

    <ul class="menu">
        <li><a class="current">Home</a></li>
        <li><a href="#">Speakers<small></small></a>
            <ul class="submenu">
                <li><a href="#">David Brancaccio</a></li>
                ...
                <li><a href="#">Ronan Tynan</a></li>
            </ul>
        </li>
        <li><a href="#">Get Tickets</a></li>
        <li><a href="#">Become a Member</a></li>
        <li><a href="#">About Us</a></li>
    </ul>
</nav>
```

Figure 8-8   A 2-tier responsive navigation menu (part 1)

Part 2 of figure 8-8 starts by presenting the CSS for the smallest viewports. Since most of the style rules are the same as the ones presented in the last figure, they aren't presented again here.

The style rule for each menu item that has a submenu starts by setting the position property to relative. Then, it has two nested style rules for the <small> element. The first rule sets the content property of the ::after pseudo-element to a right triangle symbol, while the second adds a small amount of space to the left of the triangle.

The third nested style rule applies when the user hovers over the <li> element that contains the submenu class. It selects the child element assigned to the submenu class and sets its display property to block. This displays the submenu.

The style rule for the submenu class uses absolute positioning to place the submenu relative to the item that contains the submenu. Then, it sets the top property to 0 and the left property to 100%. This aligns the submenu with the top of the menu item to its right. In other words, it positions the top left side of the submenu to the right of the item that contains it. In addition, this style rule sets the display property to none so the submenu is initially hidden.

Part 2 finishes by presenting the media query for larger screens. However, since most of the CSS is the same as the CSS presented in the last figure, it isn't shown here.

The style rule for the <small> element of each item that has a submenu sets the content of the ::after pseudo-element to a triangle symbol that points down. This overrides the mobile-first CSS of a triangle that points right.

The style rule for the submenu class overrides the mobile-first CSS and restores the top property to its default setting of auto. This aligns the top of the element with the flow of the page. Then, it sets the left property to 0, which aligns the left side of the submenu with the left side of the item that contains the submenu. In other words, it positions the submenu below the item that displays it.

The last declaration in this style rule overrides the mobile-first CSS to set the width property to 200%. This makes the submenu twice as wide as the item that contains it, which makes sense on a larger screen.

The mobile-first CSS for the submenu

```css
/* .navbar, .menu, #checkbox-toggle, #checkbox-toggle:checked, .menu-btn, and
.menu-btn::after same as figure 8-7 */

.menu li:has(.submenu) {
    position: relative;
    & small::after {
        content: "\25b6";           /* right triangle */
    }
    & small {
        margin-left: .25em;      /* add space */
    }
    &:hover > .submenu {
        display: block;
    }
}

.submenu {
    position: absolute;    /* relative to the menu item that contains it  */
    top: 0;
    left: 100%;
    background-color: var(--dark_menu);
    color: var(--light_menu);
    width: 150%;
    display: none;
}
```

The media query for the submenu for larger screens

```css
@media only screen and (min-width: 745px) {
    /* header, footer, .menu, .menu-btn, and #checkbox-toggle:checked
        same as figure 8-7 */

    /* change triangle direction */
    .menu li:has(.submenu) small::after {
        content: "\25bc";   /* down triangle */
    }

    /* change submenu position */
    .submenu {
        top: auto;              /* remove mobile-first value */
        left: 0;
        width: 200%;
    }
}
```

Figure 8-8   A 2-tier responsive navigation menu (part 2)

How to work with container queries

So far, this book has shown how to use media queries to change elements on a page based on the width of the viewport. Now, it shows how to use *container queries* to change elements on a page based on the width of the container they're in. To do that, you need to define an HTML element on a page as a container. Then, you can code queries based on characteristics of that container.

How to define a container

Figure 8-9 presents the properties that you can use to define a container. The container-type property indicates the type of container to create. This is similar to the media features for a media query. Most of the time, you'll use the inline-size value to define a container that applies styles based on the width of the container.

The container-name property lets you specify a name for the container. Then, you can use that name in your container queries to refer to a specific container.

The container property is a shorthand property that allows you to set both the container type and name. When you use this property, you must separate the two values with a slash (/).

The HTML shown in this figure defines a <section> element that contains a heading and an unordered list. Then, the first CSS example defines the <section> element as a container by setting its container-type property to inline-size. That's all you need to do to define a container. Then, you can code container queries for that container as shown in the next figure.

The second CSS example defines the <section> element as a named container by setting both the container-type and container-name properties. That way, you can include this name when you code container queries as shown in the next figure. Named containers can be useful when you have more than one container on a page.

The third CSS example shows how to use the shorthand container property to define a named container. This example is another way to define the same container as the second CSS example.

When working with containers, inline-size usually means width, but in some cases, it can mean height. For instance, the default flow of a web page is horizontal, from left to right. However, it's possible to change the flow to vertical, from top to bottom. In that case, inline-size becomes height and block-size becomes width.

Properties for defining a container

Property	Description
container-type	The type of container to create. Possible values include inline-size (width), block-size (height), or size (width or height, whichever is larger).
container-name	The name of a container. You can include this name in a container query to refer to a specific container.
container	A shorthand property for specifying the container type and name. The two values must be separated by a slash (/).

The HTML for a container

```
<section>
    <h2>Latest Articles</h2>
    <ul>
        <li>...</li>
        <li>...</li>
    </ul>
</section>
```

Define a container

```
section {
    container-type: inline-size;
}
```

Define a named container

```
section {
    container-type: inline-size;
    container-name: articles;
}
```

Another way to define a named container

```
section {
    container: inline-size / articles;    /* shorthand property */
}
```

Description

- To define a container, you code the container-type property for the element that contains other elements.

- If a page defines more than one container, you can use the container-name property to name them.

- For the purposes of this chapter, inline-size means width and block-size means height. However, it's possible to change this by changing the default right-to-left horizontal flow.

Figure 8-9   How to define a container

How to code a container query

Figure 8-10 shows how to code a container query. The top of this figure presents the basic syntax. It starts with an *@container selector* followed by one or more conditional expressions. Like media queries, each expression specifies a media feature that describes a characteristic of the container.

The first container query example applies when the unnamed container from the previous figure has a width of 500 pixels or more. This query is declared with the @container selector followed by a conditional expression that uses the min-width media feature.

The second container query example applies when the container named articles has a width of 500 pixels or more. The query is declared with the @container selector followed by the name of the named container, followed by a conditional expression that uses the min-width feature.

Within a container query, you code the CSS that adjusts the format of the web page so it's appropriate for the specified condition of the container. For instance, the first container query example selects the <ul> element in the container and sets its flex-direction property, while the second selects the <h2> element in the container and sets its text-align property.

A container query can't style the container itself. For instance, the container queries shown here apply to the <section> element presented in the previous figure. As a result, you can't use either of the container queries shown in this figure to set the background-color property of that <section> element. If you attempted to do this, nothing would happen. But, if you tried to set the background color of the <ul> or <h2> element in the container, that would work.

Chapter 6 presented relative units of measure that are based on the size of the viewport such as vw and vh. Now, this figure presents some relative units of measure that are based on the size of the container such as cpw and cph. Then, it shows an example of a container query that uses the cqw unit to set the flex-basis property of the <li> elements in a container based on the width of the container.

When you work with container queries, you need to know that they can affect how positioned elements display. In particular, they can cause a positioned element to display behind another element. When that happens, you may need to adjust the z-index of the container or the positioned elements so the page displays the way you want.

By default, most elements have a z-index of 0. However, you can set the z-index of an element to an integer that's greater than 0 to display it in front of other elements. In that case, the element with the highest z-index value displays in front of other elements.

The basic syntax of a container query

```
@container [container-name] (feature-1) [and (feature-2)]... {
    /* style rules */
}
```

A container query for the unnamed container from the last figure

```
@container (min-width: 500px) {
    ul {
        flex-direction: row;
    }
}
```

A container query for the named container from the previous figure

```
@container articles (min-width: 500px) {
    h2 {
        text-align: center;
    }
}
```

Relative units of measure based on the container

Symbol	Description
cqh	1% of the height of the container.
cqw	1% of the width of the container.
cqi	1% of the container's inline size.
cqb	1% of the container's block size.
cqmin	The smaller value of either cqi or cqb.
cqmax	The larger value of either cqi or cqb.

A container query that uses a relative unit of measure

```
@container (min-width: 500px) {
    li {
        flex-basis: 30cqw;
    }
}
```

Description

- A *container query* is similar to a media query, but it's based on the element that's defined as a container rather than the viewport. As a result, you can use relative units of measure that are based on the size of the container rather than the size of the viewport.

- To code a container query, you use the CSS @container selector.

- You can use container queries with the media queries for a page.

- A container query can't style the element that's defined as the container, but it can style the elements within the container.

- Containers can affect how positioned elements display. To address this, you can adjust the z-index of the container or the positioned elements.

Figure 8-10   How to code a container query

A web page with a container query

Figure 8-11 presents the tickets page of the Town Hall website. This web page displays tickets for each guest speaker in a container. When the container is wider, it displays two tickets per row as shown here. But when the container is narrower, it only displays one ticket per row. This happens in both the 2-column layout for large screens and the 1-column layout for small screens. That's because the container query doesn't care about the width of the viewport, only the width of the container.

The HTML shows that this page uses a <section> element as a container for a heading and a <div> element that's assigned to the tickets class. Within this <div> element, the HTML defines one link for each speaker.

The CSS that styles this page uses the <section> element as the container for the query. Then, it defines a <div> element that contains the links and assigns it to the tickets class. This allows the container query to change the flex-direction property of that class, which is necessary because a container query can't style the <section> element itself, since that element is the container.

A tickets page at screen widths that display two tickets per row

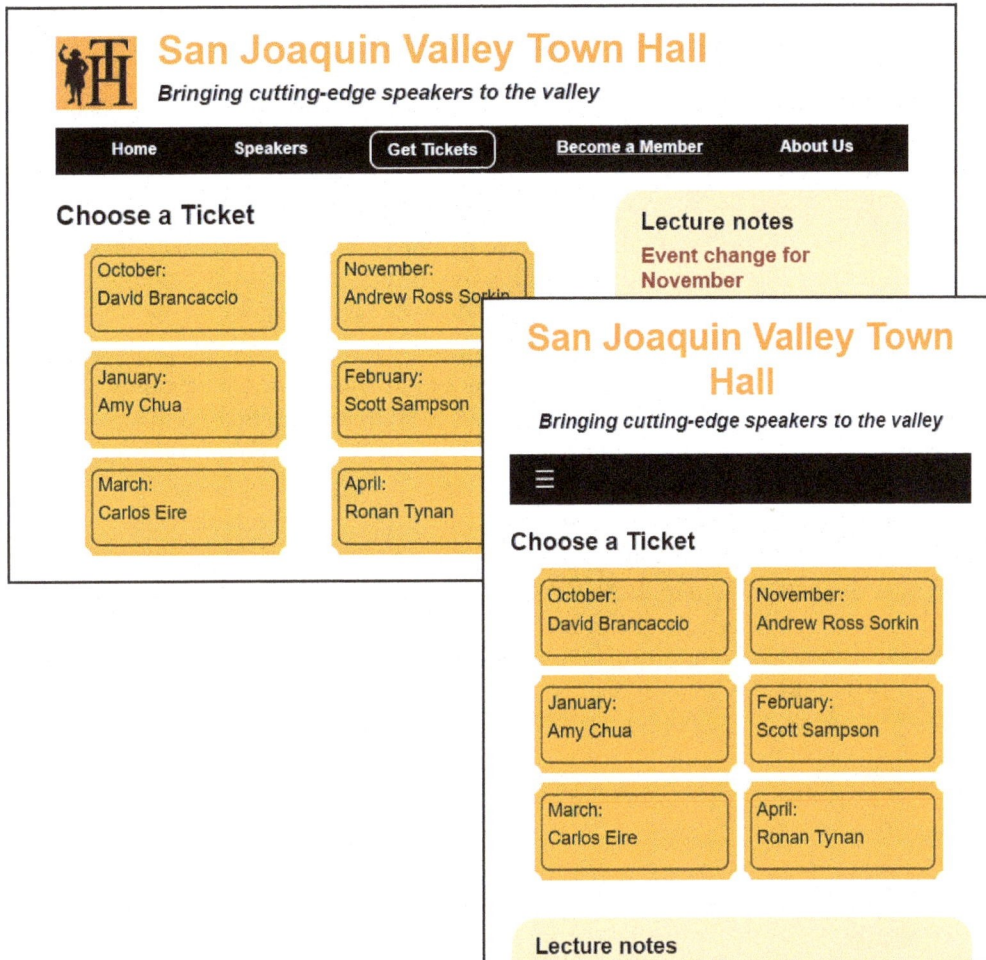

The HTML for the <section> element

```
<section>
    <h1>Choose a Ticket</h1>
    <div class="tickets">
        <a href="tickets/brancaccio">October:<br>David Brancaccio</a>
        .
        .
        <a href="tickets/tynan">April:<br>Ronan Tynan</a>
    </div>
</section>
```

Description

- This page displays one or two tickets per row, depending on the size of the
  <section> element that's the container.

Figure 8-11   A web page with a container query (part 1)

Part 2 of figure 8-11 starts by showing two style rules from the mobile-first CSS presented earlier in this chapter after they have been updated to work with container queries. Here, the style rule for the <section> element defines that element as a container named tickets.

The style rule for the menu class sets the z-index property to 1. This makes sure that the absolutely positioned menu displays in front of the container that contains the tickets, which has a z-index of 0.

Part 2 continues by showing the mobile-first style rule for the tickets class. This styles the <div> element that contains the links for the tickets. To do that, the CSS defines it as a flex container with a flex-direction property of column. As a result, it displays one ticket per row. It also adds some left and right padding.

The nested style rule for <a> elements makes the links block elements so it can add padding to all four sides and also adds a small top and bottom margin. After that, it removes the underline from the link, sets the text color to the dark menu color, sets the font size in pixels so it doesn't change as the screen size changes, and sets the font weight to normal. Finally, this style rule sets the background of the links to an image of a ticket, and sets the height of the link to make sure it's tall enough to display the background image.

The nested style rule for the :hover and :focus pseudo-classes of the <a> element sets the text color of the link to the light menu color and sets the background of the link to an image of a ticket that's a dark color. As a result, when the user hovers over a link or tabs to it, the browser displays light text on a dark background image.

Part 2 finishes by presenting the container query that applies when the width of the <section> element is 450 pixels or more. It contains a style rule for the tickets class that sets the flex-direction property to row. This overrides the flex-direction property set in the mobile-first CSS so the ticket links display horizontally. Then, it sets the flex-wrap property so the tickets wrap to multiple rows as needed, sets the justify-content property to control the spacing of the tickets, and removes the padding set in the mobile-first CSS.

The nested style rule for the <a> element uses a relative measure based on the container width to set the flex-basis property. As a result, the container displays two tickets per row at this larger size.

The updated CSS for the structural elements and navigation menu

```css
section {
    /* define the container */
    container-type: inline-size;
    container-name: tickets;

    /* rest of CSS same as figure 8-5 */
}

.menu {
    z-index: 1; /* to make sure it displays on top of container */

    /* rest of CSS same as figure 8-7 */
}
```

The mobile-first CSS for the tickets

```css
.tickets {
    display: flex;
    flex-direction: column;     /* display one ticket per row */
    padding: 0 3em;
    & a {
        display: block;
        padding: .75em;
        margin: .25em 0;
        text-decoration: none;
        color: var(--dark_menu);
        font-size: 18px;         /* fixed font size */
        font-weight: normal;
        background: url("../images/ticket.png") no-repeat;
        min-height: 94px;         /* fixed height for background image */
    }
    & a:hover, a:focus {
        color: var(--light_menu);
        background: url("../images/ticket-hover.png") no-repeat;
    }
}
```

The container query for the tickets

```css
@container tickets (min-width: 450px) {
    /* display two tickets per row */
    .tickets {
        flex-direction: row;
        flex-wrap: wrap;
        justify-content: center;
        padding: 0;
        & a {
            flex-basis: 45cqw;
        }
    }
}
```

Figure 8-11   A web page with a container query (part 2)

Perspective

Now that you've finished this chapter, you should be able to use media queries and container queries to format your web pages so they look good on devices of all screen sizes. In other words, you should be able to develop web pages that provide for responsive web design.

Terms

responsive web design (RWD)	media feature
fluid design	breakpoint
scalable image	mobile-first design
media query	desktop-first design
@media selector	container query
media type	@container selector

Exercise 8-1   Test responsive web pages

In this exercise, you'll test the responsive web pages presented in this chapter. This should help you understand how responsive web design works.

Test the Town Hall home page

1. Use Chrome to run the index.html file in this folder:

   `book_apps\ch08\town_hall`

2. Size the browser window as wide as you can. Then, narrow the window to see how the size of the text and the layout of the page change at various widths. When you're done, make the window wide again.

3. Press F12 to display the Developer Tools. Then, narrow the browser window and note that the size of the window is displayed near the upper right corner.

4. Continue to narrow the browser window to see that it has a minimum width.

5. Dock the Developer Tools on the right side of the window.

6. Continue to narrow the window. You should be able to make it much narrower.

Test a responsive navigation menu

7. Use Chrome to run the index.html file in this folder:

   `book_apps\ch08\town_hall_responsive_nav`

8. Size the browser window as wide as you can. Then, narrow the window to see how the navigation menu changes at various widths.

9. When the menu button is visible, click it to display the drop-down menu. Then, click the close button to hide the drop-down menu. When you're done, make the browser window wide again.

10. Press F12 to display the Developer Tools and click the "Toggle device toolbar" icon to display the Device toolbar.

11. Select different devices from the Dimensions drop-down list. For each device you choose, click the Rotate icon to see how the page looks in both portrait and landscape orientation.

12. When you're done, click the "Toggle device toolbar" icon again to return to the standard Developer Tools view.

Exercise 8-2   Make a responsive web page

In this exercise, you'll convert a web page that uses a desktop-first design to use a responsive design. As part of this process, you'll convert it to use a mobile-first design. When you're through, the page should look like this at different screen sizes:

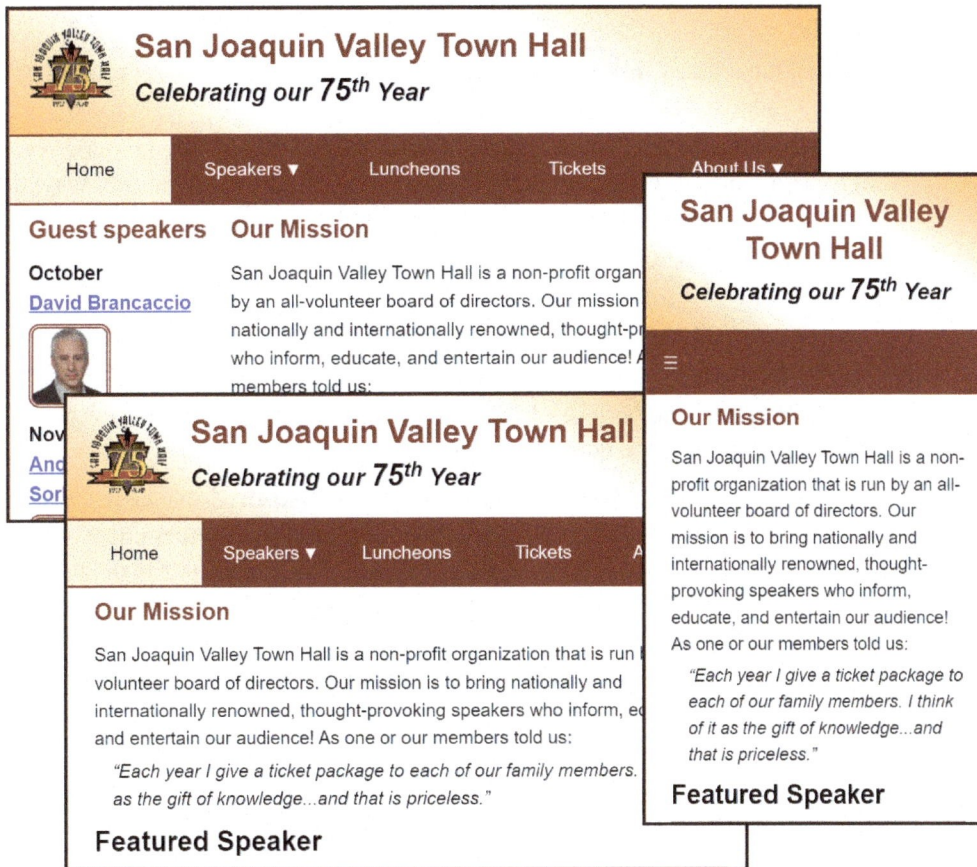

Open the HTML and CSS files for this page

1. Use your text editor to open these HTML and CSS files:

```
exercises\ch08\town_hall\index.html
exercises\ch08\town_hall\styles\main.css
```

Identify a breakpoint for the column layout and add a media query

In this section you'll move the CSS for a 2-column layout to a media query and make the existing CSS mobile-first with a 1-column layout.

2. Run the web page in Chrome. Size the browser window as wide as you can and press F12 to display the Developer Tools.

3. Narrow the window until the Guest Speakers heading in the aside wraps. Note the width of the page displayed in the upper right corner and add a media query for when the viewport width is greater than or equal to that breakpoint.

4. Copy the existing style rules for the <main>, <section>, and <aside> elements to the media query.

5. In the media query, in the main style rule, delete all the declarations. Then, set the flex-direction property to row.

6. In the media query, in the section and aside style rules, delete all properties except the flex-basis properties.

7. In the mobile-first CSS (the CSS outside the media query), in the main style rule, add a flex-direction property and set it to column-reverse. That way, the <section> element displays first.

8. In the mobile-first CSS, in the section and aside style rules, delete the flex-basis properties.

9. Refresh the web page and test it to see how the layout changes at different screen widths.

10. Add a border to the bottom of the <section> element in the mobile-first CSS. Be sure to remove the border in the media query.

Identify a breakpoint for the navigation menu and add a media query

In this section, you'll move the CSS for a horizontal navigation menu that's appropriate for larger screens to a media query, and you'll modify the default CSS for the page to display a navigation menu with a button that displays a vertical drop-down menu.

11. Reduce the width of the window until the navigation menu wraps and note the width of the page displayed in the upper right corner. Then, add a media query for when the viewport width is greater than or equal to that width.

12. In the HTML, update the <nav> element by adding a checkbox and a label. Be sure to bind the label to the checkbox and to assign the label to the menu-btn class.

13. Find the <small> elements and delete the HTML code that's between the tags.

14. In the CSS, copy the existing style rule for the menu class to the media query. Then, delete the flex-wrap property, set the position to relative, and set the width to 100%.

15. In the media query, add a style rule for the menu-btn class and set the display to none.

16. In the mobile-first CSS, modify the style rule for the navbar class so it sets the position to relative.

17. In the mobile-first CSS, delete all declarations from the style rule for the menu class. Then, set the position to absolute, the left property to 0, the background color to the accent color, the width to 40%, and the display to none.

18. In the mobile-first CSS, add style rules that format the menu button and add the hamburger symbol.

19. Add a style rule that hides the checkbox.

20. Add style rules that display the menu and change the hamburger symbol to the close (X) symbol when the checkbox is checked.

21. Refresh the web page and see how the navigation menu changes at different screen sizes. When the hamburger symbol is visible, click it to display the menu. Then, click the close symbol to hide the menu.

22. On a small screen, display the menu again. Then, make the browser window wider, and note how the navigation menu still displays vertically even when the screen gets wider.

23. To fix this, in the media query, add the style rule from the bottom of part 2 of figure 8-7 that makes sure the menu displays horizontally.

24. Refresh the web page and test this change to make sure it fixed the problem.

25. On a small screen, display the menu again and note that none of the submenus have a symbol. Hover the mouse over the Speakers menu item and note that the submenu displays below the menu item, and that it doesn't look good or work well.

26. In the CSS, copy the style rules for the list item that has a submenu and the submenu class to the media query.

27. In the media query, delete the position property from the submenu class. Then, change the nested style rule to display a down pointing triangle symbol after each submenu item.

28. In the media query, in the submenu class, delete all but the left and right properties and the nested style rule. Then, set the top property to auto.

29. In the mobile-first CSS, modify the style rule for the submenu class by adding a nested style rule that displays a right pointing triangle symbol after each item that contains a submenu.

30. In the style rule for the submenu class, set the top property to 0, the left property to 100%, the width to 100%, and delete the right property.

31. Refresh the page and test the submenus at various screen sizes.

32. Adjust the menu borders. To do that, move the style rules of the first child and last child of the menu to the media query. Then, copy the style rule for the navbar class to the media query and delete everything but the margin property in the nested style rule for <a> elements.

33. In the mobile-first CSS, set the margin for the nested <a> element in the navbar class to 0 2px 2px 0. Then, set the margin for the nested <li> <a> element in the submenu class to 2px.

34. Refresh the page and make sure the menu borders look good at various screen sizes.

Identify a breakpoint for the header section and add a media query

In this section you'll move the CSS for the scalable logo image to a media query and make the existing header mobile-first with no logo image and the headings centered.

35. Reduce the width of the window until the headings in the header start to look bad. Note the width of the page displayed in the upper right corner and add a media query for when the viewport width is equal to or greater than that width.

36. Copy the existing header style rule to the media query. Delete all the styles except the nested style rule for the image. Then, set the text-align property of the header to left and the display of the nested image to block.

37. In the mobile-first CSS, in the header style rule, set the text-align property to center, delete the declarations for the nested image, and set the display of the nested image to none.

38. Refresh the web page and see how the header looks at various screen sizes.

Test the responsive page

39. Press F12 to display the Developer Tools, and then display the Device toolbar.

40. Select different devices from the Dimensions drop-down list. For each device you choose, click the Rotate icon to see how the page looks and works in both portrait and landscape orientation.

Section 2

More HTML
and CSS skills

Section 1 presented a subset of HTML and CSS skills that you can use for building most web pages. Now, this section reviews some of these skills and presents more HTML and CSS skills that you can learn whenever you need them. To make that possible, each chapter in this section has been written as an independent module. As a result, you can read these chapters in whatever sequence you prefer.

Chapter 9

How to work
with lists and links

Chapter 3 introduced the basic skills for coding lists and links. Now, this chapter reviews those skills and presents some more skills that you may need for working with lists and links.

How to code lists

The figures that follow present the HTML skills that you need to work with lists. That includes unordered, ordered, and description lists.

How to code unordered lists

Figure 9-1 reviews the two elements for coding an *unordered list*. You use the <ul> element to create an unordered list, and you use the <li> element to create each item in the list. This is shown by the example in this figure.

In addition to containing text, a list item can contain inline elements like links and images. It can also contain block elements like headings, paragraphs, and other lists. For example, the first list item in this figure contains a <p> element that includes a link. Then, the second list item contains two <p> elements. As a result, the browser aligns the second paragraph with the first paragraph, but it doesn't display a bullet before the second paragraph because it's part of the same list item.

Elements that create unordered lists

Element	Description
ul	An unordered list.
li	A list item.

An unordered list with text, links, and paragraphs

```
<h1>San Joaquin Valley Town Hall Programs</h1>
<ul>
    <li>
        <p>Join us for a coffee hour at the
            <a href="saroyan.html">William Saroyan Theatre</a>.</p>
    </li>
    <li>
        <p>Extend the excitement of Town Hall by purchasing
            tickets to the post-lecture luncheons.</p>
        <p>This unique opportunity allows you to ask
            more questions of the speakers--plus spend
            extra time meeting new Town Hall friends.</p>
    </li>
</ul>
```

The list in a web browser

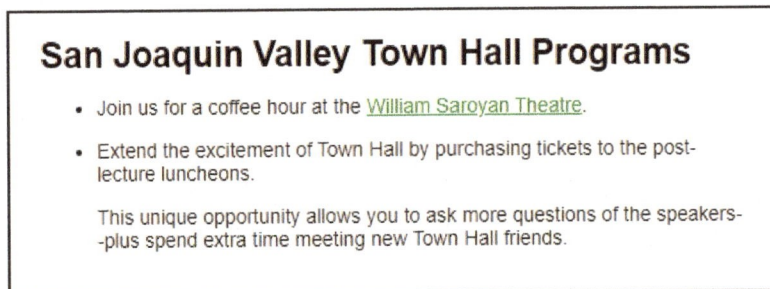

San Joaquin Valley Town Hall Programs

- Join us for a coffee hour at the William Saroyan Theatre.
- Extend the excitement of Town Hall by purchasing tickets to the post-lecture luncheons.

  This unique opportunity allows you to ask more questions of the speakers--plus spend extra time meeting new Town Hall friends.

Description

- By default, an *unordered list* is displayed as a bulleted list.
- An <li> element typically contains text, but it can also contain other inline elements such as links, as well as block elements such as paragraphs and other lists.

Figure 9-1　　How to code unordered lists

How to code ordered lists

If you want to indicate that the items in a list have a sequence, you can use an *ordered list*. To create an ordered list, you use the <ol> and <li> elements as shown in figure 9-2.

This figure shows how to use the start attribute of the <ol> element to start a list at a value other than the default. You might want to do that if one list continues from a previous list. For example, this figure shows a procedure that's divided into two parts, and each part is coded as a separate ordered list. Because the first list consists of three items, the HTML for the second list sets its start attribute to 4.

When you use the start attribute, it doesn't represent the actual value that displays. Instead, it represents the position of the item in the list. When you use decimal values to indicate the sequence of the items in a list as shown here, the position and the value are the same. However, you can also use other numbering styles for a list such as letters or roman numerals. Then, a start value of 4 represents the letter D or the roman number IV.

Elements that create ordered lists

Element	Description
ol	An ordered list. You can include the start attribute to specify the starting value for the list. The default is 1.
li	A list item.

An ordered list that continues from another ordered list

```
<h1>How to make a ham and cheese sandwich</h1>
<h2>Prepare the ingredients</h2>
<ol>
    <li>Get two pieces of bread.</li>
    <li>Cut one slice of cheese.</li>
    <li>Cut one slice of ham.</li>
</ol>
<h2>Assemble the sandwich</h2>
<ol start="4">
    <li>Place the slice of ham on top of one of the
        pieces of bread.</li>
    <li>Place the slice of cheese on top of the slice of ham.</li>
    <li>Place the second piece of bread on top of the cheese.</li>
</ol>
```

The lists in a web browser

How to make a ham and cheese sandwich

Prepare the ingredients

1. Get two pieces of bread.
2. Cut one slice of cheese.
3. Cut one slice of ham.

Assemble the sandwich

4. Place the slice of ham on top of one of the pieces of bread.
5. Place the slice of cheese on top of the slice of ham.
6. Place the second piece of bread on top of the cheese.

Description

- By default, an *ordered list* is displayed as a numbered list.
- You can use the start attribute to continue the numbering from one list to another.

Figure 9-2   How to code ordered lists

How to code nested lists

When you code a list within another list, the lists are referred to as *nested lists*. Figure 9-3 shows how nested lists work.

The HTML in this figure defines three lists. To start, it defines an unordered list that contains two list items. Within this list, the first item includes some text and a nested list that contains three items. Then, the second item includes some text and a nested list that contains three more items.

The nested lists in this example are coded just like any other unordered list. However, because they're nested within another list, the browser indents them an additional amount by default and also changes the bullet for each item to a hollow circle.

Two unordered lists nested within another unordered list

```
<h1>How to make a ham and cheese sandwich</h1>
<ul>
    <li>Prepare the ingredients
        <ul>
            <li>Get two slices of bread.</li>
            <li>Cut one slice of cheese.</li>
            <li>Cut one slice of ham.</li>
        </ul>
    </li>
    <li>Assemble the sandwich
        <ul>
            <li>Place the slice of ham on top of one of the pieces of bread.</li>
            <li>Place the slice of cheese on top of the slice of ham.</li>
            <li>Place the second piece of bread on top of the cheese.</li>
        </ul>
    </li>
</ul>
```

The lists in a web browser

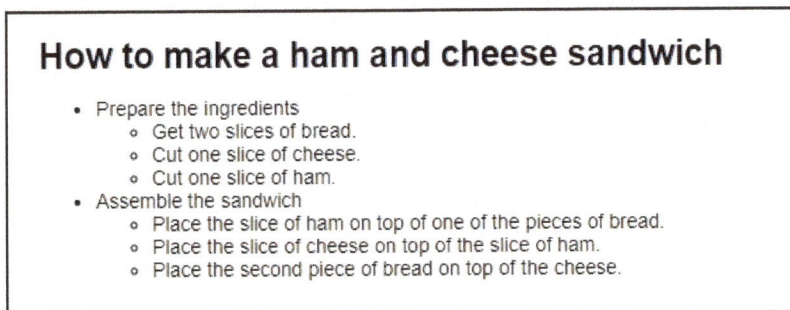

How to make a ham and cheese sandwich

- Prepare the ingredients
    - Get two slices of bread.
    - Cut one slice of cheese.
    - Cut one slice of ham.
- Assemble the sandwich
    - Place the slice of ham on top of one of the pieces of bread.
    - Place the slice of cheese on top of the slice of ham.
    - Place the second piece of bread on top of the cheese.

Description

- You can nest lists by coding one list as an item for another list.

- When you nest an unordered list within another list, the default bullet is a hollow circle.

Figure 9-3   How to code nested lists

How to code description lists

HTML also provides for *description lists*. As the name implies, a description list provides a list of terms and their descriptions.

To code a description list, you use the <dl>, <dt>, and <dd> elements presented in figure 9-4. The <dl> element defines the description list, the <dt> element defines the term, and the <dd> element defines the description for the term.

The example in this figure shows how this works. Here, the <dl> element defines a single list, and the <dt> and <dd> elements define the terms and descriptions in this list. In this example, the <dt> and <dd> elements are coded in pairs so there's one description for each term. However, you can code more than one <dd> element for each <dt> element. That's useful if you're creating a glossary and a term has more than one meaning. Conversely, you can code a single <dd> element for two or more <dt> elements. That's useful if you want to list multiple terms that have the same description.

When you code a description list, the <dt> element can only contain text and inline elements. By contrast, <dd> elements can also contain block elements. For example, they can contain paragraphs and nested lists.

Elements that create description lists

Element	Description
dl	A list that contains terms and descriptions.
dt	A term in a description list.
dd	A description in a description list.

A description list

```
<h2>Components of the internet</h2>
<dl>
    <dt>client</dt>
    <dd>A computer that accesses the web pages of a web application using a
        web browser.</dd>

    <dt>web server</dt>
    <dd>A computer that holds the files for each web application.</dd>

    <dt>local area network (LAN)</dt>
    <dd>A small network of computers that are near each other and can
        communicate with each other over short distances.</dd>

    <dt>wide area network (WAN)</dt>
    <dd>A network that consists of multiple LANs that have been connected
        together over long distances using routers.</dd>

    <dt>Internet exchange point</dt>
    <dd>Large routers that connect WANs together.</dd>
</dl>
```

The list in a web browser

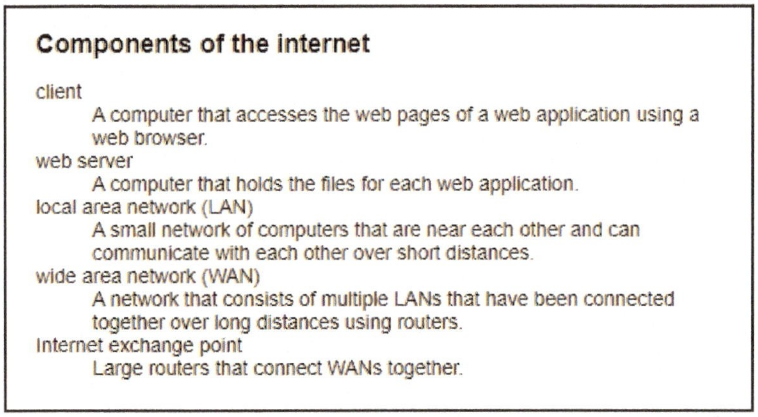

Components of the internet

client
    A computer that accesses the web pages of a web application using a
    web browser.
web server
    A computer that holds the files for each web application.
local area network (LAN)
    A small network of computers that are near each other and can
    communicate with each other over short distances.
wide area network (WAN)
    A network that consists of multiple LANs that have been connected
    together over long distances using routers.
Internet exchange point
    Large routers that connect WANs together.

Description

- A *description list* consists of terms and descriptions for those terms.
- A <dt> element can only contain text and inline elements. However, a <dd> element can contain block elements such as headings and paragraphs.
- You can use one or more <dd> elements to describe a <dt> element, and you can describe two or more <dt> elements with a single <dd> element.

Figure 9-4   How to code description lists

How to format lists

Once you use HTML to define a list, you can use CSS to format it. For example, you can change the bullets that are used for an unordered list.

How to change the bullets for a list

To change the bullets for an unordered list, you can use the properties shown in the first table in figure 9-5. For example, you can use the list-style-type property to specify one of the values listed in the second table.

By default, most browsers use a solid round bullet for a list item. However, you can specify a value of circle to display a hollow circle bullet as shown in the first list in this figure. You can specify a value of square to display a square bullet. Or, if you don't want to display a bullet, you can specify a value of none.

If these predefined bullet types aren't adequate, you can display a custom image before each item in an unordered list. To do that, you start by getting or creating the image that you want to use. For instance, you can get many images that are appropriate for lists from the internet, often for free or for a small charge. Once you have the image that you want to use, you use the list-style-image property to specify the URL for the image file as shown by the second example.

Properties for formatting unordered lists

Property	Description
list-style-type	The type of bullet that's used for the items in the list. See the table below for possible values. The default is disc.
list-style-image	The URL for an image that's used as the bullet.

Common values for the list-style-type property of an unordered list

Value	Description
disc	A solid circle. This is the default.
circle	A hollow circle.
square	A solid square.
none	No bullet.

HTML for two unordered lists

```
<h2>Popular web browsers include</h2>
<ul class="circle">
    <li>Chrome</li>
    <li>Safari</li>
    <li>Edge</li>
</ul>
<h2>Prime skills for web developers are</h2>
<ul class="star">
    <li>HTML and CSS</li>
    <li>JavaScript</li>
    <li>Python</li>
</ul>
```

CSS that changes the bullets

```
ul.circle { list-style-type: circle; }
ul.star { list-style-image: url("../images/star.png"); }
```

The lists in a web browser

Popular web browsers include

- Chrome
- Safari
- Edge

Prime skills for web developers are

★ HTML and CSS
★ JavaScript
★ Python

Figure 9-5   How to change the bullets for a list

How to change the numbering system for a list

By default, browsers use decimal values (1, 2, 3…) to number the items in an ordered list. To change that, you can use the list-style-type property as shown in figure 9-6.

In this figure, the HTML marks the two nested lists as members of the nested class. Then, the CSS uses a style rule to set the list-style-type to lower-alpha for ordered lists that are members of the nested class. As a result, the browser uses lowercase letters (a, b, c…) to order the items in the nested lists.

Note that the HTML sets the start attribute of the second nested list to 4. However, since this list uses lowercase letters, the browser doesn't start the second list at 4. Instead, it starts the second list at the letter d, the fourth letter of the alphabet.

A property for formatting ordered lists

Property	Description
list-style-type	The numbering style that's used for the items in the list. See the table below for possible values. The default is decimal.

Common values for the list-style-type property of an ordered list

Value	Example
decimal	1, 2, 3, 4, 5 ...
decimal-leading-zero	01, 02, 03, 04, 05 ...
lower-alpha	a, b, c, d, e ...
upper-alpha	A, B, C, D, E ...
lower-roman	i, ii, iii, iv, v ...
upper-roman	I, II, III, IV, V ...

HTML for an ordered list

```html
<h1>How to make a ham and cheese sandwich</h1>
<ol>
    <li>Prepare the ingredients
        <ol class="nested">
            <li>Get two slices of bread.</li>
            <li>Cut one slice of cheese.</li>
            <li>Cut one slice of ham.</li>
        </ol>
    </li>
    <li>Assemble the sandwich
        <ol start="4" class="nested">
            <li>Place the slice of ham on top of one of the pieces of bread.</li>
            <li>Place the slice of cheese on top of the slice of ham.</li>
            <li>Place the second piece of bread on top of the cheese.</li>
        </ol>
    </li>
</ol>
```

CSS that formats the list

```css
ol.nested { list-style-type: lower-alpha; }
```

The list in a web browser

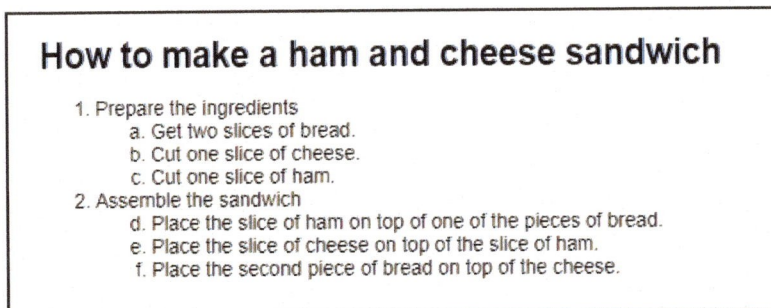

How to make a ham and cheese sandwich

1. Prepare the ingredients
   a. Get two slices of bread.
   b. Cut one slice of cheese.
   c. Cut one slice of ham.
2. Assemble the sandwich
   d. Place the slice of ham on top of one of the pieces of bread.
   e. Place the slice of cheese on top of the slice of ham.
   f. Place the second piece of bread on top of the cheese.

Figure 9-6   How to change the numbering system for a list

How to change the alignment of list items

Often, the browser aligns the items in a list the way you want. However, you may want to adjust the spacing before and after a list, and you may want to adjust the spacing before or after the list items. Beyond that, you may want to change the indentation of the items in a list or the amount of space between the bullets or numbers in a list.

Figure 9-7 shows the HTML and the CSS for a list. In this case, the CSS adjusts the space between the list items. It moves the list items so the bullets align on the left margin instead of being indented, and it increases the space between the bullets and the items.

The CSS in this figure begins by setting the margins for the unordered list and its list items to 0. Then, it sets the left padding for the <ul> element to 1em. This determines the left alignment of the items in the list. Often, you have to experiment with this value to get the alignment the way you want.

Similarly, this CSS sets the left padding for the <li> element to .25em. This determines the distance between the bullet and the text for an item. Here again, you may have to experiment with this value to set it the way you want.

For ordered lists, you can use the same techniques to adjust the indentation and to adjust the space between the number and the text in an item. However, by default, the numbers in an ordered list are right aligned. So, if the numbers vary in width, you have to adjust your values accordingly.

HTML for an unordered list

```
<h2>Popular web browsers</h2>
<ul>
    <li>Chrome</li>
    <li>Safari</li>
    <li>Edge</li>
    <li>Firefox</li>
    <li>Opera</li>
</ul>
```

CSS that aligns the list items

```
h2, ul, li {
    margin: 0;
    padding: 0;
}
h2 {
    padding-bottom: .25em;
}
ul {
    padding-left: 1em;      /* determines left alignment */
}
li {
    padding-left: .25em;    /* space between bullet and text */
    padding-bottom: .25em;  /* space after list item */
}
```

The list in a web browser

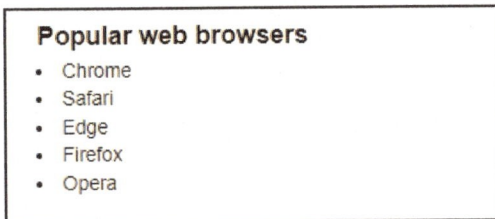

```
Popular web browsers
  • Chrome
  • Safari
  • Edge
  • Firefox
  • Opera
```

Description

- You can use margins and padding to control the indentation for the items in an ordered or unordered list and to control the space between the bullets or numbers and the text that follows.

Figure 9-7   How to change the alignment of list items

How to code links

The figures that follow present the skills you need for coding and formatting links. This includes coding a link that opens another web page in the same browser tab as well as coding other types of links.

How to link to another page

Figure 9-8 begins by reviewing some of the information presented earlier in this book about coding <a> elements to define a *link* to a resource such as another web page. Then, it presents some new skills for coding links.

To start, you code the href attribute so it specifies the URL for the link. Then, you code the content that you want to be displayed for the link. The content can be text, an image, or both text and an image.

The example in this figure shows how this works. Here, the first link displays text, and the second link displays an image. Then, the third link shows how the title attribute can be used to improve accessibility. Here, the text for the link is "TOC", but the text for the title attribute is "Review the complete table of contents". This causes the browser to display the title as a tooltip if the mouse hovers over the link, and assistive devices can read the title. As a result, if you include an image-only link, you should code the title attribute, as shown by the second link in this example.

If the user presses the Tab key, the focus moves from one link to another based on the *tab order*. By default, the tab order is the sequence in which the links and controls are coded in the HTML, which is usually what you want. However, to change that order, you can use the tabindex attribute of a link. For example, in this figure, the first link sets this attribute to 1. As a result, it's the first link in the tab order. Then, the third link sets this attribute to 2. As a result, it's the second link in the tab order. Since this attribute isn't coded for the second link, it becomes the third link in the tab order.

To set an *access key* for a link, you can code the accesskey attribute. Then, the user can press that key in combination with one or more other keys to activate the link. For example, the first link sets the accesskey attribute to the letter c. As a result, the user can activate the link by pressing a control key plus the shortcut key. For example, for the Chrome browser, Alt+c works on Windows, and Control+Option+c works on macOS.

Four attributes of the <a> element

Attribute	Description
href	The URL for the link.
title	The description that's displayed as a tooltip.
tabindex	The tab order for the link starting with 1. To take a link out of the tab order, code a negative value.
accesskey	The keyboard key that can be used in combination with other keys to activate the element. The key combination depends on the operating system and browser.

A text link, an image link, and a text link with a title attribute

```
<div>
    <a href="/orders/cart.html" accesskey="c" tabindex="1">Shopping cart</a>
    <a href="/orders/cart.html" title="Go to shopping cart">
        <img src="media/cart_animated.gif" alt="Shopping cart">
    </a>
</div>
<div>
    <a href="/books/php_toc.html" tabindex="2"
        title="Review the complete table of contents">TOC
    </a>
</div>
```

The text and image links in a web browser

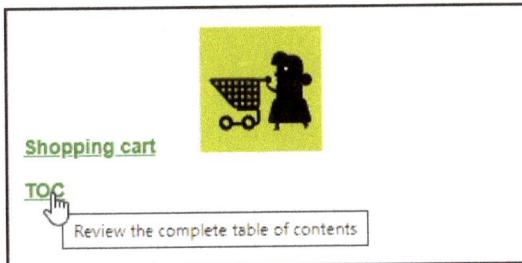

Accessibility guidelines

- If the text for a link has to be short, code the title attribute to clarify what the link does.
- Code the title attribute if a link includes an image with no text.

Description

- You use the <a> element to create a *link* that loads a resource such as another web page. The content of an <a> element can be text, an image, or text and an image.
- The *tab order* is the sequence that the browser uses to move the focus when the user presses the Tab key.
- An *access key* is a keystroke combination that can be used to activate a link.

Figure 9-8   How to link to another page

How to format links

Figure 9-9 reviews the skills for formatting a link that were introduced in chapter 4. In particular, you can use the pseudo-class selectors to change the default styles for a link. This figure also shows how to remove the underline for a text link. To do that, you can set the text-decoration property for a text link to none.

In the early days of the web, links were almost always underlined to show that they were links. Today, links are often displayed without underlining. However, it's still a good practice to avoid underlining text that isn't a link since that can be confusing to users.

It's also good to avoid the use of image-only links unless it's clear what the link does. For instance, most users recognize a shopping cart or shopping bag image because they're common to many sites. However, users may not recognize images that are specific to a single site. One exception is using a logo in the header to link to the home page of a site because that's become a website convention.

Common CSS pseudo-classes for formatting links

Name	Description
:link	A link that hasn't been visited. Blue is the default color.
:visited	A link that has been visited. Purple is the default color.
:hover	An element with the mouse hovering over it. Hover has no default color.
:focus	An element like a link or form control that has the focus. It has a thick purple border by default.
:active	An element that's currently active. Red is the default color.

The property for removing underlining

Property	Description
text-decoration	Sets the text decoration for a link. To remove all text decoration such as underlining, set this property to none.

A list of three links

```
<ul>
    <li><a href="brancaccio.html">David Brancaccio</a></li>
    <li><a href="sorkin.html">Andrew Ross Sorkin</a></li>
    <li><a href="chua.html">Amy Chua</a></li>
</ul>
```

The pseudo-class selectors that apply to the links

```
a:link {
    color: green;
}
a:hover, a:focus {
    text-decoration: none;
    font-size: 125%;
}
```

The links in a web browser with the focus on the third link

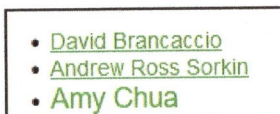

```
• David Brancaccio
• Andrew Ross Sorkin
• Amy Chua
```

Accessibility guideline

- Apply the same formatting for the :hover and :focus selectors. That way, the formatting is the same whether you hover the mouse over a link or use the keyboard to tab to the link.

Figure 9-9   How to format links

How to use a link to open a new browser tab

In most cases, when you code a link for another page, you want to display that page in the same browser tab as the current page. In some cases, though, you may want to open the next page in a new browser tab. For example, if you code a link to a page on another website, you may want to display that page in a new tab so the original website remains open in the first tab.

To open a page in a new tab, you can set the target attribute of the <a> element to _blank as shown in figure 9-10. Then, the browser opens the page for the link in a new tab.

This figure shows a link to Wikipedia that's opened in a second browser tab. However, some browsers may open a link in a completely new browser window instead of a new tab, depending on the browser settings.

A link that loads a web page into a new tab

```
<h1>Space Exploration Website</h1>
<p>To learn more about Mars,
    <a href="https://en.wikipedia.org/wiki/Mars"
        target="_blank">visit Wikipedia</a>
</p>
```

The web page in the first browser tab

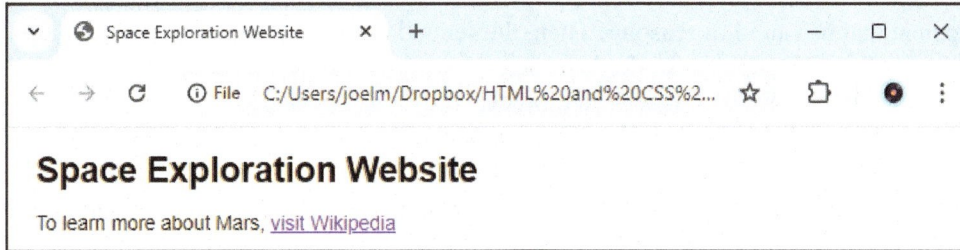

The web page in the second browser tab

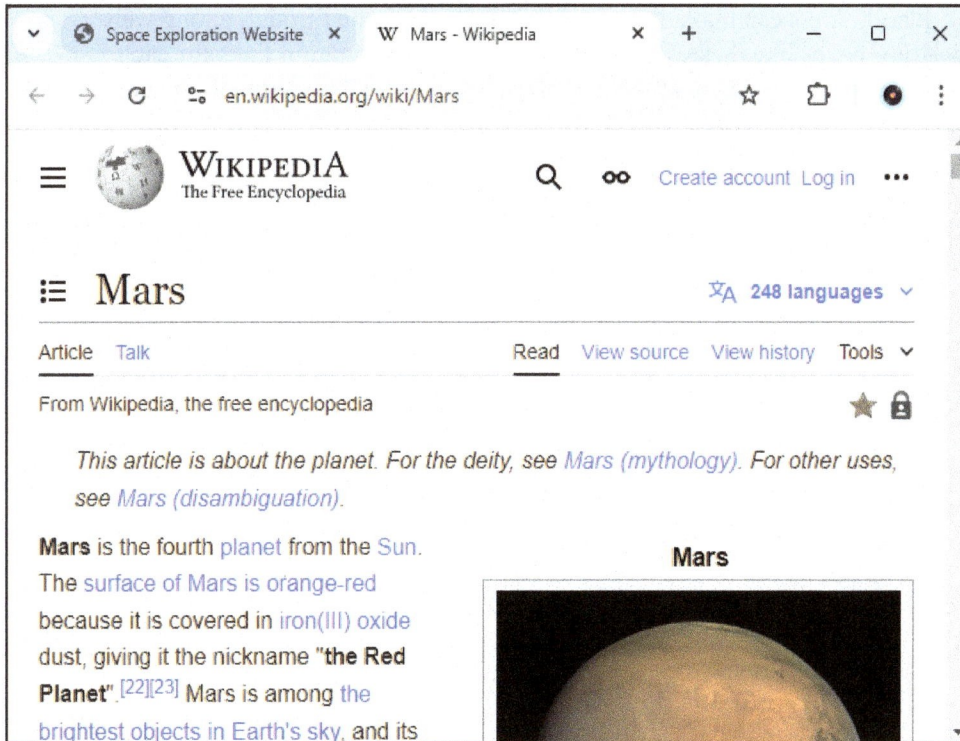

Description

- To open a link in a new browser tab, set the target attribute of the <a> element to _blank.
- Some browsers may open a link in a new browser window instead of a new tab. This depends on the browser settings.

Figure 9-10   How to use a link to open a new browser tab

How to create and link to placeholders

Besides displaying another page, you can code links that jump to a location on the same page. To do that, you first create a *placeholder* that identifies the location you want the link to jump to. Then, you code a link that points to that placeholder as shown in figure 9-11.

To create a placeholder, you code the id attribute for the element you want to jump to. In this figure, the first example shows a placeholder for an <h2> element that has an id of reason6. Then, the second example shows a link that jumps to that placeholder. To do that, the href attribute of the link uses the value of the id attribute for the placeholder, preceded by a hash character (#). When the user clicks on this link, the browser jumps to the element with that id.

Although placeholders are typically used for navigating within a single page, they can also be used to go to a location on another page. The third example in this figure shows how this works. Here, the URL for the page is coded before the placeholder. Then, when the browser displays that page, it also jumps to the element at the specified placeholder.

Placeholders can make it easier for users to navigate through a long web page. For pages like these, it's common to include navigation links for each section of the page. Then, at the end of each section, it's common to include a link to return to the top of the page.

The fourth example shows two ways to jump to the top of the page. In the first link, the href attribute is set to #top. In the second link, the href attribute is set to #. Either way, you don't need to code a placeholder for the top of the page.

A web page that provides links to topics on the same page

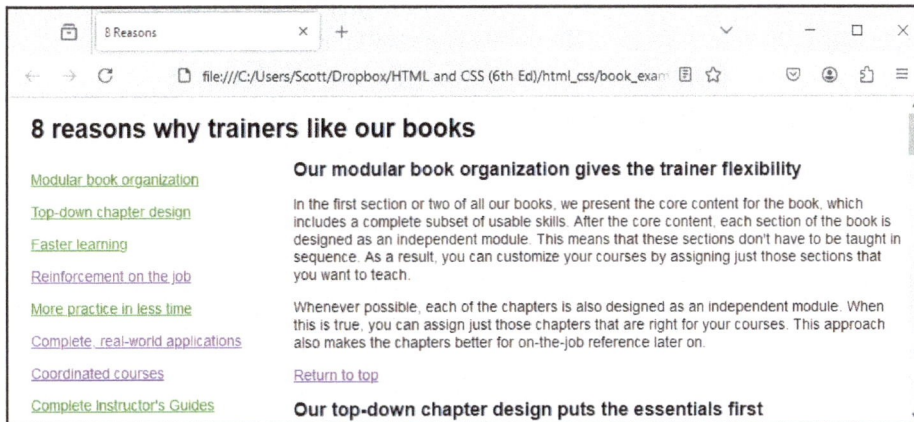

The topic that's displayed when the sixth link is clicked

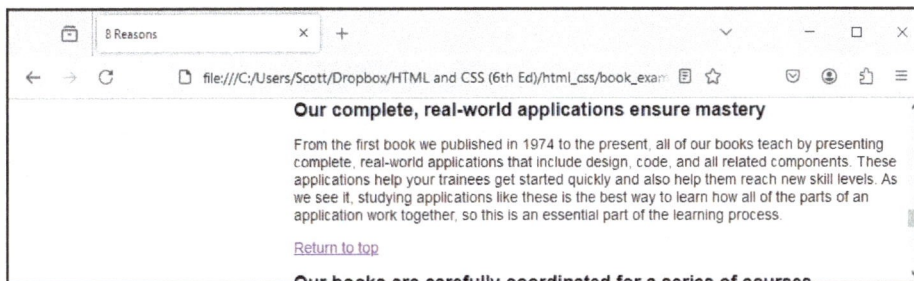

The placeholder for the sixth reason

```
<h2 id="reason6">Our complete, real-world applications ensure mastery</h2>
```

A link on the same page that jumps to the placeholder

```
<p><a href="#reason6">Complete, real-world applications</a></p>
```

A link on another page that jumps to the placeholder

```
<a href="8reasons.html#reason6">Complete, real-world applications</a>
```

Two links that jump to the top of the page

```
<p><a href="#top">Return to top</a></p>
<p><a href="#">Return to top</a></p>
```

Description

- To create a *placeholder*, code an id attribute for the element you want to jump to.
- To jump to a placeholder, set the href attribute to a hash (#) character followed by the id for the placeholder.
- To jump to a placeholder on another page, set the href attribute to the URL for the page followed by the hash (#) character followed by the id for the placeholder.
- To jump to the top of the page, set the href attribute to #top or #.

Figure 9-11   How to create and link to placeholders

How to link to a media file

If the href attribute of an <a> element points to a media file, the browser typically attempts to open it. For instance, the first example in figure 9-12 opens a PDF file in a new browser tab. The second example opens an image file and displays it. And the third example opens an MP3 audio file and plays it.

However, if the browser isn't able to open the file, it may download it instead. Then, you can use software on your device to work with that file. For instance, the fourth example downloads a PowerPoint slide show.

Chapter 13 shows how to add audio and video to your website. For other types of media, though, the <a> element can do a good job of getting the results that you want.

Common media types

Format	Description
PDF	Portable Document Format
MP3	MPEG audio
MP4	MPEG video

A link that displays a PDF file in a new browser tab

```
<a href="media/The-Road-Not-Taken.pdf" target="_blank">Open PDF</a>
```

The PDF file in a browser

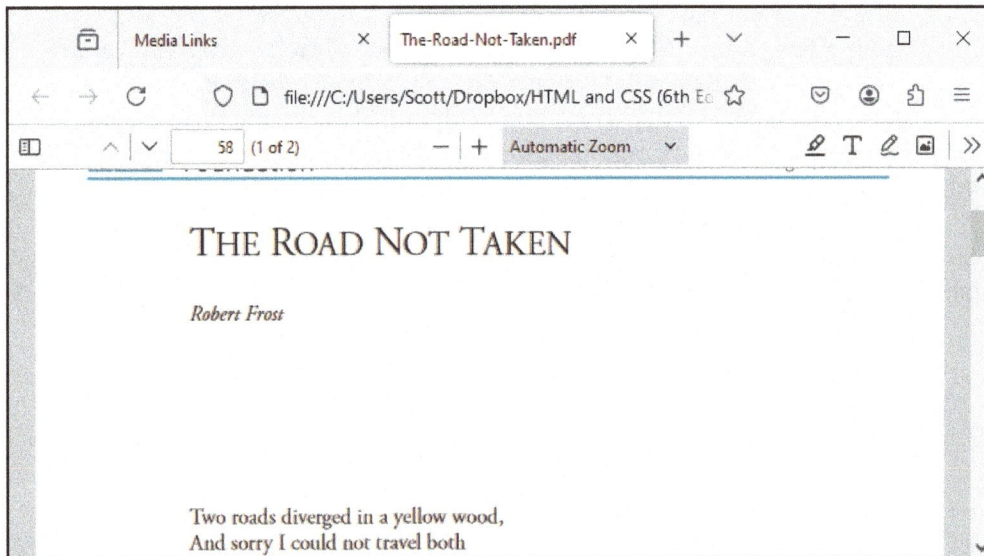

A link that displays an image

```
<a href="media/logo.jpg">View logo</a>
```

A link that plays an MP3 file

```
<a href="media/mp3_sample.mp3">Play MP3</a>
```

A link that downloads a PowerPoint slide show

```
<a href="media/powerpoint_sample.pptx">Download slide show</a>
```

Description

- When you use the <a> element to open a media file, the browser typically attempts to open the file.

Figure 9-12   How to link to a media file

How to create email and phone links

Figure 9-13 shows how to create links that start email messages and phone calls. To code a link that starts an email, you code the mailto: prefix in the href attribute of the link as shown in the first example. Beyond that, the second example shows how to add a cc and a subject to the email, and you can use a similar technique to add a bcc or body.

However, the links that start emails are designed to work with the user's default email program. As a result, if that program isn't set up correctly, the email link won't work. In addition, the mailto: prefix makes the specified email address easy for email spammers to find. One alternative is to use a form to allow a user to send an email instead of using the :mailto prefix.

To code a link that calls a phone number, you code the tel: prefix. This often works well on devices like mobile phones that are designed for making phone calls. However, on devices that aren't designed for making phone calls, the link typically lets you pick a program like Skype to make the phone call. Again, if this program isn't set up correctly, the phone link won't work.

If you can't get the :tel prefix to work correctly for your website, you can display the phone number on your web page. Then, users can manually enter the phone number into their phones.

Prefixes for coding email and phone links

Link type	Prefix
Email	mailto:
Phone	tel:

A link that starts an email

```
<a href="mailto:support@murach.com">Send us an email</a>
```

A link that starts an email with a CC address and a subject

```
<a href="mailto:support@murach.com?cc=ben@murach.com&subject=Web mail">
    Send us an email with a copy to Ben
</a>
```

A link that calls a phone number

```
<a href="tel:555-555-5555">Call us</a>
```

A web page with the links that start email messages and phone calls

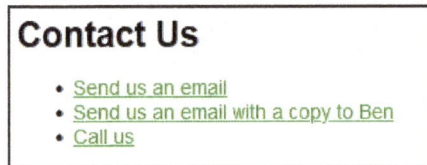

Contact Us

- Send us an email
- Send us an email with a copy to Ben
- Call us

Description

- To provide a link that starts an email message, code the email address after the mailto: prefix.
- For a link that starts an email, you can specify the subject, cc, bcc, or body fields by coding a question mark (?) after the email address followed by one or more name/value pairs where each name and value is separated by an equal sign (=) and each pair is separated by an ampersand (&).
- A link that starts an email only works correctly on devices that have a program that's configured to send email.
- To provide a link that starts a phone call on a mobile phone, code the phone number after the tel: prefix.
- A link that starts a phone call only works correctly on devices that have a program that's configured to make phone calls.

Figure 9-13   How to create email and phone links

Perspective

Now that you've completed this chapter, you should be able to define and format most types of lists and links, including nested lists. From this point on, you can use this chapter as a reference whenever you need it.

Terms

unordered list
ordered list
nested list
description list
link
tab order
access key
placeholder

Exercise 9-1   Add a speakers page

In this exercise, you'll modify the Town Hall home page to format the list on the page and to add a couple of new links. When you're through, the page should look like this:

Open the HTML and CSS files for the home page

1. Use your text editor to open these HTML and CSS files:

    ```
    exercises\ch09\town_hall\index.html
    exercises\ch09\town_hall\styles\main.css
    ```

Format a list

2. In the index.html file, find the code that defines the unordered list at the bottom of the page. Note that this unordered list is coded within the <section> element.

3. In the main.css file, add a nested style rule to the <section> element that changes the bullets in the unordered list at the bottom of the page to squares.

4. Test this to make sure it works correctly.

5. Modify the style rule so it changes the bullets to the image in the star.png file that's in the images folder.

6. Modify the style rule to change the alignment of the list so the bullets align with the rest of the text in the section element.

7. Add another nested style rule to increase the amount of space between each bullet and its text to .25em.

Add a link to a video

8. In the index.html file, find the text that says "Or meet us there!", and change it to a link that says "Or play a video!". This link should link to the sampson.mp4 file in the media folder.

9. Test this link to make sure it works correctly. Note that it plays the video on a new page. When you're done viewing the video, click the browser's Back button to return to the Town Hall home page.

Add a link to the logo so clicking it displays the home page

10. Use your text editor to open this HTML file:

    `exercises\ch09\town_hall\speakers\sampson.html`

11. Run this web page and click on the logo. Note that it doesn't display the home page.

12. In the sampson.html file, code an <a> element around the <img> element for the logo. This <a> element should link to the index.html file in the root directory. It should also include a title attribute with a value of "Go to home page".

13. Test this change to make sure it works correctly. From the sampson.html page, you should be able to click the Town Hall logo to display the home page. Also, if you hover the mouse pointer over the logo, it should display a message that says, "Go to home page".

Chapter 10

How to work with images, icons, fonts, and colors

This chapter begins by showing how to work with images and icons. Then, it shows how to get new fonts and embed them within your web pages. Finally, it presents some advanced techniques for specifying colors. For graphic designers, these features open up a whole new range of options.

How to work with images

This chapter begins by presenting some important skills for working with images. To do that, it reviews some skills that were presented in chapters 3 and 4, and it also presents new skills that bring you up to a new level of expertise.

Types of images for the web

Figure 10-1 presents the most common types of images you can use on a web page. To start, *JPEG* files are typically used for photographs and scanned images because these files can represent millions of colors, and they use a type of compression that can display complex images with a small file size.

Although JPEG files lose information when they're compressed, this loss of information isn't usually noticeable on a web page. Similarly, although JPEG files don't support transparency, you usually don't need it for any of the colors in a photograph.

In contrast, *GIF* files are typically used for simple illustrations or logos that require a limited number of colors. Two advantages of storing images in this format are (1) they can be compressed without losing any information, and (2) one of the colors in the image can be transparent.

A GIF file can also contain an *animated image*. An animated image consists of a series of images called *frames*. When you display an animated image, each frame is displayed for a preset amount of time, usually fractions of a second. Because of that, the image appears to be moving. For example, the two globes in this figure are actually two of 30 frames that are stored in the same GIF file. When this file is displayed, the globe appears to be rotating.

Unlike the GIF and JPEG formats, which were originally developed for printed materials, *PNG* files were developed specifically for the web. In particular, the PNG format was developed as a replacement for the GIF format. When compared with GIF files, PNG files provide better compression, support for millions of colors, and support for variable transparency.

WebP is a newer image format. It is an open-source format developed by Google based on technology acquired from On2 Technologies. The goal of WebP is to provide smaller images that load more quickly and also have better quality. WebP files are typically used to replace PNG and JPEG files.

AVIF (AV1 Image Format) is an even newer image format. It was developed by the Alliance for open Media, of which Mozilla is a founding member. Like WebP, it is an open-source format with the goal of providing smaller images with better quality.

When compared to the older image formats, the newer formats have better performance, and at the time of this writing, they are supported by most modern browsers. If you want to check current browser support, you can go to caniuse.com and enter the image type in the search box.

Image types

Type	Description
JPEG	A JPEG file uses a type of compression that can display complex images with a small file size. It can represent millions of colors, loses information when compressed, and doesn't support transparency.
GIF	A GIF file can represent up to 256 colors, doesn't lose information when compressed, and supports transparency on a single color.
PNG	A PNG file can represent millions of colors and supports transparency on multiple colors. Compressed PNG files are typically smaller than compressed GIF files, although no information is lost.
WebP	A newer format developed by Google. When compressed with no loss of information, WebP files are 26% smaller than PNG files. When compressed with loss of information, WebP files are 25 to 34% smaller than JPEG files. Both lossless and lossy formats support transparency.
AVIF	A newer format developed by Alliance for Open Media. An AVIF file supports up to 68 billion colors, both lossless and lossy formats, and transparency. On average, AVIF files are 50% smaller than JPEG files and 20% smaller than WebP files.

Typical JPEG images

Typical GIF images

Description

- *JPEG* (Joint Photographic Experts Group) images are commonly used for photographs and images. Although information is lost when you compress a JPEG file, the reduced quality of the image usually isn't noticeable.
- *GIF* (Graphic Interchange Format) images are commonly used for logos and small illustrations. They can also be used for *animated images* that contain *frames*.
- The *PNG* (Portable Network Graphics) format was developed specifically for the web as a replacement for GIF files.
- The *WebP* format was developed by Google as a replacement for JPEG and PNG files.
- *AVIF* (*AV1 Image Format*) was developed by Alliance for Open Media (AOMedia) as a replacement for JPEG, PNG, and WebP files.

Figure 10-1   Types of images for the web

How to align an image vertically

When you include an image on a web page, you may want to align it with the inline elements that surround it. If an image is preceded or followed by text, for example, you may want to align the image with the top, middle, or bottom of the text. To do that, you use the vertical-align property shown in figure 10-2.

To indicate the alignment you want to use, you typically specify one of the keywords listed in this figure. For example, if you specify text-bottom, the browser aligns the image with the bottom of the adjacent text. By contrast, if you specify bottom, the browser aligns the image with the bottom of the box that contains the adjacent text.

In most cases, the bottom of the text and the bottom of the box are the same. The exception is if a line height is specified for the box. In that case, the text is centered in the box. Then, if you specify bottom for the image alignment, the browser aligns the image at the bottom of the box, which is below the bottom of the text. The top and text-top keywords work similarly.

The middle keyword lets you center an image vertically with any surrounding text. This is illustrated by the example in this figure. Here, the HTML includes three paragraphs, each with an image followed by some text. Then, the CSS specifies middle for the vertical alignment of the image. As a result, the center of each image aligns with the center of the text. Otherwise, the bottom of each image would align with the bottom of the text.

Note that the CSS also specifies a right margin for the images. This creates space between the images and the text. You can also use padding and borders with images. For example, the CSS in this example adds some padding to the bottom of each image. This works just like it does for block elements.

The property for aligning images vertically

Property	Description
vertical-align	A relative or absolute value or a keyword that determines the vertical alignment of an image. See the table below for common keywords.

Common keywords for the vertical-align property

Keyword	Description
top	Aligns the top of the image box with the top of the box that contains the adjacent in-line elements.
text-top	Aligns the top of the image box with the top of the text in the containing block.
middle	Aligns the midpoint of the image box with the midpoint of the containing block.
text-bottom	Aligns the bottom of the image box with the bottom of the text in the containing block.
bottom	Aligns the bottom of the image box with the bottom of the box that contains the adjacent inline elements.

HTML for three images

```
<h2>We want to hear from you</h2>
<p><img src="images/computer.gif" alt="web address">
   <strong>Web:</strong> www.murach.com</p>
<p><img src="images/telephone.gif" alt="phone">
   <strong>Phone:</strong> 1-800-221-5528</p>
<p><img src="images/email.gif" alt="email">
   <strong>Email:</strong> murachbooks@murach.com</p>
```

CSS that aligns the images

```
img {
    vertical-align: middle;
    margin-right: .5em;
    padding-bottom: .5em;
}
```

How it looks in a web browser

We want to hear from you

Web: www.example.com

Phone: 1-800-555-1234

Email: info@example.com

Description

- If you use pixels, points, or ems to specify the value for the vertical-align property, the image is raised if the value is positive and lowered if it's negative. If you specify a percent, the image is raised or lowered based on the percent of the line height.

- You can use margins, padding, and borders with images just as you can with block elements.

Figure 10-2   How to align an image vertically

How to add an image to a figure

Figure 10-3 shows how to use the <figure> and <figcaption> elements to display an image as part of a figure that has a caption. Here, the HTML defines an article that contains a heading, a figure, and a long paragraph. Within the article, the <figure> element contains an <img> element that specifies an image for the figure, and it contains a <figcaption> element that specifies a caption for the figure.

After the HTML, the CSS sets the width of the figure to 50% of the width of the article. Then, it floats the figure to the left side of the article. As a result, the nested image is scalable, it and caption float with it.

The CSS also sets the width of the image to 100% of the figure. As a result, the image is scalable and occupies the entire width of the figure, and the caption displays below the image.

By default, a <figcaption> element is an inline element, not a block element. To make sure the caption displays on its own line on its own line, you can use CSS to change it to a block element as shown in this figure. In this case, the caption in this figure would display on its own line anyway since the image above it occupies 100% of the width. However, it's still a good practice to include it in case you change the width of the image later.

The CSS in this figure also formats other aspects of the caption. For example, it sets the caption to bold with a bottom border and some space after it.

In this figure, the HTML specifies the image before the <figcaption> element. As a result, the browser displays the caption below the image. But you can change that by coding the <figcaption> element before the image in the HTML.

Although you can get the same results without using these HTML elements, it's better to use them because they're semantic. That way, it's easy to tell that the image is used as a figure and that the caption applies to the image.

A web page that uses <figure> and <figcaption> elements

Fossil Threads in the Web of Life

What's 75 million years old and brand spanking new? A teenage Utahceratops! Come to the Saroyan, armed with your best dinosaur roar, when Scott Sampson, Research Curator at the Utah Museum of Natural History, steps to the podium. Sampson's research has focused on the ecology and evolution of late Cretaceous dinosaurs and he has conducted fieldwork in a number of countries in Africa.

Scott Sampson is a Canadian-born paleontologist who received his Ph.D. in zoology from the University of

Scott Sampson and friend

Toronto. His doctoral work focused on two new species of ceratopsids (horned dinosaurs) from the Late Cretaceous of Montana, as well as the growth and function of certopsid horns and frills.

The HTML for the <article> element

```
<article>
    <h1>Fossil Threads in the Web of Life</h1>
    <figure>
        <img src="images/sampson_dinosaur.jpg" alt="Scott Sampson">
        <figcaption>Scott Sampson and friend</figcaption>
    </figure>
    <p>What's 75 million years old and brand spanking new?...</p>
</article>
```

The CSS for the figure and its caption

```
figure {
    float: left;
    width: 50%;
    margin-right: 1.5em;
    & img {
        width: 100%;
    }
    & figcaption {
        display: block;
        font-weight: bold;
        border-bottom: 2px solid black;
        margin-bottom: .5em;
    }
}
```

Description

- The <figure> element contains all elements for a figure such as an image or a table.
- The optional <figcaption> element provides a caption that describes a figure. It can be coded anywhere within the <figure> element.
- By default, the <figcaption> element is an inline element, but you can use CSS to change it to a block element.

Figure 10-3   How to add an image to a figure

How to change an image based on viewport size

When you use responsive web design, you want the images that you use to be the right sizes for the varying viewport sizes. For example, you may want to use large, high-resolution images for larger viewports, but not for devices with smaller viewports.

If you want to use the same image in various viewport sizes but you want to reduce its file size for smaller viewports, you can use the <img> element as shown in the first part of figure 10-4. To make it work, you use the srcset attribute to specify a set of images, and you use the sizes attribute to specify when each image should be used.

In this figure, the first example uses the srcset attribute to specify a set of three images with widths of 800, 600, and 400 pixels. Then, the sizes attribute indicates that if the viewport has a minimum width of 1200 pixels, the browser should display the image so it's 800 pixels wide. And if the viewport has a minimum width of 800 pixels, the browser should display the image so it's 600 pixels wide. However, the last item doesn't specify a condition. Instead, it only specifies a width of 75vw, which is 75% of the viewport width. As a result, the browser uses this width if none of the other conditions are met. In this case, the image would be 75% as wide as the viewport.

In this example, the image widths in the srcset attribute are the same as the widths of the images for the first two conditions on the sizes attribute. However, that doesn't have to be the case. For example, you could set the sizes attribute like this:

```
sizes="(min-width: 1200px) 1000px, (min-width: 800px) 600px, 75vw">
```

Then, if the screen has a minimum width of 1200px, the browser would still display the image that's 800 pixels wide because it's the largest image that's less than or equal to 1000 pixels.

The second example shows how you can use different images for different viewport sizes. That lets you use images with fewer details and smaller file sizes for smaller viewports. To do that, you use the <picture> element. In this example, the <picture> element contains three <source> elements followed by an <img> element. Then, the media attributes for each <source> element provide the minimum width for a viewport, and the srcset attribute identifies the image for that width. But if none of the <source> elements contain a media attribute for the current viewport size, the browser displays the image that's identified by the <img> element.

The third example in this figure shows that you can also use the <picture> element to display an image in another format if the first formats aren't supported by the browser. Here, the <source> element provides an image file in AVIF format. But if the browser doesn't support that format, it can use the PNG file that's provided by the <img> element.

Attributes of the <img> element

Attribute	Description
srcset	The URL of one or more images, each followed optionally by a space and the image width. Multiple images must be separated by commas.
sizes	A media condition, followed by a space and the size of the image. Multiple conditions must be separated by commas, and the media condition must be omitted for the last item in the list.

Change the size of an image based on viewport size

```
<img src="images/mountains_medium.png" alt="mountains"
    srcset="images/mountains_large.png 800w,
            images/mountains_medium.png 600w,
            images/mountains_small.png 400w"
    sizes="(min-width: 1200px) 800px, (min-width: 800px) 600px, 75vw">
```

Elements for adding multiple image resources

Element	Descrption
picture	Contains one or more <source> elements and an <img> element.
source	The source for an image.
img	The image that's used if none of the images on the <source> elements can be used.

Attributes of the <source> element

Attribute	Description
media	A media condition that determines when the image should be used.
srcset	The URL of the image. This is required.
type	The optional MIME type for the image to be loaded.

Change the image based on viewport size

```
<picture>
    <source media="(min-width: 960px)" srcset="images/mountains_far.png">
    <source media="(min-width: 768px)" srcset="images/mountains_mid.png">
    <source media="(min-width: 460px)" srcset="images/mountains_close.png">
    <img src="images/mountains_mid.png" alt="">
</picture>
```

Display an image in an alternative format

```
<picture>
    <source srcset="images/mountains_small.avif">
    <img src="images/mountains_small.png" alt="">
</picture>
```

Description

- The <img> element allows you to change the size of an image based on viewport size without the need for media queries.

- The <picture> element allows you to change an image based on viewport size without the need for media queries. It also allows the browser to use a different format for an image if it doesn't support the first format.

Figure 10-4   How to change an image based on viewport size

How to use Scalable Vector Graphics

Scalable Vector Graphics (SVG) is an XML-based markup language that you can use to create scalable images within a web page. That usually helps the web page load faster, which improves the user experience. Besides that, SVG images are scalable, so they work well for responsive web design.

To illustrate, figure 10-5 shows the code for a logo as an <svg> element in the HTML for a web page. To create code like this, you typically use a graphics editor and then export the SVG code into the HTML for a web page.

One drawback of using SVG images is that they are hard to create unless you have a background in graphics design. In addition, SVG isn't suitable for complex images like photographs. As a result, SVG is typically used for relatively simple images and diagrams.

The Murach logo as a Scalable Vector Graphic

The code that's exported from Adobe Illustrator

```
<svg viewBox="0 0 247.83 76.96">
    <rect class="cls-1" width="80" height="76.96"/>
    <path class="cls-2" d="M28.19,18.36H18.13s3.29.09,3.29,4.33V52l10.53-
    3.33V22.47C31.95,19.88,30.58,18.37,28.19,18.36Z"/>
    <path class="cls-2" d="M40.33,21.9H30.26s3.29.08,3.29,4.32V55.52l10.53-
    3.33V26C44.09,23.43,42.72,21.91,40.33,21.9Z"/>
    <path class="cls-2" d="M52.42,25.53H42.36s3.29.08,3.29,4.32V59.14l10.53-
    3.33V29.64C56.18,27.05,54.81,25.54,52.42,25.53Z"/>
</svg>
```

Benefits of SVG

- SVG images loads faster than other types of images.
- SVG images are scalable so they work well for responsive web design.

Drawbacks of SVG

- Without a background in graphics design, SVG images can be hard to create.
- SVG isn't suitable for complex images like photographs.

Description

- *Scalable Vector Graphics* (*SVG*) is an XML-based format for defining two-dimensional vector graphics.
- SVG is an open standard developed by the World Wide Web consortium.
- You typically create SVG images by using a graphics editor like Adobe Illustrator. Then, you typically export the image as an <svg> element and paste that element into your HTML.

Figure 10-5   How to work with Scalable Vector Graphics

How to get and edit images

At this point, you've learned the most important HTML and CSS skills for working with images. If someone else is responsible for getting and sizing the images that you use, that may be all you need to know. But if you need to get and size your own images, here are some other skills that you'll need.

How to get images

For many websites, you'll use your own photos and create most of the graphic images that you use. But sometimes, you may need to get images or icons from another source. The easiest way to do that is to copy or download them from another website. To help you with that, figure 10-6 lists several of the websites you can use to get images.

Although many of the images that you find on the web are available for free, most require a Creative Commons license. The types of licenses that are available and the conditions required by these licenses are summarized in this figure. This shows that all of the licenses require attribution, which means that you must give credit to the author of the image or the website that provided the image. The other license conditions determine how an image can be shared, whether it can be used for commercial purposes, and whether you can derive new images from the existing image.

Stock photos are special images that are typically produced in a studio. For example, the two images in this figure are stock photos. You often must pay for these types of images, and they can be expensive.

You may occasionally want to link to an image that's available from another site by coding an absolute URL for that image. This is known as "hot linking". However, it is highly discouraged unless you have an agreement with the other site.

You can also use AI (artificial intelligence) to generate some types of images. However, the results can be inconsistent and creating the prompt to get the image you want isn't always easy. To find an AI image generator, you can search the internet.

Creative Commons license conditions

Conditions	Description
Attribution	You can use the image and images derived from it as long as you give credit as requested by the author or website providing the image.
Share Alike	You can distribute the image based on the license that governs the original work.
Non-Commercial	You can use the image and images derived from it for non-commercial purposes only.
No Derivative Works	You can use only the original image and not images derived from it.

Two stock photos

Popular websites for images

freeimages.com

pixabay.com

unsplash.com

rawpixel.com

Popular websites for stock photos

istockphoto.com

gettyimages.com

A popular search engine for stock photos

everystockphoto.com

Description

- Many of the images available from the web are available under a Creative Commons license. The license can restrict the use of an image to one or more of the conditions. The most common condition is attribution.

- Stock photos are typically produced in studios and can be purchased for a one-time fee.

- Some types of images can be generated by AI (artificial intelligence). To find an AI image generator, you can search the internet.

Figure 10-6   How to get images

When to use an image editor

When you use images on a web page, you want them to be the right size and format. If they aren't, you can use an image editor like the one in figure 10-7 to make the adjustments.

Today, one of the most popular image editors is Adobe Photoshop, which is currently available as a monthly subscription. Adobe also offers a product called Photoshop Elements that you can purchase for a moderate price. Beyond that, there are many other image editors that range from free to expensive.

One important use of an image editor is to size images and save them in the right format so they load as quickly as possible while still looking sharp and clear. In this figure, you can see some of the controls for doing that.

When you edit an image, you can change its resolution by setting the PPI (pixels per inch). If you decrease the PPI, the file size of the image becomes smaller, which helps the image load more quickly. However, that may also make the image less sharp and clear, especially on devices like mobile phones that have screens with high resolution. Because of that, you should try to save your images with a resolution that makes them look sharp and clear on screens with high resolution while also loading efficiently on screens with lower resolution.

You can also use an image editor to work with an animated image like a GIF file that contains frames. For instance, you can specify whether the frames are shown only once or repeated. You can also set the timing between the frames.

If an image has a transparent color, you can save it with *transparency*. To understand how this works, you need to know that an image is always rectangular. This is illustrated by the first starburst image in this figure. Here, the area that's outside the starburst is white, and it isn't transparent. As a result, the browser displays the white even when it displays the image over a colored background. By contrast, the area outside the second starburst image is transparent. Because of that, the browser displays the background color behind it.

An image editor as it is used to change the size of an image

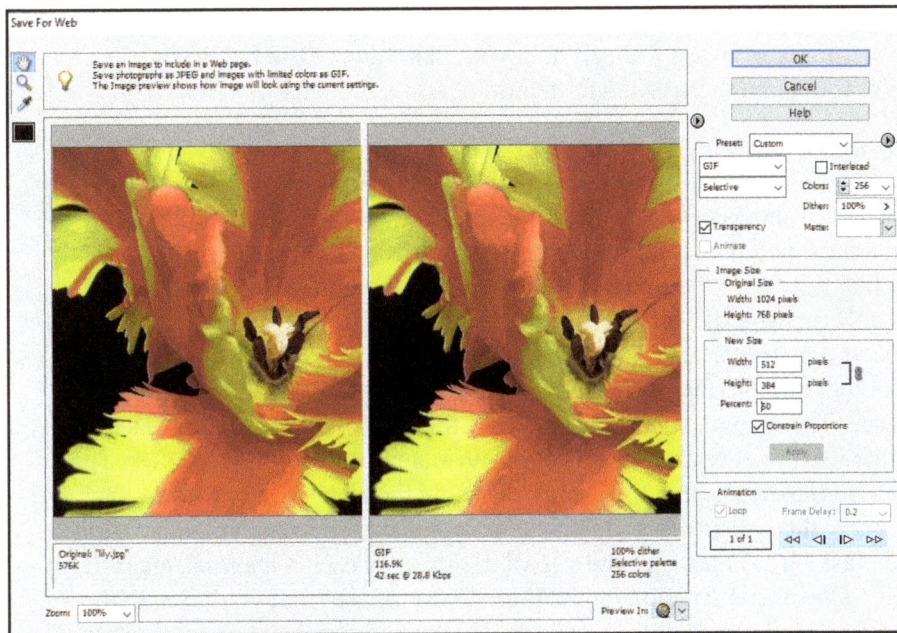

Typical editing operations

- Change the size, image type, or resolution of an image.
- Control the animation of an animated GIF file.
- Save an image with transparency.

Three popular image editors

- Adobe Photoshop (the industry standard)
- Adobe Photoshop Elements (an inexpensive editor)
- GIMP (a free editor)

An image without transparency and with transparency

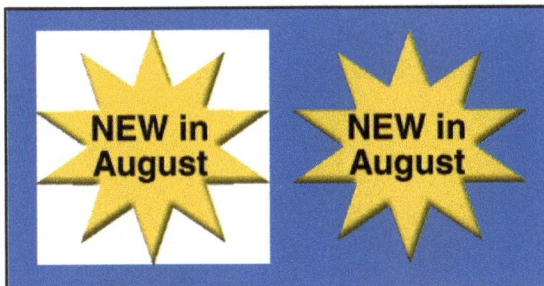

Description

- If an image with *transparency* is displayed over a colored background, the background color shows through the transparent portion of the image.

Figure 10-7   When to use an image editor

How to work with icons

You can get icons from the web. To help with that, figure 10-8 begins by presenting four popular websites for icons.

How to work with Font Awesome icons

This figure also shows how to work with icons in the Font Awesome library, one of the most popular icon libraries. To do that, you start by adding a <link> element like the one in the first example to import the Font Awesome library from a CDN (content delivery network).

Once you've imported the Font Awesome library, you can use the class attribute of an inline element such as a <span> element to identify the icon you want to use. The second example shows how this works for three icons. Here, each of the <span> elements uses its class attribute to specify three classes. The first class (fas) indicates that the element is using the standard Font Awesome library. Then, the second class identifies the icon for each <span> element. And the third class (fa-lg) identifies the size of the icon for each <span> element. If you refer to the documentation on the Font Awesome website, you'll see that other classes are also available for controlling the appearance of an icon.

How to work with favicons

In chapter 3, you learned that *favicons* are special-purpose icons. Like other icons, a favicon should always be stored in a file with the ico extension.

To create an icon from scratch, you can use an online icon generator like the favicon.io website, you can purchase a program like Axialis IconWorkshop, or you can download a free program like IrfanView. Or, if you're using Photoshop, you can get the plugin that lets you save files with the ico extension.

Although they can be larger, most favicons are 16 pixels wide and 16 pixels tall. In fact, some image converters automatically convert an image to 16x16 pixels. Then, to provide the favicon for a web page, you code a <link> element in the head section like the one shown at the bottom of this figure. When you do that, the browser displays the favicon in the tab for the page, and it may also display the favicon in the address bar and for any bookmarks for the page.

Popular websites for icons

fontawesome.com

glyphicons.com

glyphish.com

flaticon.com

HTML that imports the Font Awesome library from a CDN

```
<link rel="stylesheet" href="https://cdnjs.cloudflare.com/ajax/libs/font-
awesome/6.6.0/css/all.min.css">
```

HTML that uses three Font Awesome icons

```
<h2>We want to hear from you</h2>
<ul>
    <li><span class="fas fa-globe fa-lg"></span>
        <strong>Web:</strong> www.example.com</li>
    <li><span class="fas fa-phone-square fa-lg"></span>
        <strong>Phone:</strong> 1-800-555-1234</li>
    <li><span class="fas fa-envelope-square fa-lg"></span>
        <strong>Email:</strong> info@example.com</li>
</ul>
```

The icons in a web browser

We want to hear from you

🌐 **Web:** www.example.com

📞 **Phone:** 1-800-555-1234

✉ **Email:** info@example.com

Three tools for creating favicons

- The favicon.io website
- Axialis IconWorkshop
- Photoshop plugin

Add a favicon to the browser tab

```
<link rel="icon" href="images/favicon.ico">
```

Description

- You can get many icons under a Creative Commons license.
- To add a Font Awesome icon to the HTML for a web page, you code a class that identifies the library and a class that identifies the icon. You can also include classes that control the icon's size, style, and more.
- A *favicon* is typically 16 pixels wide by 16 pixels tall with an extension of ico.

Figure 10-8   How to work with icons

How to work with fonts

In the early days of the internet, web designers were frustrated by the limited number of fonts that were available for a website. In fact, web pages had been limited to the fonts that were available to each browser. That's why the font families throughout this book have been set to a series of fonts like Arial, Helvetica, and sans-serif. Then, each browser uses the first font in the series that's available to it.

These days, you can use CSS to *embed fonts* within your web pages. That way, you know the fonts are available to the browser. You can also use third-party services like Google Fonts and Adobe Fonts to embed fonts within your pages.

How to embed fonts in a web page

Figure 10-9 shows how to use the @font-face selector to embed fonts in a web page. To start, you can look in the folders that are listed in this figure. In most cases, you'll see that you already have access to many fonts that most browsers don't have access to. Then, you can copy the font that you want into one of the folders for your website. In this example, the TrueType font named HARNGTON.TTF has been copied to a folder named fonts in the root folder of the website.

Once you've copied the font to a folder, you code a CSS style rule for the @font-face selector that names and locates the font. In this example, the font-family property gives the font a name of MyHarrington, and the src property points to the TrueType font in the fonts folder.

After the font has been imported, you can code CSS style rules that apply the new font to HTML elements. In this example, the second style rule applies the font to <h1> elements. To do that, the font-family property is set to the name of the embedded font.

Where to find the fonts on your computer

Windows
```
C:\Windows\Fonts
```

macOS
```
System/Library/Fonts
```

A heading that uses an embedded font

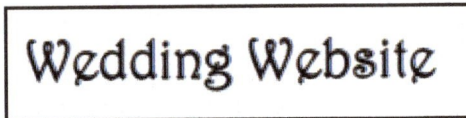

Wedding Website

The HTML for the heading
```
<h1>Wedding Website</h1>
```

The CSS that imports the font
```
@font-face {
    font-family: MyHarrington;
    src: url("fonts/HARNGTON.TTF");
}
```

The CSS that uses the font
```
h1 {
    font-family: MyHarrington;
}
```

How to import a font

1. Copy the file for the font family into a folder for your website.
2. In the CSS for the page, code a style rule for the @font-face selector.
3. Use the font-family property to specify a name for the font family.
4. Use the src property and the url() function to specify the location of the file for the font family.

Description

- CSS provides an @font-face selector that you can use to import a font family.
- Once you've imported a font family, you can use it just as you would use any other font family.
- A TrueType font typically has an extension of ttf.

Figure 10-9   How to embed fonts in a web page

How to use fonts from Google and Adobe

Figure 10-10 shows how you can use services available from Google and Adobe to import fonts for your web page. Google Fonts is a free service that lets you select fonts from hundreds of font families. Adobe Fonts also lets you select fonts from hundreds of font families. This service isn't free, but it's included as part of the Adobe Creative Cloud subscription.

Because the exact procedures for getting fonts from these services change often, they aren't presented in this chapter. However, you can start by searching for Google Fonts or Adobe Fonts and visiting the website for the service you want to use. Then, you can follow the instructions to get a link to one or more of the fonts.

When you select a Google font, the Google Fonts service provides you with a <link> element that points to the font, as shown by the first example in this figure. Then, after you include that element in the head of a web page, you can use the font-family property to apply that font to elements on the page.

In the first example, the <link> element imports the font family named "Big Shoulders Stencil Display". Then, the CSS applies this font family to the <h1> element. However, the list of font families also includes the generic sans-serif font in case something goes wrong and the Google font isn't available.

The technique for using an Adobe font is similar. However, to use an Adobe font, you have to create an Adobe Font project that includes that font. Then, the Adobe Fonts service provides you with a <link> element that points to the font in that project.

In the second example, the <link> element imports a font named alex-brush. Then, the CSS applies this font family to the <h1> element. In addition, the CSS adds the cursive font to the list of font families in case something goes wrong with the Adobe font.

A heading that uses a Google font

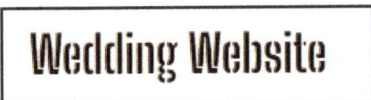

Wedding Website

The HTML that imports the font
```
<link rel="stylesheet" href="https://fonts.googleapis.com/css2?
family=Big+Shoulders+Stencil+Display:wght@100..900&display=swap">
```

The HTML for the heading
```
<h1>Wedding Website</h1>
```

The CSS that uses the font
```
h1 {
    font-family: "Big Shoulders Stencil Display", sans-serif;
}
```

A heading that uses an Adobe font

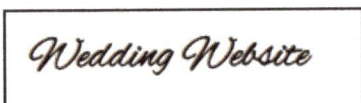

Wedding Website

The HTML that imports the font
```
<link href="https://use.typekit.net/uzs8kai.css" rel="stylesheet">
```

The HTML for the heading
```
<h1>Wedding Website</h1>
```

The CSS that uses the font
```
h1 {
    font-family: "alex-brush", cursive;
}
```

Description

- Google Fonts is a free service that lets you select and use any of the fonts in Google's extensive collection of font families.

- Adobe Fonts is a service that's part of the Adobe Creative Cloud subscription. It lets you select and use the fonts in Adobe's extensive collection of font families.

- You can locate the websites for these services by searching for Google Fonts or Adobe Fonts.

- When you use a font from Google or Adobe, you should add one or more other fonts to the font family list in case something has changed and the Google or Adobe font isn't available.

Figure 10-10 How to use fonts from Google and Adobe

How to work with colors

Chapter 4 showed how to use color names and RGB (red, green, and blue) values to specify colors. Now, the next three figures present some advanced skills for working with colors.

How to use functions to specify colors

Figure 10-11 begins by showing three functions that you can use to work with three color specifications that CSS provides. These color specifications go beyond using RGB values to specify a color.

First, you can use *RGBA* values. This works like RGB values, but with a fourth *alpha* parameter that provides an opacity value. If you set this value to 0, the color is fully transparent so anything behind it shows through. On the other hand, if you set this value to 1, the color is fully opaque and nothing shows through. Typically, you set the alpha parameter to a value between 0 and 1 so the color is partially transparent.

Second, you can use *HSL* (hue, saturation, and lightness) values. In this color specification, *hue* specifies the color, *saturation* specifies the intensity of the color, and *lightness* specifies the brightness of the color. To specify hue, you provide the number of degrees from 0 through 359. Then, you provide a number from 0 through 100 that represents the saturation percent with 100 being the most intense and 0 being the most muted. Last, you provide a number from 0 through 100 that represents the lightness percent with 50 being normal, 100 being white, and 0 being black.

Third, you can use *HSLA* values. This works like HSL values, but with a fourth alpha parameter that provides an opacity value between 0 and 1.

The example in this figure shows how these values work. In this example, the second and third style rules specify the same hue, but the saturation and lightness percents make them look quite different. This should give you some idea of the many color variations that CSS offers. As always, when working with colors, please keep accessibility in mind. In particular, when displaying text, you want to make sure there is enough contrast between the colors to make the text easy for everyone to read.

The color specifications presented here and in chapter 4 use the *standard RGB (sRGB) color space* or the HSL color space. However, in recent years CSS has introduced color specifications that use color spaces like CIELAB and OkLab. These color spaces attempt to cover the entire range of human color perception. While these new color specifications are beyond the scope of this book, you can find information about them by searching the web.

Three more functions for specifying color

```
rgba(red, green, blue, alpha)
hsl(hue-degrees, saturation%, lightness%)
hsla(hue-degrees, saturation%, lightness%, alpha)
```

Four parameters for these functions

Parameter	Description
alpha	A number from 0 to 1 with 0 being fully transparent and 1 being fully opaque.
hue-degrees	A number of degrees ranging from 0 to 359 that represents the color.
saturation%	A percent from 0 to 100 with 0 causing the hue to be ignored and 100 being the full hue.
lightness%	A percent from 0 to 100 with 50 being normal lightness, 0 being black, and 100 being white.

CSS that uses functions to specify color

```
h1 { color: rgba(0, 0, 255, .2); }        /* transparent blue */
h1 { color: hsl(120, 100%, 25%); }        /* dark green */
h1 { color: hsl(120, 75%, 75%); }         /* pastel green */
h1 { color: hsla(240, 100%, 50%, .5); }   /* semi-transparent blue */
```

The colors in a browser

San Joaquin Valley Town Hall

San Joaquin Valley Town Hall

San Joaquin Valley Town Hall

San Joaquin Valley Town Hall

Description

- *RGBA* enhances the RGB specification by providing an alpha value that controls opacity.
- *HSL* (hue, saturation, and lightness) lets you specify the number of degrees for the hue of a color. Then, you can enhance the hue by specifying saturation and lightness percents.
- *HSLA* enhances the HSL specification by offering an alpha value that controls opacity.

Figure 10-11   How to use functions to specify colors

How to use relative colors

When you code a specific color name or RGB or HSL value, you create an *absolute color*. However, you can also create a *relative color* that's based on another color. This makes it possible to create a palette of colors that are relative to a base color. Figure 10-12 presents two ways to do this.

The first way to create a relative color is to use the syntax presented at the top of this figure. This syntax uses a color function like rgb() or hsl() with the from keyword followed by a color. This is the color that the new color is based on, and it can be any valid color value, such as a color name, a hex value, a color in a CSS variable, or a color specification like RGB, RGBA, HSL, or HSLA.

After the from keyword and the color, you code an output value for each *color channel*. For instance, for the rgb() function, you code output values for the red, green, and blue channels. Similarly, for the hsl() function, you code output values for the hue, saturation, and lightness channels.

Optionally, you can code an output value for the alpha channel to provide an opacity value. If you do, the alpha value must be separated from the other channel values by a slash. If you don't specify an alpha channel, the relative color inherits the opacity value of the base color.

When you code the output value for a channel, you can use the original value, provide a completely new value, or calculate a new value that's based on the original value. The first three examples show how this works. All of these examples use the color named rebeccapurple after the from keyword.

In the first example, the hsl() function specifies the letter h to use the original hue. Then, it provides new values for saturation and lightness.

In the second example, the rgb() function specifies the letters r, g, and b to use the original red, green, and blue values. Then, it changes the opacity by specifying a slash followed by a new alpha value.

In the third example, both of the hsl() functions use the calc() function to create a new hue value that's relative to the original hue. Then, both functions specify the letters s and l to use the original saturation and lightness values.

The second way to create a relative color is to use the color-mix() function to mix two colors. This function uses the in keyword followed by a color space such as sRGB (not RGB) or HSL. Then you code the colors that you want to mix. Again, these colors can be any valid color value, such as a color name, a hex value, a color in a CSS variable, or a color specification like RGB, RGBA, HSL, or HSLA.

With the color-mix() function, you can code optional percent values for one or both of the colors you specify. This indicates how much of each color to add to the new color. If you omit both percents, each color defaults to 50%. If you code a percent for the first color, the percent for the second color is 100% minus the specified percent. Similarly, if you code a percent for the second color, the percent for the first color is 100% minus the specified percent.

The fourth example shows how this works. The first color-mix() function doesn't specify any percents, so the colors are mixed at 50% each. In the second function, the percent of the first color is 30%, so the percent of the second color is 70%. Conversely, in the third function, the percent of the second color is 30%, so the percent of the first color is 70%.

Syntax to create a color from another color

```
color-function(from color channel1 channel2 channel3 / alpha)
```

Change saturation and lightness

```
h1 { color: hsl(from rebeccapurple h 20% 60%); }
```

Change alpha (opacity)

```
h1 { color: rgb(from rebeccapurple r g b / .65); }
```

Change hue based on current hue value

```
h1 { color: hsl(from rebeccapurple calc(h - 150) s l); }
h1 { color: hsl(from rebeccapurple calc(h - 250) s l); }
```

The colors in a browser

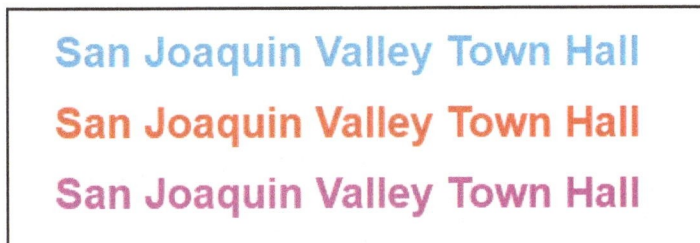

San Joaquin Valley Town Hall

San Joaquin Valley Town Hall

San Joaquin Valley Town Hall

San Joaquin Valley Town Hall

Syntax for the color-mix() function

```
color-mix(in color-space, color [percent], color [percent]);
```

Change the color mix

```
h1 { color: color-mix(in srgb, rebeccapurple, aqua); }
h1 { color: color-mix(in hsl, rebeccapurple 30%, orange); }
h1 { color: color-mix(in hsl, rebeccapurple, orange 30%); }
```

The colors in a browser

San Joaquin Valley Town Hall

San Joaquin Valley Town Hall

San Joaquin Valley Town Hall

Description

- You can use the rgb() and hsl() functions to create a new color from an existing color. Then, you can use the letters like r, g, b, h, s, and l to access the original value of each *channel* of the color, and you can use the calc() function to calculate a new value for each channel. The alpha parameter is optional and must be preceded by a slash.
- You can use the color-mix() function to mix two colors into a new color.

Figure 10-12   How to use relative colors

How to work with light and dark color schemes

On most devices, users can change their settings to indicate whether they prefer a light or dark color scheme. Then, you can use this setting to provide different formatting based on that user preference.

One way to provide formatting for light and dark color schemes is to use the light-dark() function presented in figure 10-13. Before you can use this function, you must enable it. To do that, you can use the CSS shown by the first example in this figure. This CSS sets the color-scheme property to "light dark". In most cases, it makes sense to set this property in the :root pseudo-class since that enables the light-dark() function throughout your CSS. Once you've enabled the light-dark() function, you can use it with any property that accepts a color value.

The syntax in this figure shows how to use the light-dark() function after you enable it. To do that, you pass two colors to this function. These colors can be any valid color value, such as a color name, a hex value, a color in a CSS variable, or a color specification like RGB, RGBA, HSL, or HSLA.

The first color you pass to the light-dark() function is the color that's used if the user prefers a light color scheme or has no preference. The second color you pass is the color that's used if the user prefers a dark color scheme.

The example after the syntax shows how this works. Here, the style rule for the <body> element uses the light-dark() function to set the background-color and color properties. Specifically, it sets the background color to aliceblue (a light color) for the light color scheme and rebeccapurple (a dark color) for dark color schemes. Then, it reverses the colors for the color property.

The screens in this figure show how an <h1> element on this web page looks in a browser in both color schemes. When the user prefers a light color scheme or has no preference, the browser displays the heading with dark text on a light background. However, when the user prefers a dark color scheme, the browser displays the heading with light text on a dark background.

To test your web page, you can use Chrome's Developer Tools to emulate a browser that prefers a light or dark color scheme. To do that, follow the steps presented in this figure.

Enable the light-dark() function

```
:root {
    color-scheme: light dark;
}
```

Syntax for the light-dark() function

```
property: light-dark(light-color, dark-color);
```

Define light and dark color schemes

```
body {
    background-color: light-dark(aliceblue, rebeccapurple);
    color: light-dark(rebeccapurple, aliceblue);
}
```

The light color scheme in a browser

San Joaquin Valley Town Hall

The dark color scheme in a browser

San Joaquin Valley Town Hall

How to use Chrome to emulate a preference for a dark color scheme

1. Display the Developer Tools for a web page.

2. In the Developer Tools toolbar, click on the Customize icon (three vertical dots next to the Close icon).

3. Select "More tools"→Rendering.

4. In the Rendering panel, scroll down to "Emulate CSS media feature prefers-color-scheme" and select "prefers-color-scheme: dark".

5. When you're done testing the dark color scheme, select "No emulation" to exit the dark color scheme.

Description

- Users can indicate whether they prefer a light or dark color scheme via operating system or user agent settings.

- You can use the light-dark() function to provide a light or dark color scheme based on a user's preferences.

- The light-dark() function returns the first color if the user prefers a light color scheme or has not set a preference. It returns the second color if the user prefers a dark color scheme.

- The light-dark() function can be used with any property that accepts a color value.

Figure 10-13   How to work with light and dark color schemes

Perspective

Now that you've finished this chapter, you should have the HTML and CSS skills that you need to work with the images, icons, fonts, and colors on a web page. However, if you need to edit your own images, you'll also need to get an image editor and learn how to use it.

Terms

JPEG
GIF
animated image
frame
PNG
WebP
AVIF
Scalable Vector Graphics (SVG)
transparency
favicon
embedded font

RGBA
alpha
HSL
HSLA
hue
saturation
lightness
standard RGB (sRGB) color space
absolute color
relative color
color channel

Exercise 10-1   Use <figure> and <picture> elements

In this exercise, you'll modify a speaker page of the Town Hall website to add <figure>, <figcaption>, and <picture> elements. When you're done, the page should look like this:

1. Use your text editor to open these HTML and CSS files:

```
exercises\ch10\town_hall\speakers\sampson.html
exercises\ch10\town_hall\styles\main.css
```

2. In the HTML file, enclose the <img> element at the top of the article in a <figure> element. Then, below the <img> element, add a <figcaption> element with text that says, "Scott Sampson and Friend".

3. In the CSS file, add nested style rules for the <figure> and <figcaption> elements to the style rule for the <section> element.

4. Move the float and margin declarations from the <img> style rule to the <figure> style rule.

5. Style the <figcaption> element to give it bold text and a bottom border.

6. In the <img> style rule, change the percent in the clamp() function from 50% to 100%.

7. Run the page to make sure it works correctly.

8. In the HTML file, enclose the <img> element in a <picture> element. Then, add a <source> element before the <img> element that displays the image named sampson_dinosaur.avif.

9. Test this change in Chrome to be sure the image is still displayed. To see which file the browser used, you can right-click the image and then select "Open image in new tab". The name of the file should display in the URL that's displayed for the tab.

Exercise 10-2   Size an image for different viewports

In this exercise, you'll display an image with different file sizes in different viewport sizes. When you're done, the page should look like this:

1. Use your text editor to open this file:

   exercises\ch10\students\index.html

2. View this page in your browser.

3. Narrow the browser window until the image is cut off.

4. In the HTML file, add a srcset attribute to the <img> element that provides for three images: students_750.png, students_550.png, and students_400.png. (The numbers indicate the widths of the images.) Then, change the src attribute so the default image is students_550.png.

5. Add a sizes attribute to the <img> element. If the minimum viewport width is 800 pixels, the image with a width of 750 pixels should display. If the minimum viewport width is 600 pixels, the image with a width of 550 pixels should display. And if neither of these conditions is met, the image with a width of 400 pixels should display.

6. Test the page in your browser to see that the image that displays depends on the width of the browser window.

Exercise 10-3   Use an embedded font

In this exercise, you'll enhance the header of the website with an embedded font. When you're done, the page should look like this:

1. Use your text editor to open these files:

   ```
   exercises\ch10\town_hall\index.html
   exercises\ch10\town_hall\styles\main.css
   ```

2. View the page in a browser.

3. Choose a font file on your system and copy it into the styles folder.

4. Use CSS to apply the font to the <h2> element in the header. Then, refresh the web page in the browser to make sure the new font is applied.

5. Navigate to the speaker page for Scott Sampson to see that the new font is applied to all the pages of the website.

Chapter 11

How to work with tables

In the early days of the web, tables were used to lay out pages. But now, it's a best practice to lay out pages by using CSS flexbox or grid layout as shown in this book. As a result, you should only use tables to display tabular data as shown in this chapter.

Basic HTML skills for defining tables

This chapter begins by presenting some basic HTML skills for defining tables. But first, it introduces the structure of an HTML table.

An introduction to tables

Figure 11-1 presents a simple table and points out its various components. To start, a table consists of one or more *rows* and *columns*. In this figure, the table consists of three columns and six rows.

In HTML, you begin by defining the rows for a table. Then, within each row, you define a *cell* for each column.

Within each row, a table can contain two different kinds of cells. *Header cells* identify what's in the columns and rows, and *data cells* contain the actual data. For example, the first row of the table in this figure contains three header cells that identify the contents of each column. In contrast, the next four rows contain data cells. The last row starts with a header cell that identifies the contents of the data cell at the end of that row.

In broad terms, a table starts with a *header* that consists of one or more rows. Then, the *body* consists of one or more rows that present the data. Last, the *footer* provides summary data that consists of one or more rows. In this figure, the header and footer contain a single row, but the body contains four rows.

A table with basic formatting

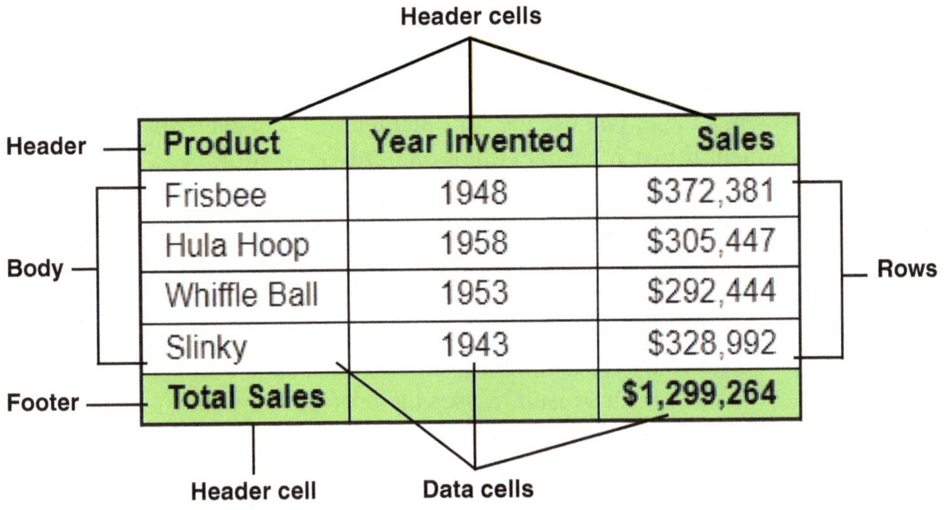
Header cells
Header
Body
Footer
Rows
Header cell
Data cells
Product
Year Invented
Sales
Frisbee
1948
$372,381
Hula Hoop
1958
$305,447
Whiffle Ball
1953
$292,444
Slinky
1943
$328,992
Total Sales
$1,299,264

Description

- A *table* consists of *rows* and *columns* that intersect at *cells*.
- *Header cells* identify the data in a column or row.
- *Data cells* contain the data of the table.
- The *header* for a table can consist of one or more rows.
- The *body* of a table typically consists of one or more rows.
- The *footer* for a table can consist of one or more rows.

Figure 11-1 An introduction to tables

How to define a table

Figure 11-2 presents four HTML elements for coding tables. The <table> element defines the table itself. Then, you code the other elements within the <table> element.

To define each row in a table, you use the <tr> element. Within each row, you code one <th> element for each header cell or one <td> element for each data cell. The table in this figure shows how that works. For the first row, this figure uses three <th> elements to define a header cell for each column. Then, the following rows use <td> elements to define a data cell for each column. Finally, the last row uses a header cell for the first column to describe the data cell in the third column.

This table doesn't include any formatting beyond the font that's set by the CSS for the <body> element. As a result, it uses the browser's default formatting. This centers the content in the header cells and displays them in bold, and it left aligns the content in the data cells. In addition, it sets the width of the cells based on the data they contain, with each cell in a column being as wide as the widest cell.

Four elements for coding tables

Element	Description
table	A table that contains rows.
tr	A table row.
th	A table header cell.
td	A table data cell.

The HTML for a table

```
<table>
    <tr>
        <th>Product</th>
        <th>Year Invented</th>
        <th>Sales</th>
    </tr>
    <tr>
        <td>Frisbee</td>
        <td>1948</td>
        <td>$372,381</td>
    </tr>
    <tr>
        <td>Hula Hoop</td>
        <td>1958</td>
        <td>$305,447</td>
    </tr>
    <!-- Other rows go here -->
    <tr>
        <th>Total Sales</th>
        <td></td>
        <td>$1,299,264</td>
    </tr>
</table>
```

The table in a browser with no CSS formatting

Product	Year Invented	Sales
Frisbee	1948	$372,381
Hula Hoop	1958	$305,447
Whiffle Ball	1953	$292,444
Slinky	1943	$328,992
Total Sales		$1,299,264

Description

- By default, a browser automatically determines the width of each column in a table based on its content.
- By default, a browser centers the content of a <th> element and applies bold to it.
- By default, a browser left aligns the content of a <td> element.

Figure 11-2 How to define a table

How to add a header and footer

Figure 11-3 presents three elements for grouping the rows of a table into headers and footers. It also presents the element for grouping rows in the body of the table, which is typical when you use a header and footer. Although these elements don't change the default appearance of a table, they make it easier to format a table with CSS later.

To code a header, you code the row or rows that make up the header between the opening and closing tags of the <thead> element. Similarly, you code a footer by coding the row or rows that make up the footer within a <tfoot> element. And you code a body by coding the row or rows for the body within a <tbody> element.

The example in this figure shows how this works. Here, the first row of the table is coded within a <thead> element, the last row of the table is coded within a <tfoot> element, and the remaining rows are coded within a <tbody> element. In addition, each data cell in the first column has an attribute that makes it a member of the left class. Although this class isn't used in this figure, the next figure uses it to format the table.

Elements for adding a header and a footer

Element	Description
thead	A table header that contains one or more rows.
tbody	A table body that contains one or more rows.
tfoot	A table footer that contains one or more rows.

A table with a header, body, and footer

```
<table>
    <thead>
        <tr>
            <th class="left">Product</th>
            <th>Year Invented</th>
            <th>Total sales</th>
        </tr>
    </thead>
    <tbody>
        <tr>
            <td class="left">Frisbee</td>
            <td>1948</td>
            <td>$372,381</td>
        </tr>
        <!-- Other rows go here -->
    </tbody>
    <tfoot>
        <tr>
            <th class="left">Total Sales</th>
            <td></td>
            <td>$1,299,264</td>
        </tr>
    </tfoot>
</table>
```

The table in a browser with no CSS formatting

Product	Year Invented	Sales
Frisbee	1948	$372,381
Hula Hoop	1958	$305,447
Whiffle Ball	1953	$292,444
Slinky	1943	$328,992
Total Sales		$1,299,264

Description

- The <thead>, <tbody>, and <tfoot> elements make it easier to style a table with CSS.
- You can code the <thead>, <tbody>, and <tfoot> elements in any sequence, and the header will always be displayed first and the footer last.

Figure 11-3   How to add a header and footer

Basic CSS skills for formatting tables

Now that you know the basic HTML skills for defining a table, you're ready to learn the basic CSS skills for formatting a table.

How to format a table

Figure 11-4 presents some common CSS properties for formatting tables. However, you can also use many of the other CSS properties that you've already learned to format a table.

By default, a table doesn't include borders. In many cases, though, you may want to add a border around a table to make it stand out on the page. You may also want to add borders around the cells or rows in a table to help identify the columns and rows.

To do that, you can use any of the border properties presented in chapter 5. In this figure, for example, the first style rule uses the border property to add a solid black border around the table, and the second style rule adds a solid black border around each cell in the table.

By default, a browser displays a small amount of space between the cells of a table. To remove that space, you can set the border-collapse property to collapse. Then, the borders between adjacent cells collapse to a single border as shown in the first table in this figure. When you do that, if two adjacent cells have different borders, the most dominant border displays. For example, if one border is wider than the other, the wider border displays.

When you use the padding property with tables, it works much as it does with the box model. That is, it specifies the amount of space between the contents of a cell and the outer edge of the cell. In this figure, for example, the second style rule adds .2em above and below the contents of each cell, and it adds .7em of space to the left and right of the contents.

To complete the formatting, the second style rule right aligns all the <th> and <td> elements and then overrides that formatting by left aligning the elements in the left class. Then, the third style rule applies a background color to the header and footer using a selector for the <thead> and <tfoot> elements. Finally, the fourth style rule changes the font weight for the footer to bold.

Properties for formatting tables

Property	Description
border-collapse	Determines whether to add some space between the borders of adjacent cells. By default, the separate keyword separates the borders and adds space between them. By contrast, the collapse keyword collapses the borders into a single border.
border-spacing	The space between cell borders when they aren't collapsed.
padding	The space between the cell contents and the outer edge of the cell.
text-align	The horizontal alignment of text.
vertical-align	The vertical alignment of text.

The CSS for the table from figure 11-3

```
table {
    border: 1px solid black;
    border-collapse: collapse;
}
th, td {
    border: 1px solid black;
    padding: .2em .7em;
    text-align: right;
    &.left {
        text-align: left;
    }
}
thead, tfoot {
    background-color: lightgreen;
}
tfoot {
    font-weight: bold;
}
```

The table in a browser

Product	Year Invented	Sales
Frisbee	1948	$372,381
Hula Hoop	1958	$305,447
Whiffle Ball	1953	$292,444
Slinky	1943	$328,992
Total Sales		$1,299,264

The same table without collapsed borders

Product	Year Invented	Sales
Frisbee	1948	$372,381
Hula Hoop	1958	$305,447
Whiffle Ball	1953	$292,444
Slinky	1943	$328,992
Total Sales		$1,299,264

Figure 11-4   How to format a table

How to use structural pseudo-classes to format a table

Chapter 4 presented some of the pseudo-classes available from CSS. Now, figure 11-5 presents some of the *structural pseudo-classes*, which select elements based on their position in the HTML document. In other words, they select elements based on their structural location. When you use these pseudo-classes to format your tables, you can you can avoid the use of class selectors as shown in the previous figure. This is generally considered a good practice because it simplifies your HTML and separates the formatting from the content.

The example in this figure shows how this works. Here, the first style rule uses the first-child pseudo-class to select the first <th> and the first <td> element in each row. Then, it left aligns the data in these cells. In other words, it left aligns the contents of the cells in the first column. As a result, you don't need to use the left class as shown in the previous two figures.

The second style rule uses the nth-child selector to select the second <th> and <td> element in each row. Then, the style rule centers the content in those cells. In other words, it centers the contents of the cells in the second column.

The third style rule uses the nth-child selector to select all even rows in the body of the table. Then, it applies silver as the background color for these rows. The table of n values presented in this figure shows that the n value could also be coded as 2n to select all even rows or 2n+1 to select all odd rows. If you wanted to apply a different color to every third row, you could use other combinations of n values to get that result.

The syntax for the CSS structural pseudo-class selectors

Syntax	Description
:nth-child(n)	nth child of parent
:nth-last-child(n)	nth child of parent counting backwards
:nth-of-type(n)	nth element of its type within the parent
:nth-last-of-type(n)	nth element of its type counting backwards

Typical n values

Value	Meaning
odd	Every odd child or element
even	Every even child or element
n	The nth child or element
2n	Same as even
3n	Every third child or element (3, 6, 9, ...)
2n+1	Same as odd
3n+1	Every third child or element starting with 1 (1, 4, 7, ...)

The CSS for a table

```
th:first-child, td:first-child {      /* all cells in first column */
    text-align: left;
}
th:nth-child(2), td:nth-child(2) {    /* all cells in second column */
    text-align: center;
}
tbody tr:nth-child(even) {            /* even rows in table body */
    background-color: silver;
}
```

The table in a browser

Product	Year Invented	Sales
Frisbee	1948	$372,381
Hula Hoop	1958	$305,447
Whiffle Ball	1953	$292,444
Slinky	1943	$328,992
Total Sales		$1,299,264

Description

- In CSS, *structural pseudo-classes* select elements based on their position in the HTML document.
- Structural pseudo-classes let you format a table without using classes or ids.

Figure 11-5   How to use structural pseudo-classes to format a table

More skills for working with tables

This chapter has already presented the most common skills for working with tables. However, you may also need to use some of the skills that follow.

How to add a table to a figure

For accessibility, a table should have a caption above or below it that summarizes what's in the table. One way to do that it to add the table to a figure that has a caption. This works similarly to adding an image to a figure that has a caption as shown in chapter 10.

The HTML for figure 11-6 defines a <figure> element that contains a <figcaption> element followed by a <table> element. Then, the CSS formats these elements. By default, a <figcaption> element is an inline element, not a block element. As a result, if you want to make sure the caption displays on its own line, it's a good practice to change it to a block element by setting its display property.

In this example, the <figcaption> element comes before the <table> element in the HTML. As a result, the browser displays the caption above the table. If you want to display the caption below the table, you can do that by coding the <figcaption> element after the <table> element.

The CSS for the <table> element uses the margin property to set the top and bottom margins to .5em and the right and left margins to auto. This centers the table within the figure.

It's also possible to use the <caption> element to provide a caption for a table as shown later in this chapter. However, if you're presenting a table as a figure, it's a better practice to use the <figure> and <figcaption> elements. That way, it's easier for a screen reader to tell that the table is part of the figure and that the caption applies to the figure.

A figure with a caption and a table

Total Sales by Product		
Product	**Year Invented**	**Sales**
Frisbee	1948	$372,381
Hula Hoop	1958	$305,447
Whiffle Ball	1953	$292,444
Slinky	1943	$328,992
Total Sales		**$1,299,264**

The HTML

```
<figure>
    <figcaption>Total Sales by Product</figcaption>
    <table>
        <!-- The code for the table goes here -->
    </table>
</figure>
```

The CSS

```
figure {
    border: 1px solid black;
    width: 450px;
    padding: 1em;
}
figcaption {
    display: block;
    font-weight: bold;
    text-align: center;
    font-size: 120%;
    padding-bottom: .25em;
}
table {
    border-collapse: collapse;
    border: 1px solid black;
    margin: .5em auto;
}
```

Description

- The <figure> element contains all elements for a figure such as an image or a table.
- The optional <figcaption> element provides a caption that describes a figure. It can be coded anywhere within the <figure> element.
- By default, the <figcaption> element is an inline element, but you can use CSS to change it to a block element.

Figure 11-6   How to add a table to a figure

How to merge cells in a column or row

To make a table easier to read, it often makes sense to merge some of the cells. This is illustrated in figure 11-7. Here, four cells are merged in the first row so the Sales header cell spans the last four columns of the table. Also, two cells are merged in the first column, so the Product header cell spans the first two rows.

To merge cells, you use the colspan and rowspan attributes of the <th> and <td> elements. To merge cells so a cell in a row spans two or more columns, you use the colspan attribute. To merge cells so a cell in a column spans two or more rows, you use the rowspan attribute. The value you use for these attributes indicates the number of cells to merge.

The example in this figure shows how these attributes work. Here, the table header consists of two rows and five columns. However, the first header cell spans two rows. As a result, the second row doesn't include a <th> element for the first column. Then, the second header cell spans the remaining four columns of the row. As a result, this row includes only two <th> elements: the one that defines the cell that contains the Product header, and the one that defines the cell that contains the Sales header.

The CSS in this figure uses structural pseudo-classes to format the merged cells. The first style rule selects the first header cell in the first row (the merged "Product" cell) and vertically aligns it with the bottom of the cell. Then, the second style rule selects the second header cell of the first row (the merged "Sales" cell) and centers it horizontally. In this case, it would be possible to reduce code duplication and improve readability by using nested classes like this:

```
tr:first-child {
    & th:first-child {
        vertical-align: bottom;
    }
    & th:nth-child(2) {
        text-align: center;
    }
}
```

Attributes of the \<th> and \<td> elements

Attribute	Description
colspan	The number of columns that a cell spans. The default is 1.
rowspan	The number of rows that a cell spans. The default is 1.

A table with merged cells

Product	Sales			
	North	South	West	Total
Frisbee	$55,174	$73,566	$177,784	$306,524
Hula Hoop	$28,775	$39,995	$239,968	$308,738
Whiffle Ball	$27,688	$24,349	$168,228	$220,265
Slinky	$23,082	$24,858	$129,619	$177,559
Sales Totals	$134,719	$162,768	$715,599	$1,013,086

The HTML for the table

```
<table>
    <thead>
        <tr>
            <th rowspan="2">Product</th>
            <th colspan="4">Sales</th>
        </tr>
        <tr>
            <th>North</th>
            <th>South</th>
            <th>West</th>
            <th>Total</th>
        </tr>
    </thead>
    <tbody>
        <!-- Other rows go here -->
    </tbody>
    <tfoot>
        <tr>
            <th>Sales Totals</th>
            <td>$134,719</td>
            <td>$162,768</td>
            <td>$715,599</td>
            <td>$1,013,086</td>
        </tr>
    </tfoot>
</table>
```

The CSS for the merged cells

```
tr:first-child th:first-child { vertical-align: bottom; }  /* Product */
tr:first-child th:nth-child(2) { text-align: center; }     /* Sales */
```

Figure 11-7   How to merge cells in a column or row

How to make a table accessible

Because tables are difficult for visually-impaired users to decipher, HTML provides some elements and attributes that can improve accessibility. These elements and attributes are summarized in figure 11-8, and they can be read by screen readers.

First, it's important to provide a caption that summarizes what each table contains. To do that, you can use the <caption> element as shown in this figure or the <figure> and <figcaption> elements described earlier in this chapter.

Second, you can use the headers attribute of the <td> element to identify one or more header cells that the data cell is associated with. To identify a header cell, you code the value of the cell's id attribute. In this figure, the example defines an id attribute for each of the three <th> elements in the header as well as the <th> element in the footer. Then, each <td> element includes a headers attribute that names the associated <th> element or elements.

For instance, the headers attribute for the first <td> element in each row of the body specifies the header cell that contains the Product header because the content of these cells are product names. Similarly, the <td> element in the footer includes a headers attribute that specifies two <th> elements. The first one specifies the header cell in the third column of the first row (the Sales header), and the second one specifies the header cell in the first column of the last row (the Total Sales header).

Third, you can use the scope attribute to identify whether a cell is associated with a column, a row, or a group of merged cells. Although you can use this attribute with both <td> and <th> elements, it's typically used with the <th> element as shown in this figure. Here, this attribute indicates that each <th> element in the header row is associated with a column. In addition, this attribute indicates that the <th> element in the footer row is associated with a row.

Even if you use these attributes, a table can be difficult for a visually-impaired person to interpret. So besides coding these attributes, you should try to keep your tables simple. That's good for the visually-impaired, and it's usually good for everyone else too.

An element for a caption

Element	Description
caption	A caption that describes the contents of the table.

Two attributes of the <th> and <td> elements

Attribute	Description
headers	One or more header cells that describe the content of the cell.
scope	A keyword that indicates whether a cell is associated with a column or row. You can use the col and row keywords to refer to columns and rows, and you can use the rowgroup keyword to refer to merged cells.

The HTML for a table that provides for accessibility

```
<table>
<caption>Total sales by product for famous toys</caption>
<thead>
    <tr>
        <th id="hdr_product" scope="col">Product</th>
        <th id="hdr_year" scope="col">Year Invented</th>
        <th id="hdr_sales" scope="col">Sales</th>
    </tr>
</thead>
<tbody>
    <tr>
        <td headers="hdr_product">Frisbee</td>
        <td headers="hdr_year">1948</td>
        <td headers="hdr_sales">$372,381</td>
    </tr>
    <tr>
        <td headers="hdr_product">Hula Hoop </td>
        <td headers="hdr_year">1958</td>
        <td headers="hdr_sales">$305,447</td>
    </tr>
    <!-- Other rows go here -->
</tbody>
<tfoot>
    <tr>
        <th id="hdr_total" scope="row">Total Sales</th>
        <td></td>
        <td headers="hdr_sales hdr_total">$1,299,264</td>
    </tr>
</tfoot>
</table>
```

Accessibility guidelines

- Provide captions for tables. One way to do that is to use the <caption> element.
- Use the attributes of the <th> and <td> elements to associate data cells with header cells.

Figure 11-8   How to make a table accessible

How to make a table responsive

When you develop responsive websites, you may need to modify your tables so they work on all screen sizes. To do that, you can start by making the width of each table and its columns fluid.

For example, you can make the width of the <table> element fluid by setting its width to a percent of the container. Then, you can make the widths of the columns fluid by setting the widths of the column header cells like this:

```
th:first-child { width: 45%; }
th:nth-child(2) { width: 35%; }
th:nth-child(3) { width: 20%; }
```

Then, the width of the table and its columns adjust automatically.

In addition to making a table fluid, you may want to use media queries to change how it's formatted for large or small screens. Figure 11-9 shows how to use a media query for small screens with widths that are 640 pixels or less to format the table presented in figure 11-4.

To format this table, the first style rule in the media query sets the display property of the <th> and <td> elements to block. That way, the browser displays each cell in the table on a separate line. Then, the next two style rules use the display property to not display the header and the first two cells of the footer.

The next style rule formats the first data cell in each row of the table body. These data cells contain the names of the products, and this style rule formats these cells to make them look like a header that identifies the second and third data cells in the row.

The next two style rules turn off borders that are no longer needed. This includes the border for the table as well as the bottom border for all of the data cells in the table body. Because the cells in the table body are defined with a one-pixel border on each side, removing the bottom border causes the cells to have just one border between them (the top border).

The next three style rules use the ::before pseudo-element to add information to the some of the data cells. That way, you don't lose the header content for the cells.

There are two things to note about these style rules. First, no header content is added to the data cell that contains the product name. However, you could add header content to that cell if you decided it would be helpful to the user. Second, there are three <td> elements for each row in the body, but only two <td> elements in the footer since the footer starts with a <th> element. That's why the third nth-of-type() pseudo-class selects the second data cell in the footer.

Finally, the last style rule positions the pseudo-elements that contain the header content so they're left aligned. To do that, this style rule uses absolute positioning to position the pseudo-elements, and it uses the left property to indent them so they align with the content in the other cells.

This figure starts by formatting a table so it works for large screens. Then, it uses a media query to adjust the formatting for smaller screens. In other words, it uses a desktop-first design. This approach often works well with tables that contain multiple columns. However, it's also possible to use a mobile-first design with a table.

A table formatted for large and small screens

Product	Year Invented	Sales
Frisbee	1948	$372,381
Hula Hoop	1958	$305,447
Whiffle Ball	1953	$292,444
Slinky	1943	$328,992
Total Sales		$1,299,264

Frisbee	
Year Invented:	1948
Sales:	$372,381
Hula Hoop	
Year Invented:	1958
Sales:	$305,447
Whiffle Ball	
Year Invented:	1953
Sales:	$292,444
Slinky	
Year Invented:	1943
Sales:	$328,992
Total Sales:	**$1,299,264**

The media query for small screens

```css
@media only screen and (max-width: 640px) {
    th, td { display: block; }

    /* hide header and first 2 cells of footer */
    thead { display: none; }
    tfoot th, tfoot td:nth-of-type(1) { display: none; }

    /* style first data cell like a header */
    tbody td:first-child {
        font-weight: bold;
        background-color: lightgreen;
    }

    /* turn off borders that are no longer needed */
    table { border: none; }
    tbody td { border-bottom: none; }

    /* add headings to data cells in body and footer */
    tbody td:nth-of-type(2)::before { content: "Year Invented: "; }
    tbody td:nth-of-type(3)::before { content: "Sales: "; }
    tfoot td:nth-of-type(2)::before { content: "Total Sales: "; }

    /* position headings so they're left aligned */
    td::before {
        position: absolute;
        left: 30px;
    }
}
```

Description

- To make a table responsive, you can make the table and column widths fluid and use media queries to reformat the table so it works for all screen sizes.
- To reformat a table so it works on small screens, you can change the <th> and <td> elements to block elements so each cell starts on a new line. Then, you can hide other elements and use pseudo-elements to add information to the data cells.

Figure 11-9   How to make a table responsive

Perspective

Now that you've finished this chapter, you should have the HTML and CSS skills you need for defining and formatting tables that display tabular data. This includes being able to make tables accessible and responsive.

Terms

table	data cell
row	header
column	footer
cell	body
header cell	structural pseudo-class

Exercise 11-1   Add a table to the luncheons page

In this exercise, you'll enhance the luncheons page of the Town Hall website by adding a table to it that looks like this:

Open and view the web page

1. Use your text editor to open the following files:

    ```
    exercises\ch11\town_hall\luncheons.html
    exercises\ch11\town_hall\styles\main.css
    ```

2. View the web page in your browser. It should display everything but the table.

Add the HTML for the table

3.   In the HTML file, add the table to the page. To start the table, you may want to copy the HTML from one of the book examples into the file. Then, you can modify that code and add the data for the new table. To quickly add new rows to the table, you can copy and paste earlier rows.

4.   View the page in a browser. It should display the content for the table, but it won't format the table correctly.

Add the CSS for the table

5.   In the CSS file, add the style rules that format the table. To start the CSS for the table, you may want to copy the CSS from one of the book examples into the file.

6.   View the table in a browser. The CSS should format the table, but it shouldn't format it correctly for this exercise.

7.   Modify the CSS so it formats the table correctly. You can use classes or structural pseudo-elements to select the elements you need to style. The borders should use the accent color variable for their color.

8.   View the table in a browser to make sure it's formatting the table correctly.

Add the table to a figure and provide for accessibility

9.   Add the table to a figure. To do that, code a <figure> element around the <table> element, and code a <figcaption> element above the table but within the <figure> element.

10.   Copy the <h2> element that says "The luncheon schedule" into the <figcaption> element.

11.   Add the attributes for user accessibility.

Chapter 12

How to work with forms

To create dynamic web pages, you use HTML to create forms that let the user enter data into controls. Then, the user can click on a button to submit the data to a web server for processing. In this chapter, you'll learn how to code forms and the controls they contain.

How to get started with forms

A *form* contains one or more *controls* such as labels, text boxes, and buttons. This chapter starts by showing how to define and format a form that contains some controls.

How to define a form

Figure 12-1 shows how to define a form that contains six controls: two labels, two text boxes, and two buttons. To start, you code the <form> element. On the opening tag for this element, you code the action and method attributes.

The action attribute specifies the URL for the application that processes the data in the form when it's submitted to the server. In this example, the action attribute specifies an HTML file named submit.html that's in the same folder as the HTML file that contains the form. As a result, submitting the form in this example just displays the HTML file named submit.html. However, if you edited this attribute to be a URL for an application running on a web server, that application would process the data for the form. Often, these URLs just point to the folder that contains the application like this:

```
action="/register_user"
```

The method attribute specifies the HTTP method used to send the data in the form to the web server. By default, the HTTP method is set to GET. Then, the web browser uses the HTTP GET method to send the data to the server. This appends the data to the URL that's specified in the action attribute.

However, you typically want to set the method attribute to POST. Then, the web browser uses the HTTP POST method to send the data to the server. This stores the data in the body of the HTTP request, which is more secure than appending the data to the URL.

Within the <form> element, this example contains three <div> elements. These <div> elements group related controls. In addition, since all of the controls in this example are inline elements, this starts a new line after each text box and after the second button.

The first <div> element contains a <label> element that labels an <input> element for a text box. To associate the label to the text box, this example uses the for attribute of the label and the id attribute of the text box. This shows that the label is for the text box. In general, it's a good practice to label controls like text boxes so the user can easily determine what kind of data it stores.

The first <input> element sets the name attribute to email. If you submit the form to an application that's running on the server, that application can use the name attribute to get the data from the control. In this example, the <input> element uses the same value for its id attribute and its name attribute. That's a common convention. Then, client-side code such as JavaScript can use the id attribute to work with the control, and server-side code can use the name attribute to get data from the control.

The second <div> element works much like the first one, except that it stores the data for a first name instead of an email address. Finally, the third <div> element stores two buttons where the first button submits the form to the server.

A form with six controls and no formatting

```
Email: johndoe@example.com
First name: John
Submit  Reset
```

Attributes of the <form> element

Attribute	Description
action	The URL for the application that processes the data in the form.
method	The HTTP method for submitting the form data. Valid values include GET and POST. The default value is GET, but you typically want to use POST.

Attributes common to most <input> elements

Attribute	Description
type	The type of control. For example, a value of text defines a text box.
id	An id that can be referred to by client-side code.
name	A name that can be referred to by client-side or server-side code.
value	The value that's stored in the control.

The HTML for the form

```html
<form action="submit.html" method="post">
    <div>
        <label for="email">Email:</label>
        <input type="text" id="email" name="email" value="johndoe@example.com">
    </div>
    <div>
        <label for="first_name">First name:</label>
        <input type="text" id="first_name" name="first_name" value="John">
    </div>
    <div>
        <button type="submit">Submit</button>
        <button type="reset">Reset</button>
    </div>
</form>
```

Description

- A *form* contains one or more *controls* like labels, text boxes, and buttons.
- It's a good practice to use <label> elements to label all text boxes. The for attribute of a label should correspond with the id attribute of the control that it labels.
- By default, the <label>, <input>, and <button> elements are inline elements.
- When a form is submitted, the data in the controls is sent to the server as part of the HTTP request.

Figure 12-1   How to define a form

How to format a form

Figure 12-2 shows a form and the CSS that formats it. On this form, the labels appear above the text boxes instead of to the left of them. To do that, the first style rule sets the display property for both <label> elements on the form to block. This is a common practice that works well for devices with small screens such as mobile devices.

The second style rule applies to <input> elements such as text boxes. This style rule sets the width of the text boxes to 100% of the width of its containing element. This is another common modern design practice that works well for small screens such as mobile devices because it allows the user to use either thumb to select a text box.

In addition, the second style rule sets the font-size property to inherit. As a result, the text boxes inherit the font size that's set for the <body> element. Without this setting, the default font size for a text box is smaller than the font size of the labels, which isn't usually what you want.

The third and fourth style rules apply to <button> elements. Here, the third style rule begins by setting the font-size to inherit. As a result, the buttons inherit the font size that's used by other <input> elements such as text boxes, which is usually what you want. In addition, the third style rule sets the color and background color of the two buttons on the form, and the fourth style rule changes the color and background color if the mouse hovers over the button.

The fifth style rule sets the border-radius property for the text boxes and buttons to 5px. As a result, the text boxes and buttons use the same rounded corners.

As you review the CSS in this figure, note that much of it sets the padding, margins, borders, and colors for the controls on a form. This works just as it does for setting the padding, margins, borders, and colors for other elements that use the box model.

The form after it has been formatted

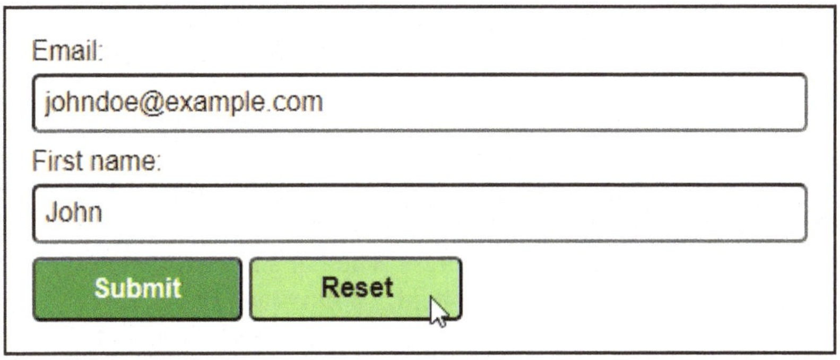

The CSS for the form

```css
label {
    display: block;
    margin-bottom: .25em;
}
input {
    width: 100%;
    font-size: inherit;
    padding: .4em;
    margin-bottom: .5em;
}
button {
    font-size: inherit;
    font-weight: bold;
    width: 8em;
    padding: .5em;
    color: white;
    background-color: forestgreen;
}
button:hover {
    color: black;
    background-color: lightgreen;
}
input, button {
    border-radius: 5px;
}
```

Description

- You can use CSS to format the controls on a form.

Figure 12-2   How to format a form

How to get started with controls

Now that you have a general idea of how forms and controls work, you're ready to learn more details for working with some of the most common controls.

How to use labels and text fields

You can use a *text field*, also known as a *text box*, to allow a user to enter text such as a name, password, phone number, or address. Then, it's usually a good practice to use a label to identify each text field. To illustrate, figure 12-3 presents the HTML and CSS for a web page that has two labels and three text fields.

The HTML shows how to set some of the attributes of the <input> element for text fields. To start, all three <input> elements use the type attribute to set the type of text field. Here, the first <input> element creates a text field, the second creates a password field, and the third creates a hidden field.

A *password field* works much like a text field, except that it displays bullet characters instead of the actual characters entered by the user. Unlike text and password fields, a *hidden field* isn't displayed by the browser. As a result, the user can't enter data into it. Instead, its value is set by the value attribute. Then, you can use client-side or server-side code to work with that value.

The first <input> element has an autofocus attribute. As a result, when the page loads, the browser automatically moves the focus to that textbox so it's ready for the user to begin entering text.

The second <input> element uses a placeholder attribute to display some information to the user about entering a password. In addition, it uses a maxlength attribute to limit the password to a maximum of 20 characters.

The HTML also shows how to use the for attribute of the <label> element to associate a label with a control. To do that, you set the for attribute of the <label> element to the same value as the id attribute of the control.

The CSS shows how to format labels and text fields. Here, the first style begins by using the display property to make the <label> elements block elements. As a result, each label starts on a new line. Then, it adds a small margin below each label.

The second style rule begins by setting the width of the <input> elements to 100% of its container element. Then, it sets the font-size property to inherit. As a result, the browser uses the same font size for text fields as it does for text in the <body> element, which is the same font size it uses for labels. Next, it sets a small amount of padding between the text in the field and its border. Finally, it sets a bottom margin for each text field that's displayed below the border.

Attributes of the `<input>` element for text fields

Attribute	Description
type	The type of text field. Valid values include text, password, and hidden.
maxlength	The maximum number of characters that the user can enter in the field.
autofocus	A Boolean attribute that tells the browser to move the focus to the field.
placeholder	Placeholder text that is removed when the user's cursor enters the control.

An attribute of the `<label>` element

Attribute	Description
for	The id for the label's control.

The HTML for labels and text fields

```
<label for="username">Username:</label>
<input type="text" id="username" name="username" autofocus>

<label for="password">Password:</label>
<input type="password" id="password" name="password"
       placeholder="Enter your password" maxlength="20">

<input type="hidden" name="productid" value="widget">
```

The CSS for the labels and text fields

```
label {
    display: block;
    margin-bottom: 0.25em;
}
input {
    width: 100%;
    font-size: inherit;    /* inherit from <body> element */
    padding: .2em;
    margin-bottom: 1em;
}
```

How it looks in a browser

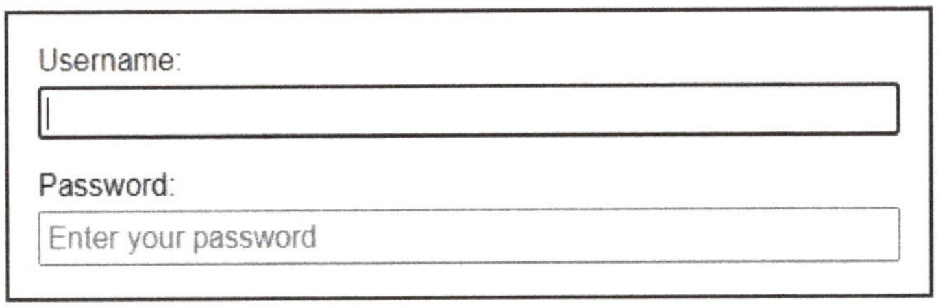

Username:
(text field)
Password:
Enter your password

Description

- You can use the `<input>` element to create *text fields*, *password fields*, and *hidden fields*.
- You should use a `<label>` element for each visible text field. Each label should use the for attribute to associate it with a text field.
- By default, an `<input>` element doesn't inherit the font size from the body, but you can change that by setting the font-size property to inherit.

Figure 12-3   How to use labels and text fields

How to use text areas

The previous figure showed how to use labels and text fields to get a single line of text from a user. Now, figure 12-4 shows how to use a *text area* that get multiple lines of text from a user. For example, if you were defining a control for a blog that allows users to leave comments on a post, you would probably want to use a text area to get the comments from the users.

Text areas provide two features to make it easier to get longer blocks of text. First, text areas automatically wrap text so that users don't have to manually enter new lines. This makes it easier for users to enter text, and it also makes it easier to display the input. Second, text areas allow you to drag the corner of the control to resize the input box. This allows users to view what they've written without having to scroll the box if their input gets too long.

By default, a text area doesn't inherit the font family or font size from the <body> and <input> elements. However, the CSS in this figure styles the text area by setting its font-family and font-size properties to inherit. As a result, the text area in this figure inherits the font family and font size from the <body> and <input> elements, which is usually what you want.

The HTML for a label and a text area

```
<label for="comments">Comments:</label>
<textarea id="comments" placeholder="Add your comments here."></textarea>
```

The CSS for the text area

```
textarea {
    width: 100%;
    padding: .2em;
    height: 5em;
    font-family: inherit;    /* inherit from <body> element */
    font-size: inherit;      /* inherit from <input> element */
}
```

How it looks in a browser

Comments:

Add your comments here.

Description

- A *text area* lets a user enter multiple lines of text.
- By default, a <textarea> element doesn't inherit font family or size from the <body> and <input> elements, but you can change that by setting the font-family and font-size properties to inherit.

Figure 12-4   How to use text areas

How to use radio buttons and check boxes

Figure 12-5 shows how to use *check boxes* and *radio buttons*. Although check boxes work independently of each other, radio buttons are typically set up so the user can select only one radio button from a group of buttons. In the example in this figure, for instance, you can select only one of the three radio buttons. However, you can select or deselect any combination of check boxes.

To create a radio button, you set the type attribute of the <input> element to radio. Then, to create a group, you set the name attribute for all of the radio buttons in the group to the same value. In this figure, all three radio buttons have their name attribute set to crust. That way, the user can only select one of these radio buttons at a time. However, each of these buttons has a different value. That way, the client-side or server-side code can get the value of the selected button.

To create a check box, you set the type attribute of the <input> element to checkbox. Then, you set the name attribute so you can access the control from your client-side and server-side code. When you submit the form to the server, the browser submits a name/value pair for the check box only if it's selected.

If you want the browser to select a check box or radio button when it loads the page, you can code the checked attribute. In this figure, for example, the code selects the first radio button. As a result, when the browser loads the page, it selects the first radio button.

In this figure, labels provide the text that labels the radio buttons and text boxes. For instance, the first label provides the text for the first radio button. To do that, the HTML sets the for attribute for the first label to the id attribute of the first radio button. As a result, the text for the first radio button is "Thin Crust".

When radio buttons and check boxes are coded this way, the user can click on the label to select or deselect a button or check box. This makes the buttons and check boxes more accessible to users with motor disabilities who may find it difficult to click on the smaller button or box.

This also makes it easier for assistive devices such as screen readers to read the text associated with a control. If you don't use labels in this way, the assistive devices have to scan the text before and after a control and guess which text is associated with the control.

To code a label after a control as shown in this figure, the label should be an inline element. Since labels are inline elements by default, you don't need to use CSS to change the display property for a label. However, you may want to use the margin-right property of the <input> element to add some space between the radio button or check box and its label.

Attributes of the <input> element for radio buttons and check boxes

Attribute	Description
type	The type of control, either radio or checkbox.
checked	A Boolean attribute that causes the control to be checked when the page loads.

The HTML for radio buttons and check boxes

```
<h3>Crust:</h3>
<div>
    <input type="radio" name="crust" id="crust1" value="thin" checked>
    <label for="crust1">Thin Crust</label>
</div>
<div>
    <input type="radio" name="crust" id="crust2" value="deep">
    <label for="crust2">Deep Dish</label>
</div>
<div>
    <input type="radio" name="crust" id="crust3" value="hand">
    <label for="crust3">Hand Tossed</label>
</div>

<h3>Toppings:</h3>
<div>
    <input type="checkbox" name="topping1" id="topping1" value="pepperoni">
    <label for="topping1">Pepperoni</label>
</div>
<div>
    <input type="checkbox" name="topping2" id="topping2" value="mushrooms">
    <label for="topping2">Mushrooms</label>
</div>
<div>
    <input type="checkbox" name="topping3" id="topping3" value="olives">
    <label for="topping3">Black Olives</label>
</div>
```

CSS that adds space between the control and its label

```
input { margin-right: .5em; }
```

How it looks in a browser

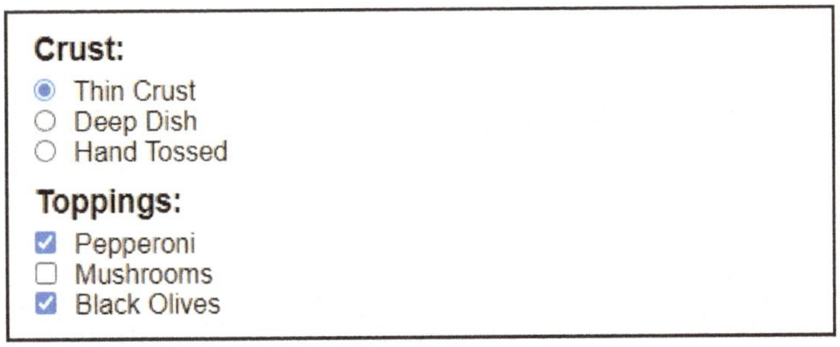

Crust:
◉ Thin Crust
○ Deep Dish
○ Hand Tossed

Toppings:
☑ Pepperoni
☐ Mushrooms
☑ Black Olives

Description

- For radio buttons and check boxes, it's common to code the <input> element before the <label> element.

Figure 12-5   How to use radio buttons and check boxes

How to group controls

In many cases, you want to group related controls on a form to make it easy for users to see how they're related. To do that, you can use the <fieldset> and <legend> elements as shown in figure 12-6.

To start, you code the <fieldset> element around any controls that you want to group. Then, if you want to provide a legend for the controls, you code the <legend> element within the <fieldset> element to specify the text for the legend.

The example in this figure illustrates how this works. Here, the first group of controls contains radio buttons that allow the user to select the crust, and the second group of controls contains check boxes that allow the user to select toppings.

By default, a browser displays a black border around each group. In addition, it displays the legend for each group across the top left part of this border. However, if you want to change the appearance of this border or its legend, you can use CSS as shown in this figure.

The first style rule shown in this figure changes the formatting for the <legend> element. First, it sets the font size so it's slightly larger than the font size for the <body> element. In addition, it sets the font weight of the legend to bold.

The second style rule changes the formatting for the <fieldset> element. To do that, it sets its padding and its bottom margin. However, if you wanted, you could change the border around the element too.

The third style rule only applies to radio buttons and check boxes. To make that possible, it uses attribute selectors to only select <input> elements that have their type attribute set to radio or checkbox. This technique is often useful when you only want to apply a style to certain types of <input> elements such as radio buttons and check boxes but not to other types of <input> elements such as text boxes.

HTML that uses <fieldset> and <legend> elements

```
<fieldset>
   <legend>Crust</legend>
   <div>
      <input type="radio" name="crust" id="crust1" value="thin">
      <label for="crust1">Thin Crust</label>
   </div>
   <!-- the next two div elements are the same as the previous figure -->
</fieldset>

<fieldset>
   <legend>Toppings</legend>
   <div>
      <input type="checkbox" name="topping1" id="topping1" value="pepperoni">
      <label for="topping1">Pepperoni</label>
   </div>
   <!-- the next two div elements are the same as the previous figure -->
</fieldset>
```

The CSS

```
legend {
    font-size: 1.2em;
    font-weight: bold;
}
fieldset {
    padding: .5em 1em;
    margin-bottom: 0.5em;
}
input[type="radio"], input[type="checkbox"] {
    margin-right: .5em;
}
```

How it looks in a browser

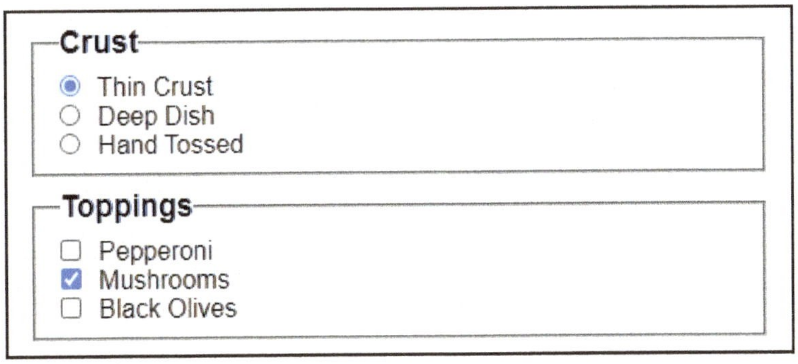

Description

- The <fieldset> element groups controls.
- The <legend> element can be coded within a <fieldset> element to label the group of controls.
- If you want to disable all of the controls within a <fieldset> element, you can code the disabled attribute for the <fieldset> element.

Figure 12-6   How to group controls

How to use drop-down lists

The first example in figure 12-7 shows how to use a *drop-down list*. To display the list of options, the user can click the control or the arrow at the right side of the control. Then, the user can select one option from that list.

To define a drop-down list, you can use a <select> element with a name attribute that provides the name that's sent to the server. Then, between the opening and closing tags, you code two or more <option> elements where each element has a value attribute that provides the value that's sent to the server. You also code the text that's displayed for each option as the content of the <option> element. This text is often the same as or similar to the value for the option. In this figure, for example, the first option has a value of "thin" and text of "Thin Crust".

When a drop-down list is first displayed, the first option in the list is selected by default. If that's not what you want, you can code the selected attribute for the option you want to be selected. In this example, the "Stuffed Crust" option has the selected attribute, so it's selected by default even though it's the third option in the list.

This figure also presents some typical CSS for the label and drop-down list presented in this figure. This CSS is the same as the CSS for the labels and text boxes presented in figure 12-3. As a result, the drop-down list in this figure looks much like the text boxes presented in figure 12-3, except that it drops down to allow the user to select an option.

Attributes of the <option> element

Attribute	Description
value	The value of the selected option that's sent to the server for processing.
selected	A Boolean attribute that causes the option to be selected when the page loads.

The HTML for a label and a drop-down list

```
<label for="crust">Crust Style:</label>
<select id="crust" name="crust">
    <option value="thin">Thin Crust</option>
    <option value="deep">Thick Crust</option>
    <option value="stuffed" selected>Stuffed Crust</option>
</select>
```

The CSS

```
label {
    display: block;
    margin-bottom: 0.25em;
}
select {
    width: 100%;
    font-size: inherit;
    padding: .2em;
    margin-bottom: .5em;
}
```

How it looks in a browser

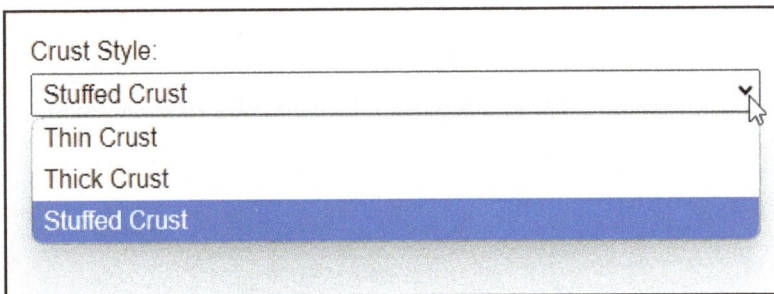

Description

- A *drop-down list* lets the user select an option from a list that drops down from a text field.
- You can use the <select> element to define a drop-down list.
- You can use the <option> element to define the options in a drop-down list.

Figure 12-7   How to use drop-down lists

How to use list boxes

While a drop-down list is useful for selecting a single option, it doesn't support selecting multiple options. To do that, you need to use a *list box* as shown in figure 12-8. A list box differs from a drop-down list in that it displays two or more options, and it can allow a user to select more than one option.

To code a list box, you start by defining a <select> element with a name attribute. Then, you code the size attribute to indicate the number of options that are displayed at one time. In this example, the HTML sets the size attribute to 4. However, the list contains seven options. As a result, the browser adds a scroll bar to the list.

By default, the user can select only one option from a list box. But you can change that by adding the multiple attribute to the <select> element. Then, to select multiple options, the user can hold down Ctrl in Windows or Command in macOS and click on multiple options.

This figure also presents some CSS for a list box that works like the CSS for the drop-down list in the previous figure. However, this CSS also provides some padding for the <option> elements. To get this padding to work the way you want, you may need to experiment with it.

Attributes of the <select> element

Attribute	Description
size	The number of items to display in the control. The default value is 1.
multiple	A Boolean attribute that determines whether multiple items can be selected.

The HTML for a label and a list box

```
<label for="multi-topping">Choose one or more toppings: </label>
<select id="multi-topping" name="toppings" multiple size="4">
    <option value="pepperoni">Pepperoni</option>
    <option value="sausage" selected>Sausage</option>
    <option value="mushrooms">Mushrooms</option>
    <option value="olives">Black olives</option>
    <option value="onions">Onions</option>
    <option value="bacon">Canadian bacon</option>
    <option value="pineapple">Pineapple</option>
</select>
```

The CSS

```
label {
    display: block;
    margin-bottom: 0.25em;
}
select {
    width: 100%;
    padding: .2em;
    margin-bottom: .5em;
    font-size: inherit;
}
option {
    padding: .2em;
}
```

How it looks with two toppings selected

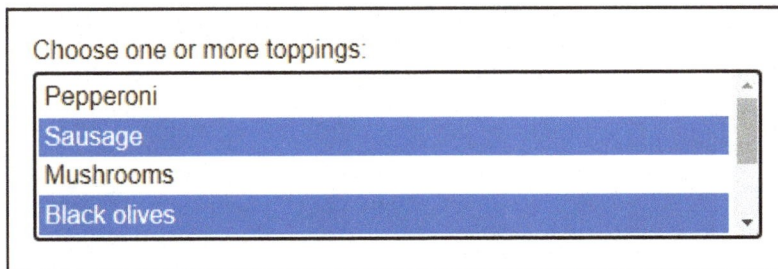

Description

- A *list box* displays two or more options and lets the user select one or more of those options.
- You can use the <select> element to define a list box.
- To select multiple items, the user can hold down the Ctrl or Command key.

Figure 12-8   How to use list boxes

How to define buttons

When coding forms, you can use a *button* to allow a user to perform an action. For example, users typically click on a button to submit a form to the server. However, you can use buttons for other purposes too such as running some client-side code.

To define a button, you can use the <button> element or the <input> element. For instance, the first example in figure 12-9 uses the <button> element to define a button. Then, the second example uses the <input> element to define the same button. To do that, it sets the type attribute button.

So, which element should you use to define buttons? In general, using the <button> element works better with screen readers and makes your code easier to read and format. As a result, for new HTML, it's generally considered a good practice to use the <button> element. However, you may want to use <input> elements if you're maintaining old code that already uses them consistently.

When you create a button, you can use the type attribute to define three types of buttons. You can create a *submit button* that submits the form to the server when it's clicked. You can create a *reset button* that sets the inputs for a form to its default values when it's clicked. And you can create a *JavaScript button* that runs some JavaScript when it's clicked.

By default, a <button> element has a type attribute of button, which creates a JavaScript button. As a result, you only need to set the type attribute if you want to create a submit or reset button as shown in the third and fourth examples. However, reset buttons aren't common because users don't typically want to reset the inputs for an entire form and clicking the button by accident can cause the user frustration.

If necessary, a <button> element can contain other elements. For example, the fifth example shows a <button> element that contains an <img> element as well as some text. Adding images to buttons used to be more common in the early days of the internet. These days, it's more common to use colors to make buttons more visually interesting as shown in the next figure.

If you need to disable a button, you can add the disabled attribute to the button. Then, by default, the browser grays out the button as shown in this figure. However, you can also use CSS to change the appearance of a disabled button as shown in the next figure.

Two attributes of the <button> element

Attribute	Description
type	The type of button. Valid values include submit, reset, or button. The default is button.
disabled	A Boolean attribute that disables the button.

A <button> element

```
<button id="start_button">Start Slide Show</button>
```

An <input> element

```
<input id="start_button_old" type="button" value="Start Slide Show">
```

A button that submits a form

```
<button type="submit">Submit</button>
```

A button that resets the controls on a page

```
<button type="reset">Reset</button>
```

A button that uses an image

```
<button><img src="images/addtocart.png"> Add To Cart</button>
```

A disabled button

```
<button id="stop_slide_show" disabled>Stop Slide Show</button>
```

How it looks in a browser

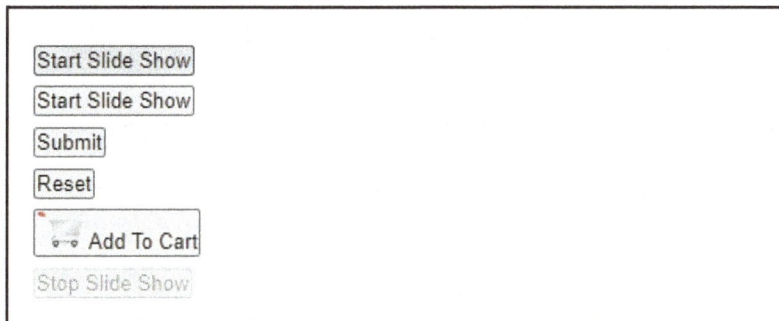

Description

- Buttons perform actions such as submitting a form to the server or running client-side code.

Figure 12-9   How to define buttons

How to format buttons

Now that you know how to use HTML to define buttons, you're ready to learn how to use CSS to format them. One common technique for formatting buttons is to create a base class that applies to all of the buttons for your website. This base class sets all of the properties that are the same for all buttons like the font size, color, background color, padding, and border. Then, you can create other classes that create variations for other types of buttons.

In figure 12-10, the button-styled class is the base class for a button. It starts by changing the cursor to a finger pointer. As a result, when the mouse hovers over a button, the cursor changes from its default shape of an arrow to the finger pointer as shown at the bottom of this figure. This base class also sets many other properties of the button. The result is a button with white text, a blue background, a black border, adequate padding, and rounded corners.

Once you've established the base class, you can create additional classes that help to visually communicate the purpose of each button. Typically, these classes set the background colors for the buttons. In this example, the button-primary class sets the background color for the button to green, and the button-danger class sets the background color to red. You can use these classes to override the default background color of blue that's set by the base class.

On a website, you can use the button-primary class for actions that move the application state forward, such as submitting a form or accepting the cookies for the site. In contrast, you can use the button-danger class for actions that move the application state backwards like canceling an action or clearing a form.

If your website uses disabled buttons, you can use the opacity property to indicate that a button is disabled. In this figure, the button-disabled class makes the button more transparent by setting the opacity property to .5, and it sets the cursor back to its default arrow. When you set the CSS for a disabled button, remember that the HTML should still include the disabled attribute. That way, the browser doesn't start an action if a user clicks the button.

These days, it's common to use hover effects to provide visual feedback to your users and make your apps feel more responsive. In this example, the :hover pseudo-class for the button-styled class changes the text and background color for the button when the mouse is hovering over it. As a result, when the user hovers the mouse over a button, the browser changes the background color to white and the text to black as shown in this figure.

The HTML for five buttons

```
<h3>Two styled buttons</h3>
<button type="submit" class="button-styled button-primary">Submit</button>
<button type="reset" class="button-styled button-danger">Cancel</button>

<h3>Two buttons with disabled styling</h3>
<button type="submit" disabled
        class="button-styled button-primary button-disabled">Submit</button>
<button type="reset" disabled
        class="button-styled button-danger button-disabled">Cancel</button>

<h3>A button with a hover effect</h3>
<button class="button-styled">Start Slide Show</button>
```

The CSS

```
.button-styled {
    cursor: pointer;
    color: white;
    background-color: blue;
    font-weight: bold;
    font-size: inherit;
    padding: 0.5em;
    border-radius: 10px;
    &:hover {
        color: black;
        background-color: white;
    }
}
.button-primary {
    background-color: green;
}
.button-danger {
    background-color: red;
}
.button-disabled {
    opacity: 0.5;
    cursor: default;
}
```

How it looks in a browser

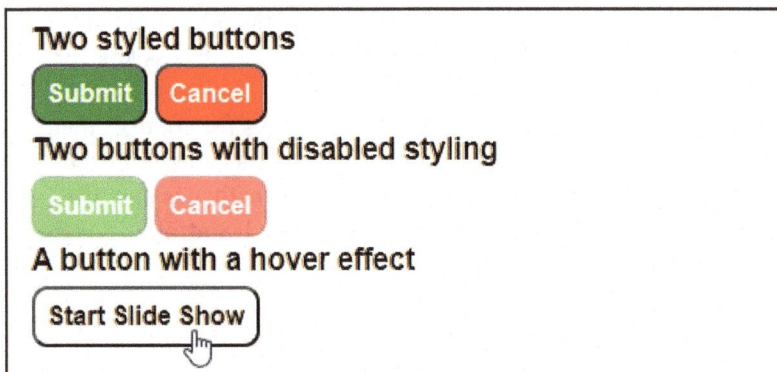

Figure 12-10   How to format buttons

Other skills for working with forms and controls

Now that you know how to code a form and some of the most common controls, you're ready to learn some other skills for working with forms and controls. That includes setting the tab order, assigning access keys, and using controls for specific types of input.

How to set the tab order and assign access keys

Figure 12-11 shows how to set the *tab order* of the controls on a form and how to assign an *access key* to a control. By default, when the user presses the Tab key on a web page, the focus moves from one control to another in the sequence that the controls appear in the HTML. The tab order doesn't include labels but does include links created by <a> elements.

To change the default tab order, you can code tabindex attributes. To remove a control from the tab order, you can set its tabindex attribute to a negative number such as -1. To set the tab order, you can set the tabindex attribute for the first control to 1, the second control to 2, and so on. In this figure, both examples use the tabindex attribute of the controls to reverse the default tab order.

If you don't set the tabindex value for a control, the control still receives the focus when the user presses the Tab key. However, the focus moves in the sequence that the controls appear in the HTML, after it is done moving through the controls that have positive tabindex values.

When you provide an access key for a control, the user can press that key in combination with one or more other keys to move the focus to the control. For example, if a page is displayed in Chrome on a Windows system, the user can press Alt+Shift and the access key to move the focus to a control that has an access key.

To define an access key, you can code the accesskey attribute as shown in the first example. The value of this attribute is the key that the user should use to move the focus to the control. Here, the access key for the "First name" text box is the letter F, the access key for the "Last name" text box is the letter L, and the access key for the Email text box is the letter E. In addition, these examples use the <u> element to underline the letters for the access keys in the labels that are associated with the controls. This is a common way to identify the access key for a control.

The second example shows another way to code access keys for controls that have labels associated with them. Here, the accesskey attribute is coded for each of the labels instead of for the text boxes. Then, when the user activates one of these access keys, the browser moves the focus to the associated control.

The attributes for setting the tab order and access keys

Attribute	Description
tabindex	The tab order for the link starting with 1. To take a link out of the tab order, code a negative value like -1.
accesskey	A keyboard key that can be pressed in combination with a control key to move the focus to the control.

Three labels with access keys

```
First name:
[                                            ]

Last name:
[                                            ]

Email:
[                                            ]
```

Access keys defined for the <input> elements

```
<label for="firstname"><u>F</u>irst name:</label>
<input type="text" id="firstname" accesskey="F" tabindex="3">
<label for="lastname"><u>L</u>ast name:</label>
<input type="text" id="lastname" accesskey="L" tabindex="2">
<label for="email"><u>E</u>mail:</label>
<input type="text" id="email" accesskey="E" tabindex="1">
```

Access keys defined for the <label> elements

```
<label for="firstname" accesskey="F"><u>F</u>irst name:</label>
<input type="text" id="firstname" tabindex="3">
<label for="lastname" accesskey="L"><u>L</u>ast name:</label>
<input type="text" id="lastname" tabindex="2">
<label for="email" accesskey="E"><u>E</u>mail:</label>
<input type="text" id="email" tabindex="1">
```

Accessibility guideline

• Set a tab order and provide access keys for users who can't use a mouse.

Description

• The *tab order* for a form is the sequence in which the controls receive the focus when the user presses the Tab key. By default, the tab order is the order of the links and controls in the HTML, not including labels.

• *Access keys* are shortcut keys that the user can press to move the focus to specific controls on a form. If you assign an access key to a label, it moves the focus to the control that's associated with the label since labels can't receive the focus.

• To use an access key, you press a control key plus the access key. For Windows, press Alt+Shift plus the access key. On macOS, press Ctrl+Option plus the access key.

Figure 12-11   How to set the tab order and assign access keys

How to use the number, email, url, and tel controls

Figure 12-12 shows how to use the *number control*, the *email control*, the *url control*, and the *tel control*. These controls are good for semantics because they indicate the type of entry that the user should make. In addition, these controls can make it easier for the user to enter data into them. The example in this figure shows how to define these controls.

Here, the number control has 100 as its minimum value, 1000 as its maximum value, 100 as the step value, and 300 as its starting value. Then, when the mouse hovers over the control or the cursor enters the control, the browser displays up and down arrows on the right side of the control. That way, the user can move the number up or down by clicking on these arrows.

This example also shows the email, url, and tel controls. Although these controls look like regular text boxes, browsers can use them to make it easier for the user to enter data. For example, when you move the focus to an email control on a mobile device, it may display a keyboard that's optimized for entering email addresses.

In addition, browsers can provide automatic data validation for email and url controls as shown later in this chapter. However, they don't typically provide data validation for tel controls because the format of telephone numbers varies so much from one country to another.

The number, email, url, and tel controls

Control	Description
number	A number with min, max, and step attributes.
email	An email address.
url	A URL.
tel	A telephone number.

The HTML for number, email, url, and tel controls

```
<label for="investment">Monthly Investment: </label>
<input type="number" id="investment"
    min="100" max="1000" step="100" value="300">

<label for="email">Email:</label>
<input type="email" id="email">

<label for="link">Website:</label>
<input type="url" id="link">

<label for="phone">Phone Number:</label>
<input type="tel" id="phone">
```

How it looks in a browser

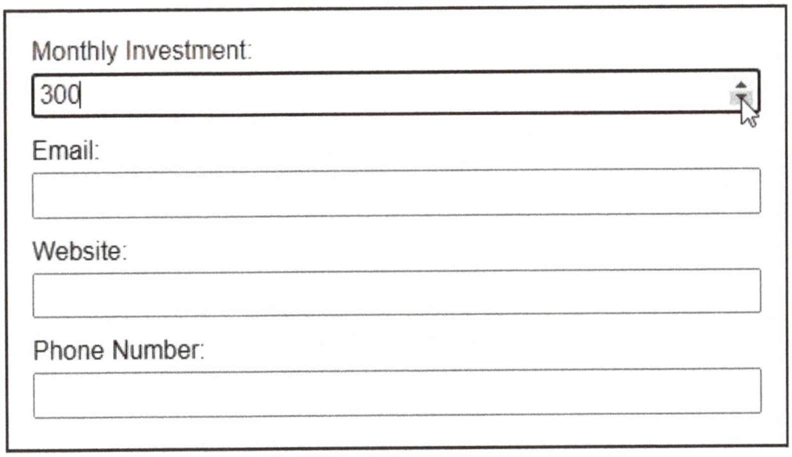

Monthly Investment:

`300`

Email:

Website:

Phone Number:

Description

- Because the number, email, url, and tel controls indicate the type of data that they accept, these controls are good for semantics. In addition, they can make it easier for users to enter data.
- Browsers can provide automatic data validation for the number, email, and url controls as shown later in this chapter. However, browsers don't provide automatic data validation for the tel control because telephone number formats vary so much from country to country.

Figure 12-12   How to use the number, email, url, and tel controls

How to use the date and time controls

Figure 12-13 shows how to use the *date and time controls*. When the user clicks on the symbol at the right of one of these controls, a panel drops down that makes it easy for the user to select a valid entry.

The examples in this figure show how to code the <input> elements for these controls. Here, the fifth control uses the datetime-local type to get the local date and time. However, you can also use the datetime type to get Coordinated Universal Time, or UTC, which is the time by which the world sets its clocks.

When you test these controls in different browsers, you may find some variations in how they look and work. In the worst case, the browser displays these controls as text boxes. Even then, it's a good practice to use these controls for semantic reasons.

HTML that uses the date and time controls

```
<label for="week">Week:</label>
<input type="week" id="week" name="week">

<label for="month">Month:</label>
<input type="month" id="month" name="month">

<label for="date">Date:</label>
<input type="date" id="date" name="date">

<label for="time">Time:</label>
<input type="time" id="time" name="time">

<label for="datetimelocal">Local date and time:</label>
<input type="datetime-local" id="datetime" name="datetime">
```

How it looks in a browser

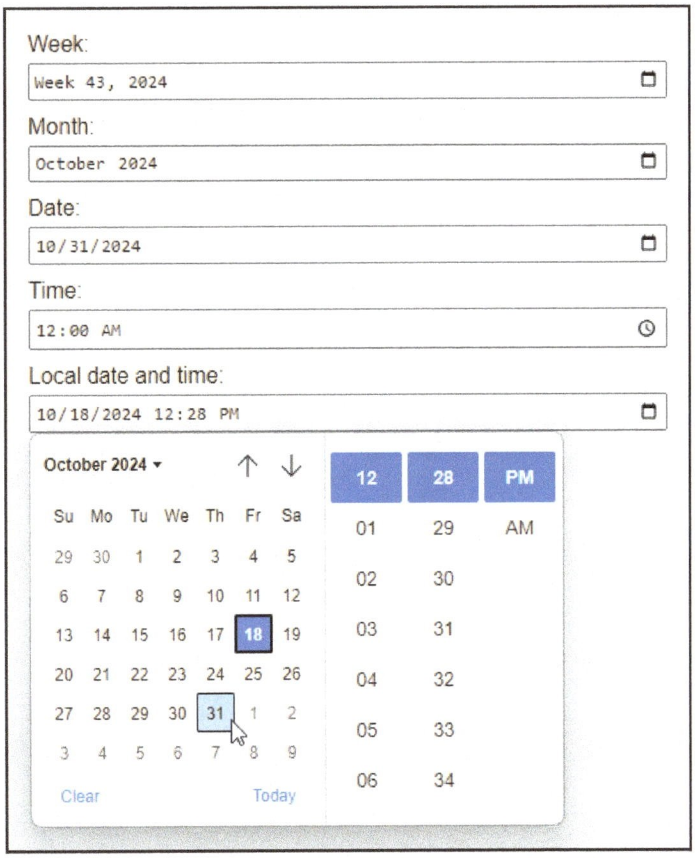

Description

- The HTML controls for dates and times make it easy for the user to enter valid data.
- The datetime-local type uses the local date and time for your computer. By contrast, the datetime type uses Coordinated Universal Time, or UTC, the time by which the world sets its clocks.

Figure 12-13   How to use the date and time controls

How to use the file upload control

Figure 12-14 shows how to use the *file upload control*. This control lets users upload one or more files to the web server. Typically, you use this control with a server-side programming language to upload the file or files from the user's computer to the web server. When you code a file upload control, you set the type attribute of the <input> element to file.

You can also set the accept attribute for a file upload control. This attribute lets you specify the types of files that are accepted. In this example, the accept attribute specifies that the control accepts JPEG and GIF types.

When you use the file upload control, you must also set the enctype attribute of the <form> element that contains it to multipart/form-data as shown in this figure. Typically, you only set the enctype attribute of a form when you're uploading files. When you set the enctype attribute, you must also set the method attribute of the <form> element to post.

Attributes of the <input> element for a file upload control

Attribute	Description
accept	The types of files that are accepted for upload. When the operating system's open dialog box opens, only files of those types are shown.
multiple	A Boolean attribute that lets the user upload more than one file.

An attribute of the <form> element

Attribute	Description
enctype	Specifies how the form data is encoded when it's submitted to the server. Can only be used when the HTTP method is POST.

A form with a file upload control

```
<form action="upload_image" method="post" enctype="multipart/form-data">
    <label for="image_file">Image: </label>
    <input type="file" id="image_file" accept="image/jpeg image/gif">

    <button type="submit">Upload</button>
</form>
```

How it looks in the browser

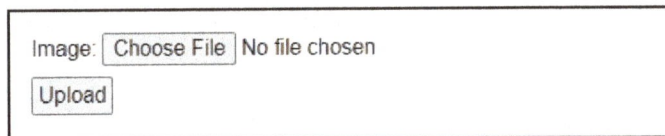

The dialog box that lets you choose a file

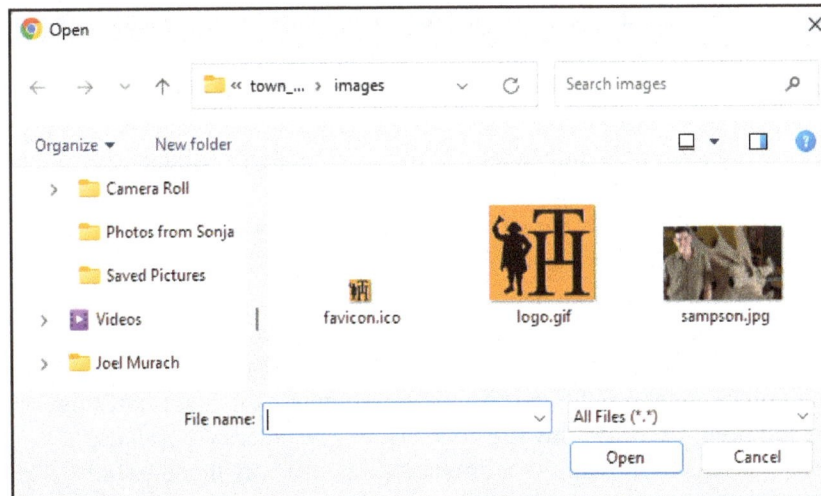

Description

- To create a *file upload control*, set the type attribute of the <input> element to file.
- The <form> element that contains the file upload control must have its enctype attribute set to multipart/form-data and its method attribute set to post.

Figure 12-14   How to use the file upload control

How to use HTML for data validation

Now, it's on to the HTML features for *data validation*. These features let the browser automatically validate some of the data that the user enters into a form without using client-side or server-side scripting languages.

The HTML attributes and CSS selectors

Figure 12-15 shows three attributes that HTML provides for data validation. By default, the autocomplete attribute is on in all modern browsers. As a result, by default, a browser uses *auto-completion* to display a list of entry options when the user starts the entry for a control. These options are based on the entries the user has previously made for controls with similar names or ids.

If you don't want the browser to use this feature, you can set the autocomplete attribute to off for an entire form or for one or more controls. For instance, you may want to turn auto-completion off for controls that accept credit card numbers or passwords. In this figure, the example turns auto-completion off for the entire form.

The required attribute causes the browser to check whether a control is empty when the user submits the form. If the control is empty, the browser displays a message like the first one shown in this figure. However, the exact wording and appearance of the message may vary from one browser to another.

When a semantic control is required, the browser may also validate it to make sure it contains valid data for that type of control. For instance, if the text in an email control isn't a valid email address, the browser may display a message like the second one shown in this figure.

If you want to stop the controls on a form from being validated, you can code the novalidate attribute for the form. Then, you can still use the required attribute to indicate which controls are required. However, the browser doesn't automatically validate any of these controls when the user submits the form. Instead, you typically use a client-side language like JavaScript to validate the controls.

Using JavaScript gives you more control over how the browser performs the data validation. For example, instead of displaying one validation message at a time as shown in this figure, which can be irritating, you can display all of the validation messages at once when the user submits the form. This is less intrusive and generally leads to a better experience for the user.

If you want, you can use the CSS pseudo-classes shown at the bottom of this figure to format required, valid, and invalid controls. For instance, you can use the :required pseudo-class to format all required controls in a way that makes it clear that they are required.

The HTML attributes for data validation

Attribute	Description
autocomplete	Set this attribute to off to tell the browser to disable auto-completion. This can be coded for a form or a control.
required	A Boolean attribute indicates that a value is required for a control.
novalidate	A Boolean attribute tells the browser that it shouldn't validate the controls on the form. You can code this attribute for the form when you want to use JavaScript to customize the way data validation works.

A form that uses some of the validation attributes

```html
<form action="submit.html" method="post" autocomplete="off">
    <label for="email">Email:</label>
    <input type="email" id="email" name="email" required>

    <label for="name">Name: </label>
    <input type="text" id="name" name="name" required>

    <button type="submit">Submit</button>
</form>
```

How it looks in a browser

When a required field isn't entered

When a field contains invalid data

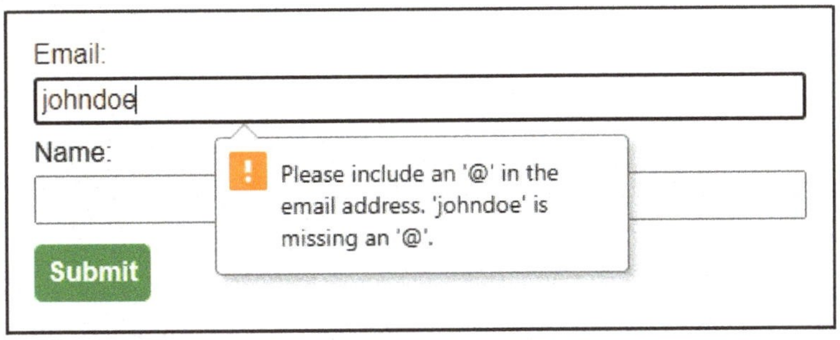

The CSS pseudo-classes for required, valid, and invalid fields

```
:required
:valid
:invalid
```

Figure 12-15   The HTML attributes and CSS selectors

How to use regular expressions

A *regular expression* (or *regex* for short) defines a pattern of characters that can be searched for in a string. As a result, regular expressions can be used for validating user entries that have a standard pattern, such as credit card numbers, zip codes, dates, phone numbers, URLs, and more. Regular expressions are supported by many programming languages including HTML.

As figure 12-16 shows, HTML provides a pattern attribute that specifies the regular expression used to validate the entry for the control. In the example, a regular expression specifies the pattern for a text box that accepts a phone number. As a result, the user must enter a phone number that has 3 digits, a hyphen, 3 more digits, another hyphen, and 4 more digits. If this control doesn't match the pattern when the user clicks the submit button, the browser displays an error message like the one shown in this figure, and it doesn't submit the form.

If you code a title attribute for a control that is validated by a regular expression, the browser displays the title when the mouse hovers over the control. The browser also displays the title at the end of its standard error message when the regular expression doesn't match as shown in this figure. However, the exact wording and appearance of the error message may vary from browser to browser.

The trick to using regular expressions is coding the regular expression that you need, and that can be difficult. Fortunately, you can often find the regular expression that you need by searching the web. Also, an AI tool like ChatGPT can often generate the regular expression you need if you give it a good prompt. In either case, once you have a regular expression, you should test it thoroughly to make sure it works the way you want.

Attributes for using regular expressions

Attribute	Description
pattern	The regular expression that provides the pattern used to validate the entry.
title	Text that displays in the tooltip when the mouse hovers over a control. This text also displays after the browser's error message when validation fails.

Regular expressions for common entries

Used for	Expression
Password (8+ alphanumeric)	[a-zA-Z0-9]{8,}
Zip code (99999 or 99999-9999)	\d{5}([\-]\d{4})?
Phone number (999-999-9999)	\d{3}[\-]\d{3}[\-]\d{4}
Date (MM/DD/YYYY)	[01]?\d\/[0-3]\d\/\d{4}
URL (starting with http:// or https://)	https?://.+
Credit card (9999-9999-9999-9999)	^\d{4}-\d{4}-\d{4}-\d{4}$

A form that uses regular expressions

```
<form action="submit.html" method="post" autocomplete="off">
    <label for="phone">Phone:</label>
    <input type="tel" id="phone" name="phone" required
        pattern="\d{3}[\-]\d{3}[\-]\d{4}"
        title="Must be 999-999-9999">
    <button type="submit">Submit</button>
</form>
```

How it looks in a browser when validation fails

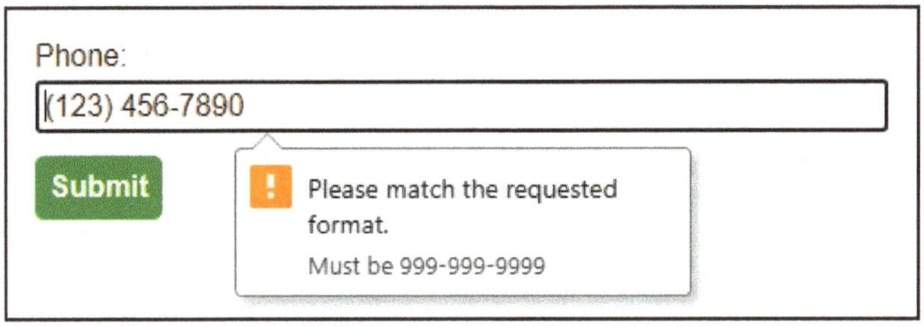

Phone:
(123) 456-7890

Submit   ! Please match the requested format.
             Must be 999-999-9999

Description

- A *regular expression* (or *regex* for short) defines a pattern that can be searched for in a string.
- To learn how to code regular expressions, you can search the web or refer to our Modern JavaScript book.

Figure 12-16   How to use regular expressions

A web page with a form

Figure 12-17 presents a web page from the Town Hall website that uses a form. This shows how you can use the skills presented in this chapter to add a form to a web page.

The web page

Part 1 shows a web page that contains a form that uses some of the controls presented in this chapter. This includes text boxes that get a first and last name, an email control that gets an email address, a drop-down list that selects the membership type, and two password fields.

This form uses a fluid and responsive design, so it works well for devices with large screens like desktop computers as well as devices with small screens like mobile phones. For example, on a small screen, the controls stretch across the width of the entire screen. This makes it easy for users to select the controls with either thumb.

For desktop users, the buttons provide visual feedback to users when the mouse hovers over them. In this figure, for example, the mouse is hovering over the first button in the first screen.

This form also uses automatic data validation. In this figure, for example, the browser is displaying a message that indicates that the second control in the first screen is required. In addition, this form changes the color of the border for required controls to red, and it changes the color of the border for valid controls to green.

A web page with a form at large and small screen sizes

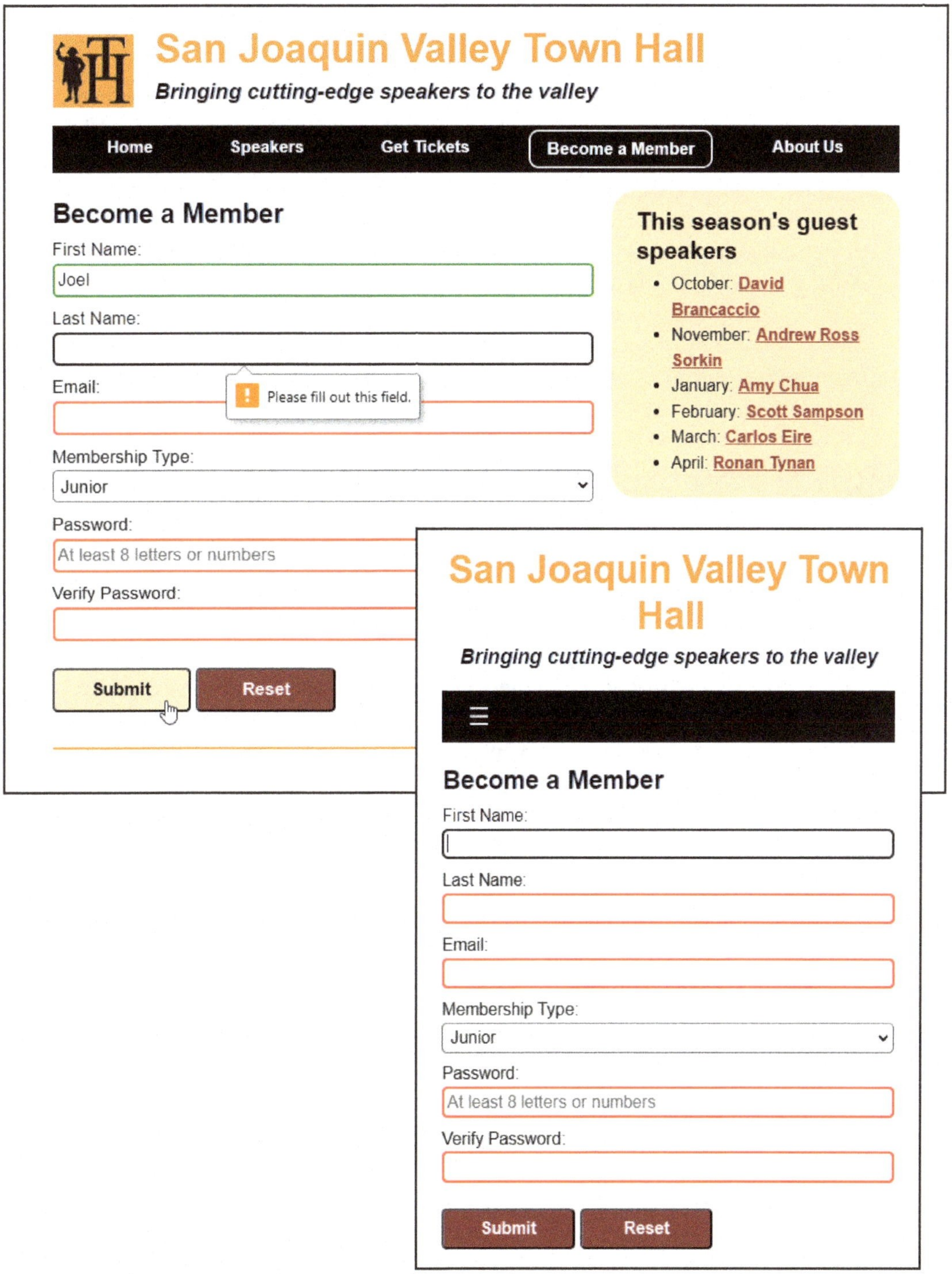

Figure 12-17   A web page with a form (part 1)

The HTML for the form

Part 2 of figure 12-17 presents the HTML for the form. The rest of the HTML for this page works like the HTML presented in earlier chapters.

The <form> element specifies that this form uses the HTTP POST method. As a result, when the user clicks the Submit button, the browser uses the HTTP POST method to send the data in the form to the server. In this figure, the action attribute is set to a URL for an HTML page named register_account.html. This page just displays a message and doesn't actually process the data. However, if you wanted to process the data, you could change the value of the action attribute to a URL for an application on the server.

Within the form, the first two controls are text boxes that get a first and last name. Then, the third control is an email control that gets the email address. The fourth control is a drop-down list that selects the membership type. And the last two text boxes are password fields that get and verify a password.

This form uses HTML data validation features. To do that, it includes the required attribute for all of the controls except the drop-down list. As a result, if the user doesn't enter values for these controls, the browser automatically displays an error message when the user submits the form. In addition, if the user doesn't enter an email address that's in a valid format, the browser displays an appropriate error message. Finally, the password control uses a regular expression to make sure that the user enters a password with at least eight alphanumeric characters.

This HTML doesn't do a complete job of data validation. For instance, HTML doesn't provide a way to make sure that the second password is the same as the first password. That's why JavaScript is commonly used to perform client-side data validation. For instance, JavaScript could check that the second password is the same as the first.

Although form data should be validated on the client with HTML or JavaScript, it should also always be validated on the server with server-side code. This code can validate data more thoroughly than is possible with HTML. As a result, the data validation on the client doesn't always have to be completely thorough. In fact, the main benefit of client-side validation is that it saves some of the round trips to the server that would be required if the validation was only done by the server. And if you use the HTML features for data validation, you will certainly save many round trips.

The HTML for the form

```
<form action="register_account.html" method="post">
    <label for="first_name">First Name:</label>
    <input type="text" name="first_name" id="first_name" autofocus required>

    <label for="last_name">Last Name:</label>
    <input type="text" name="last_name" id="last_name" required>

    <label for="email">Email:</label>
    <input type="email" name="email" id="email" required>

    <label for="membership_type">Membership Type:</label>
    <select name="membership_type" id="membership_type">
        <option value="junior">Junior</option>
        <option value="regular">Regular</option>
        <option value="charter">Charter</option>
    </select>

    <label for="password">Password:</label>
    <input type="password" name="password" id="password" required
            placeholder="At least 8 letters or numbers"
            pattern="[a-zA-Z0-9]{8,}"
            title="Must be at least 8 alphanumeric characters">

    <label for="verify">Verify Password:</label>
    <input type="password" name="verify" id="verify" required>

    <div>
        <button type="submit">Submit</button>
        <button type="reset">Reset</button>
    </div>
</form>
```

Figure 12-17   A web page with a form (part 2)

The CSS for the form

Part 3 of figure 12-17 presents the CSS for the form. By now, you should understand how most of this CSS works.

However, note that the style rule for the <button> element sets the background color to the dark color variable that's used by the rest of the web site. In addition, if the mouse hovers over the button, the nested style rule sets the background color to the light color variable that's used by the rest of the site.

Also, this CSS uses the :required and :valid pseudo-classes. To start, the fifth style rule uses the :required pseudo-class to display a 2-pixel red border around any <input> elements that are required. Similarly, the sixth style rule uses the :valid pseudo-class to display a 2-pixel green border around any valid <input> elements. For this code to work correctly, you must code the style rule for the :valid pseudo-class after the style rule for the :required pseudo-class. Otherwise, the formatting for the required controls overrides the formatting for the valid controls.

The CSS for the form

```
input, select, button {
    font-size: inherit;
    padding: .2em;
    margin-bottom: .5em;
    border-radius: 5px;
}
label {
    display: block;
}
input, select {
    width: 100%;
}
button {
    cursor: pointer;
    width: 8em;
    padding: .5em;
    margin-top: 1em;
    font-weight: bold;
    color: white;
    background-color: var(--dark);
    &:hover {
        color: black;
        background-color: var(--light);
    }
}
input:required {
    border: 2px solid red;
}
input:valid {
    border: 2px solid green;
}
```

Figure 12-17   A web page with a form (part 3)

Perspective

Now that you've completed this chapter, you should have the HTML and CSS skills you need to define and format a form. You should also be able to use the HTML features for data validation.

Terms

form	reset button
control	JavaScript button
label	tab order
text box	access key
text field	number control
password field	email control
hidden field	url control
text area	tel control
check box	date and time controls
radio button	file upload control
drop-down list	data validation
list box	auto-completion
button	regular expression
submit button	regex

Exercise 12-1   Create a form for ordering tickets

In this exercise, you'll enhance the tickets page of the Town Hall website by adding a form that looks like this:

Open the HTML and CSS files

1. Use your text editor to open the HTML and CSS files for the tickets page:

   ```
   exercises\ch12\town_hall\tickets.html
   exercises\ch12\town_hall\styles\main.css
   ```

2. Review the HTML file. Note that it includes a <form> element with a <fieldset> element that contains some labels and controls.

3. Review the CSS file. Note that it contains some basic styles for <label>, <input>, <select>, <fieldset>, and <legend> elements.

4. View the web page in a browser at different widths.

Enhance the Member Information controls

5. In the HTML file, set the autofocus to the text box for the first name.

6. Modify the controls for the email, first name, and last name so they are required.

7. Modify the control for the phone number so it uses a placeholder value of 999-999-9999. Then, use the pattern attribute to create a regular expression that validates that the user input follows this format, and use the title attribute to describe the required format.

Add the Ordering Information controls

8. Define a <fieldset> element and add a legend that says "Ordering Information".

9. Add a label and a drop-down list for the order type. The drop-down list should have three options: Base Package, Upgraded Package, and Premium Package.

10. Add a label and number control for the number of tickets. Specify a starting value of 1, a minimum value of 1, and a maximum value of 100.

Add the Payment controls and a submit button

11. Define a <fieldset> element and add a legend that says "Payment Information".

12. Add a label and a drop-down list for the card type. The drop-down list should have three options: Visa, Master Card, and Discover.

13. Add a label and text field for a credit card number. The text box should be required and use a pattern that accepts 16 digits. Add a placeholder that says "16 digits" and a title that says the entry must be 16 digits.

14. Add a label and a drop-down list for the expiration month. The drop-down list should include options for every month of the year.

15. Add a drop-down list for the expiration year. This list should include options for the current year plus the next two years.

16. Below the <fieldset> element, add a submit button for the form.

Style the form

17. In the CSS file, add a style rule that uses the :required and :invalid pseudo-classes of the <input> element to set a 2-pixel solid red border.

18. Add a style rule that uses the :valid pseudo-class of the <input> element to set a 1-pixel solid black border.

19. Set the width of the <select> elements for the expiration month and year to 49%. To do that, you can code a style rule that selects these elements by id.

20. Add a style rule for the <button> element that sets the
    - font weight to bold
    - width to 10em
    - background color to the accent color variable
    - text color to white
    - border radius to 6px
    - padding to .5em
    - bottom margin to 1em

21. Add a style rule for the :hover pseudo-class of the <button> element that sets the:
    - background color to the gradient color variable
    - text color to black
    - cursor to a pointer

22. View the page in a browser and adjust the CSS until it looks the way you want.

Test the form

23. View the page and click the submit button without entering any information. You should get an error message for the first required control in the form.

24. Fill out the form but enter invalid data, such as an email address without an @ sign, a phone number with 7 digits, "two" as the number of tickets, and a credit card number with 10 digits.

25. Click the submit button and review the error messages you get. Correct each invalid entry and click the submit button again until all the errors are corrected.

26. When you correct all of the entries and click the submit button, the browser should display a confirmation page. On that page, note that all the data you entered is at the end of the URL, after the ? character.

27. In the HTML, set the method attribute of the form to post. Then, view the tickets page again and submit valid data again. This time, note that the data isn't at the end of the URL.

Chapter 13

How to add audio and video

This chapter shows how to add audio and video to a website. But first, it starts with a brief introduction to the media types.

An introduction to media on the web

Before you learn how to include media files in your web pages, you need to be familiar with the media types that are used for audio and video. In addition, you may need to know how to convert one media type to another.

Common media types for audio and video

When most of us think about *media types*, a short list comes to mind: MP3, MP4, and maybe even FLAC. However, there are dozens of media types for both audio and video. Some of them are summarized in figure 13-1.

These media types are containers that contain content that can be played by *media players*. For example, an MP4 file contains a video track, which is what the user sees, and an audio track, which is what the user hears.

In addition to the content itself, a media type can contain metadata, such as the title of the video, an image related to the video, the length of the video, and digital rights management information.

Common media types for audio

Type	Description
MP3	One of the most widely-used media types for audio. It commonly uses an extension of mp3.
AAC	Advanced Audio Coding is an audio format that Apple uses with many of its products. AAC was designed to deliver better quality audio than MP3 at the same size.
FLAC	An audio format that compresses audio without losing quality. It's used to store high-quality audio files while reducing the size by up to 50%.
Ogg	Ogg is an open-source, open-standard media type. The audio stream of the Ogg media type is technically referred to as Vorbis.
WebM	A file format that usually has an extension of webm. It currently has native support in all modern browsers except Safari.

Common media types for video

Type	Description
MP4	One of the most widely used media types for video. It commonly uses an extension of mp4.
Ogg	Ogg is an open-source, open-standard media type. The video stream of the Ogg media type is technically referred to as Theora.
WebM	A file format that usually has an extension of webm. It currently has native support in all modern browsers except Safari.

Description

- A *media type* is a container for several components, including an encoded video track, one or more encoded audio tracks, and metadata.
- To play a media type, a browser requires a *media player* for that type.

Figure 13-1   Common media types for audio and video

How to convert from one media type to another

Occasionally, you may need to convert a media file from one type to another. For instance, you may need to convert an uncompressed video or audio file that you've captured on a digital recording device into a compressed format.

To do that type of conversion, you can use one of the many free or commercial software packages that are available. For instance, at the time of this writing CloudConvert, which is shown in figure 13-2, lets you convert a file from just about any media type to the type that you need for free. However, you rarely need to do that because modern browsers can recognize most media types and treat them as the MP3 or MP4 type.

A tool for converting audio and video files

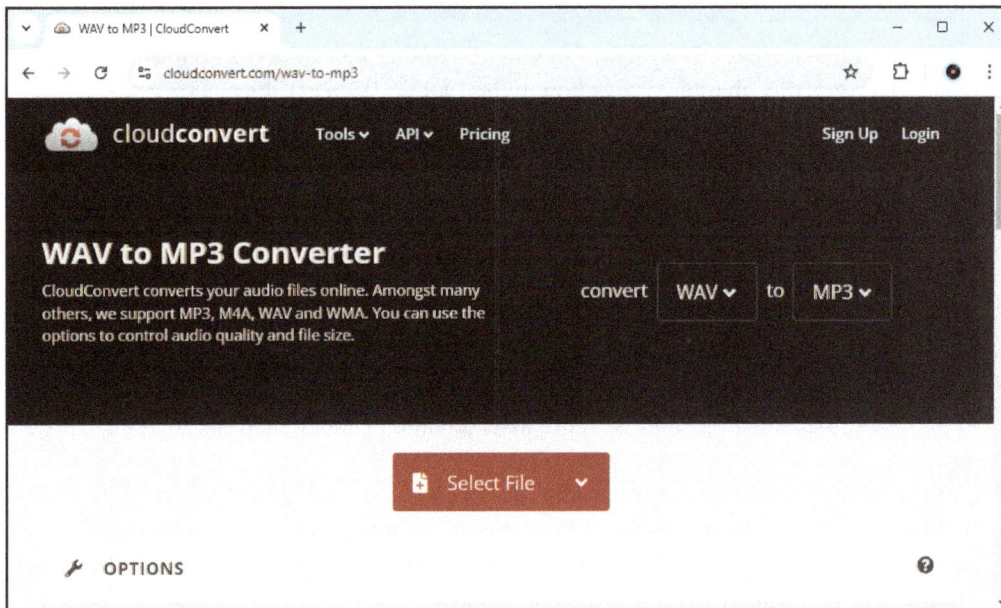

Description

- If necessary, you can use a free converter like CloudConvert to convert a file from one media type to another.
- Modern browsers can identify most media types and treat them as the MP3 or MP4 type.

Figure 13-2   How to convert from one media type to another

How to add audio and video

At this point, you're ready to learn how to add audio and video to your web pages. To start, this chapter shows how to add audio because it's easier than adding video and most of the skills for adding audio also apply to video.

How to add audio

To add audio to a web page, you can use the <audio> element as shown in figure 13-3. In the example, the <audio> element uses the src attribute to specify the URL for an MP3 file, and it uses the controls attribute to display controls for playing the audio. As a result, when the browser displays the web page, it displays the controls that allow the user to play the audio.

The <audio> element in this figure also sets the preload attribute to auto. This tells the browser that the author of the web page thinks it should automatically preload the entire media file, even if the user isn't likely to play that file. Although this can improve the responsiveness of the page, it also uses more bandwidth. If that's not what you want, you can set the preload attribute to metadata or not code the preload attribute at all.

When adding audio to a web page, remember that users with hearing disabilities might not be able to hear the audio. As a result, if the audio presents essential content, you should make your web page accessible by also including a transcript that accurately describes the audio.

Common attributes for media elements

Attribute	Description
src	The URL for the media file.
controls	Display the controls for playing the media.
loop	Loop the media so it plays again when it reaches the end. The default is false.
preload	One of three possible values that tell the browser whether to preload any data. Valid values include none (the default), metadata (only preload metadata like dimensions and track list), or auto (preload the entire media file).

HTML that adds audio

```
<audio controls src="media/sjv_welcome.mp3" preload="auto"></audio>
```

How it looks in a browser

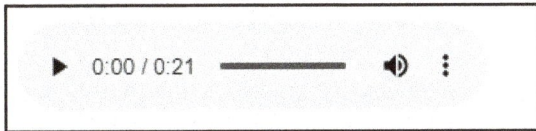

```
▶  0:00 / 0:21  ━━━━━━━━━  ◀))  ⋮
```

Description

- You can use the <audio> element to add an audio file to a web page.
- For accessibility, you should provide a transcript that accurately describes the contents of any audio files that present essential content.

Figure 13-3   How to add audio

How to add video

Adding video to a website works much like adding audio. However, you use a <video> element instead of an <audio> element, and you can specify some additional options like the ones described in figure 13-4.

The HTML in this figure adds a heading and an MP4 video file to the web page. The <video> element uses the src attribute to specify the URL for the video file. Then, it includes the controls attribute to display the controls for playing a video. Next, it uses the poster attribute to display an image instead of the video.

As a result, when the browser loads the page, it displays the image with the video controls over it. Then, if the user clicks the play button, the browser plays the video instead of displaying the image.

The CSS in this figure sets the width of the video to 100%. As a result, the browser adjusts the width of the video so it takes up 100% of its container while still maintaining its aspect ratio.

The third code example shows a <video> element that automatically plays a video. To do that, this <video> element includes the autoplay attribute and the muted attribute. In most modern browsers, this causes the video to automatically begin playing with the audio muted. If you don't include the muted attribute, most modern browsers won't automatically play the video, unless the user has explicitly enabled that feature. That's because most people find it intrusive when a video automatically plays with the audio on. However, it's less intrusive when a video automatically plays with the audio muted.

More common attributes for media elements

Attribute	Description
poster	The URL for an image that's displayed until the video file starts playing.
muted	Causes the video to begin playing with the audio muted. The default is false.
autoplay	Automatically plays the video when the page loads. The default is false. For most browsers, this only works when the audio is muted.

HTML for a heading and a video with a poster

```
<h1>Welcome to the San Joaquin Valley Town Hall</h1>
<video src="media/sjv_speakers_sampson.mp4" controls
        poster="images/poster.png"></video>
```

The CSS

```
video {
    width: 100%;
}
```

How it looks in a browser

A <video> element that automatically plays a video

```
<video src="media/sjv_speakers_sampson.mp4" controls
        autoplay muted></video>
```

Description

- You can use the <video> element to add a video file to a web page.
- For accessibility, videos should provide both captions and transcripts that accurately describe their content.

Figure 13-4   How to add video

How to provide alternate source files

If you're only using MP3 and MP4 files, you can use the skills presented in the previous two figures to add audio and video to your web pages. However, if you want to use media formats that aren't common, it's a best practice to provide alternate media files as shown in figure 13-5. To do that, you use the <audio> or <video> element to provide the settings for the audio or video. Then, you use one or more nested <source> elements to provide the URLs for the alternate source files.

The first example shows how this works for audio. Within the <audio> element, the <source> elements provide two possible source files. When the browser reads these elements, it starts at the top and tries to load the audio from the first <source> element, which is for an Ogg file. If this operation fails, the browser attempts to load the second source file, which is for an MP3 file.

The second example shows how this works for video. Within the <video> element, the <source> elements provide three possible source files (WebM, Ogg, and MP4). As a result, if the browser can't load the WebM or Ogg files, it attempts to load the MP4 file.

An <audio> element with two formats

```
<audio controls>
    <source src="media/sjv_welcome.ogg">
    <source src="media/sjv_welcome.mp3">
</audio>
```

A <video> element with three formats

```
<video controls autoplay muted>
    <source src="media/sjv_speakers_sampson.webm">
    <source src="media/sjv_speakers_sampson.ogv">
    <source src="media/sjv_speakers_sampson.mp4">
</video>
```

Description

- You can use multiple <source> elements to specify multiple media files for an <audio> or <video> element. If you provide multiple media formats, the browser uses the first format it supports.

Figure 13-5   How to provide alternate media formats

Perspective

Now that you know how to add audio and video to your web pages, please remember that users with hearing disabilities might not get the benefit of this audio or video unless you also provide transcripts or captions for your content. Fortunately, there are many good resources available from the web to help you with that.

Terms

media type
media player

Exercise 13-1   Add a video to a speaker page

In this exercise, you'll add a video to a speaker page on the Town Hall website. When you're through, the page should look like this:

San Joaquin Valley Town Hall
Celebrating our 75th Year

| Home | Speakers ▼ | Luncheons | Tickets | About Us ▼ |

Guest speakers

October
David Brancaccio

November
Andrew Ross Sorkin

January
Amy Chua

February
Scott Sampson

Fossil Threads in the Web of Life

0:00 / 0:31

February
Scott Sampson

What's 75 million years old and brand spanking new? A teenage Utahceratops! Come to the Saroyan, armed with your best dinosaur roar,

Add a video

1. Use your text editor to open the HTML and CSS files for this speaker page:

    `exercises\ch13\town_hall\speakers\sampson.html`
    `exercises\ch13\town_hall\styles\main.css`

2. In the HTML file, add the video with a poster in place until the video starts. To do that, you need to use a document-relative path that goes up one level, and you need to use the correct file names. For instance, the URL in the src attribute should be:

    `../media/sampson.mp4`

    and the URL in the poster attribute should be:

    `../images/sampson_dinosaur.jpg`

3. In the CSS file, add a style rule that formats the video so it's as wide as 90% of its container.

4. View this page and test it to make sure that the poster displays correctly and that the video plays correctly.

Debug the drop-down menu

5. In the navigation menu, hover the mouse over the Speakers item to drop it down, and move the mouse over the items that cover the video. Note that the drop-down menu disappears when you move the mouse over the items that cover the video.

6. In the CSS file, add the following declaration to the submenu class:

    `z-index: 1;`

7. Test the drop-down menu for the Speakers item again and confirm that it no longer disappears when you move the mouse over the items that cover the video.

Chapter 14

How to use grid layout

So far, this book has shown how to use the CSS flexbox layout module to lay out the structural elements of a web page. Now, this chapter shows to use another CSS module called grid layout. This module provides a way to develop complex page layouts that use a grid to lay out elements in columns and rows.

How to get started with grid layout

Grid layout, or *grid*, lets you lay out content in rows and columns. This chapter starts by presenting the basic skills that you need to do that.

An introduction to grid layout

The diagram in figure 14-1 shows the basic components of a grid layout. Here, the *grid container* consists of four rows and four columns. These rows and columns are referred to as *grid tracks*. The lines on either side of a grid track are referred to as *grid lines*. Two adjacent row and column grid lines form a *grid cell*. And two or more grid cells that form a rectangle can be combined into a *grid area*.

In simple layouts, each cell contains a *grid item*, which is a structural element like a <nav> or <div> element. In more complex layouts, you define the grid areas that contain the grid items. Although it's not shown here, grid tracks can also have space between them.

The components of a grid layout

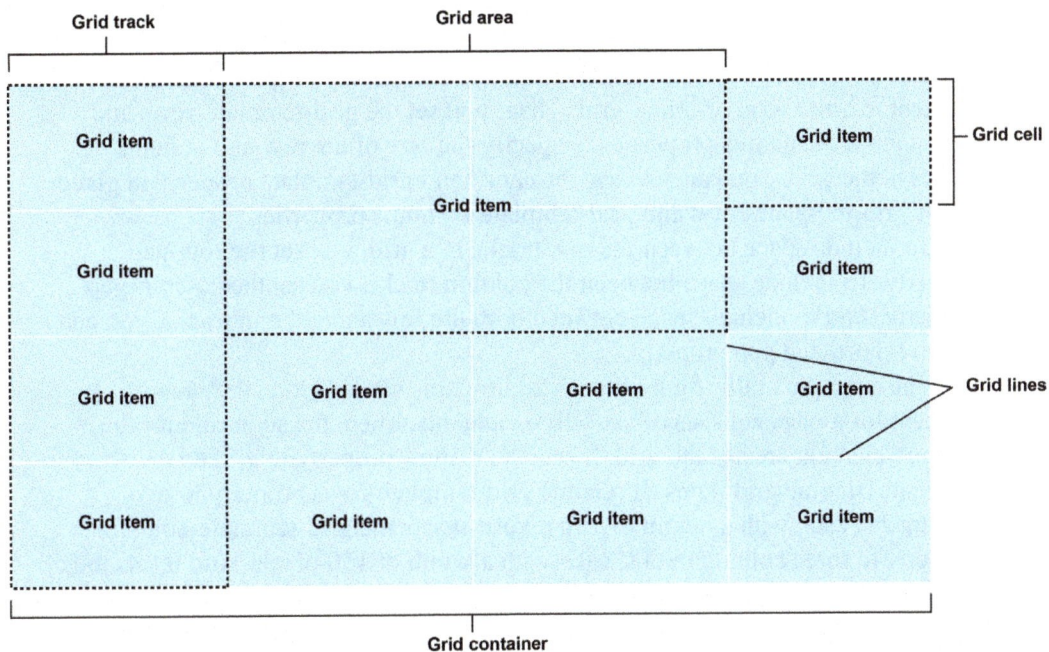

Grid layout terms

- A *grid* consists of a *grid container* with one or more columns and rows.
- A *grid track* is a column or row within the grid container.
- The lines on each side of the grid tracks are called *grid lines*.
- A *grid cell* is the space made up by two adjacent row and column grid lines.
- A *grid area* is a rectangular area that consists of one or more grid cells.

Description

- *Grid layout*, or *grid*, is a 2-dimensional layout module in CSS that allows you to lay out web pages in a more efficient and familiar way.
- In a simple layout, each cell of a grid can contain a *grid item*. In a more complex layout, you define *grid areas* that contain the grid items.

Figure 14-1   An introduction to grid layout

How to create a grid

To create a simple grid, you can set the properties presented in figure 14-2. To start, you create a grid container by setting the display property of a block element to either grid or inline-grid. Then, you set the grid-template-rows and grid-template-columns properties to specify the size of the row and column tracks in the grid. You can also use the shorthand grid-template property in place of the grid-template-rows and grid-template-columns properties.

To include space between the row tracks in a grid, you set the row-gap property. To include space between the column tracks, you set the column-gap property. And to include space between both the row and column tracks, you can use the shorthand gap property.

The example in this figure illustrates how this works. Here, the <main> element for a page consists of six <div> elements. Then, the style rule for the <main> element creates the grid. It sets the display property to grid so the <div> elements become grid items. It sets the grid-template-rows property to two row tracks, each with a height of 100 pixels. It sets the grid-template-columns property to three column tracks, each with a width of 150 pixels. And it sets the gap property to 15 pixels between rows and 20 pixels between columns.

If you want to set the sizes of the row and column tracks at the same time, you can use the shorthand grid-template property to do that. In that case, you specify the heights of the row tracks first, followed by a forward slash, followed by the widths of the column tracks. For example, if you wanted to use the grid-template property here, you'd code it like this:

```
grid-template: 100px 100px / 150px 150px 150px;
```

Similarly, if you want the row and column tracks to have the same space between them, you can code a single value that represents the space between the row and column tracks like this:

```
gap: 15px;
```

The layout in this figure shows that this HTML and CSS causes the browser to display a grid with two rows and three columns for a total of six cells. Then, the browser places the six <div> elements in these cells. The important thing to notice here is the order in which the <div> elements display. Specifically, they display from left to right and from top to bottom according to the order that they appear in the HTML.

Properties for creating a grid

Property	Description
display	The type of grid used for an HTML element: grid (rendered as a block) or inline-grid (rendered as inline content).
grid-template-rows	The size of the row tracks in a grid.
grid-template-columns	The size of the column tracks in a grid.
grid-template	The shorthand property for setting the grid-template-rows and grid-template-columns properties.
row-gap	The space between row tracks.
column-gap	The space between column tracks.
gap	The shorthand property for setting the row-gap and column-gap properties.

The HTML for a 2 row by 3 column grid

```
<main>
    <div>Div 1</div>
    <div>Div 2</div>
    <div>Div 3</div>
    <div>Div 4</div>
    <div>Div 5</div>
    <div>Div 6</div>
</main>
```

The CSS for laying out the grid

```
div { background-color: #C8DFEE; }
main {
    display: grid;
    grid-template-rows: 100px 100px;
    grid-template-columns: 150px 150px 150px;
    gap: 15px 20px;
}
```

The resulting layout

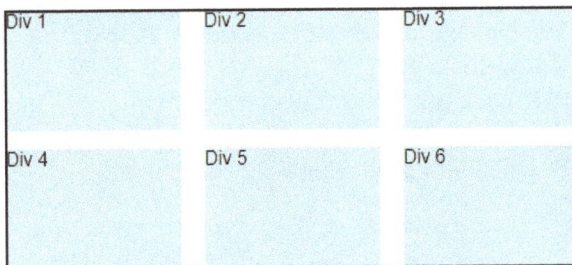

Description

- To define the rows and columns in a grid, you code one value for each row or column, and you separate the values with spaces. If you use the grid-template property, you code a forward slash (/) between the row and column values.
- By default, the grid items are laid out from left to right and top to bottom.

Figure 14-2   How to create a grid

How to set the size of grid tracks

To set the size of the row and column tracks in a grid, you can use any of the fixed or relative units of measure presented in chapter 4. However, when you use grid layout, you can also use the fr unit of measure to specify the size of grid tracks. This unit of measure represents a fraction of the space that's available within a container, and it's particularly useful for allocating space for fluid layouts.

When you use grid layout, you can use functions to specify the size of the grid tracks. For example, you can use the repeat() function to repeat one or more track sizes the specified number of times, and you can use the minmax() function to specify the minimum and maximum sizes for a grid track.

All of the examples in figure 14-3 use the same HTML presented in the previous figure. In addition, they all set the background color of the <div> elements as shown in the previous figure.

After that, all of the examples in this figure use the CSS that's presented in the first example. However, they change the value of the grid-template property to show how you can use it to set the size of the grid tracks.

The second example in this figure shows how the repeat() function and the fr unit of measure work. Here, the first repeat() function indicates that the grid contains two row tracks, each with a height of 50 pixels. Then, the second repeat() function indicates that the grid contains three column tracks, each with a width of 1fr. As a result, each of the columns occupies 1/3 of the available space within the container.

If, for example, the width of the container is 640 pixels, 600 pixels are left after the two 20-pixel column gaps are subtracted from 640. Then, the 600 is divided by 3, which makes each column track 200 pixels wide.

The third example shows that you can combine fixed, relative, and fractional track sizes. Here, the first column track is fixed at 100 pixels. The second column track is set to 30% of the container width, or 192 pixels using a 640-pixel container. And the third column track is set to 1fr, which represents the remaining space within the container, or 308 pixels.

The fourth example shows that you can use the repeat() function to repeat more than one column track. Specifically, this repeat() function repeats two column tracks that are 50 pixels and 1fr wide two times. This creates four column tracks. The first and third column tracks are 50 pixels wide, and the second and fourth column tracks each take half of the remaining space in the container.

A unit of measure for specifying the size of grid tracks

Unit	Description
fr	Represents a fraction of the available space in a grid container. Use this unit when building fluid grid layouts or layouts with a combination of fixed and fluid tracks.

Functions for specifying the size of one or more grid tracks

Function	Description
repeat(repeat, track-list)	Repeats one or more track sizes in the track list. The repeat value can be a positive integer or the keywords auto-fill or auto-fit.
minmax(min, max)	The minimum and maximum size for a grid track.

CSS used by all the examples in this figure

```
main {
    display: grid;
    grid-template: /* grid template specification goes here */;
    gap: 20px;
}
```

Use the repeat() function with integers and fractional widths

```
grid-template: repeat(2, 50px) / repeat(3, 1fr);
```

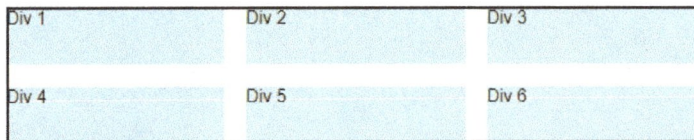

Div 1   Div 2   Div 3

Div 4   Div 5   Div 6

Use mixed units

```
grid-template: repeat(2, 50px) / 100px 30% 1fr;
```

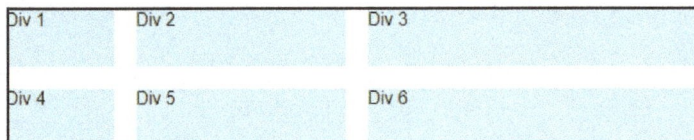

Div 1   Div 2   Div 3

Div 4   Div 5   Div 6

Repeat two columns

```
grid-template: repeat(2, 50px) / repeat(2, 50px 1fr);
```

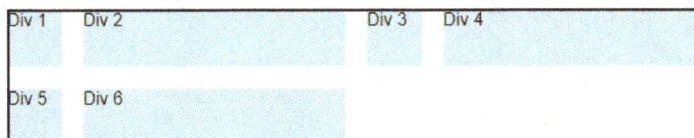

Div 1 Div 2   Div 3 Div 4

Div 5 Div 6

Description

- You can define the size of a grid track using any unit of measure, including the fr unit.

Figure 14-3   How to set the size of grid tracks (part 1)

Instead of specifying the number of tracks to create with the repeat() function, you can use the auto-fit or auto-fill keywords to determine how many tracks to create. Part 2 of this figure presents two examples that show how you can use these keywords with the minmax() function.

In both examples, the minmax() function causes the column tracks to be a minimum of 75 pixels wide and a maximum of 1 fr wide. That means that when the container becomes narrower than 550 pixels (6 column tracks at 75 pixels plus 5 column gaps at 20 pixels), the grid items roll over to the next row track. For instance, since the container in the first layout in each example is just 360 pixels wide, two of the grid items roll over to the second row track. In this case, the auto-fit and auto-fill keywords work the same way.

However, when the width of the container can hold more than the number of minimum-width columns, the auto-fit and the auto-fill keywords don't work the same way. In that case, the auto-fit keyword expands the width of the columns to fit the width of the container. This is shown by the second layout in the auto-fit example. By contrast, the auto-fill keyword expands the columns until the width of the container is large enough to hold another minimum-width column, and then provides the space for that column. This is shown by the second layout in the auto-fill example.

Since this is hard to visualize, you may want to run the example for this figure. That should help you decide which of these keywords you need to use. In most cases, the auto-fit keyword will probably give you the result that you're looking for.

Use auto-fit and the minmax() function with the repeat() function

`grid-template: repeat(2, 50px) / repeat(auto-fit, minmax(75px, 1fr));`

The layout when the container is 360 pixels wide

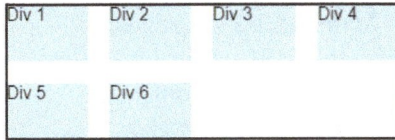

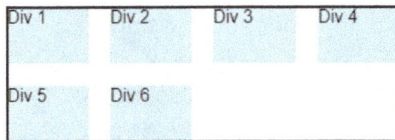

Div 1   Div 2   Div 3   Div 4

Div 5   Div 6

The layout when the container is 720 pixels wide

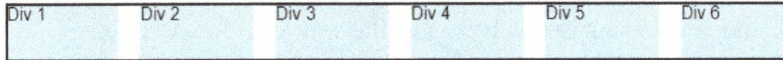

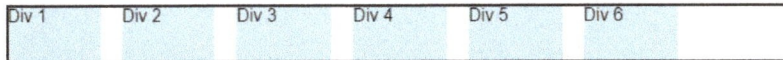

Div 1   Div 2   Div 3   Div 4   Div 5   Div 6

Use auto-fill and the minmax() function with the repeat() function

`grid-template: repeat(2, 50px) / repeat(auto-fill, minmax(75px, 1fr));`

The layout when the container is 360 pixels wide

Div 1   Div 2   Div 3   Div 4

Div 5   Div 6

The layout when the container is 720 pixels wide

Div 1   Div 2   Div 3   Div 4   Div 5   Div 6

Description

- The minmax() function is typically used within the track list of the repeat() function to specify the minimum and maximum sizes of one or more tracks.
- When you use the auto-fit keyword on the repeat() function, as many grid items as will fit in the available space will be placed in the container and they will be expanded when necessary to fit the available space.
- The auto-fill keyword works like the auto-fit keyword until the available space is large enough to provide for additional minimum-sized grid items. Then, this keyword provides enough space for those items.

Figure 14-3   How to set the size of grid tracks (part 2)

The properties for aligning grid items and tracks

When you create a grid, you can align the column tracks it contains horizontally if the tracks are less than the width of the container, and you can align the row tracks vertically if the tracks are less than the height of the container. You can also align the grid items within a grid area both horizontally and vertically.

The first two tables in figure 14-4 present the properties for horizontal and vertical alignment. To align grid tracks, you can set the justify-content and align-content properties. These properties can have any of the values shown in the third table in this figure.

First, the start, end, and center values work just the way you'd expect. For horizontal alignment, start aligns the grid tracks at the left side of the container, end aligns the grid tracks at the right side of the container, and center aligns the grid tracks in the center of the container. Similarly, for vertical alignment, start aligns the grid tracks at the top of the container, end aligns the grid tracks at the bottom of the container, and center aligns the grid tracks in the center of the container. The default for both horizontal and vertical alignment of grid tracks is start.

If you specify stretch for the justify-content property, the grid tracks are stretched so they extend from the left side of the container to the right side of the container. If you specify stretch for the align-content property, the grid tracks are stretched so they extend from the top of the container to the bottom of the container.

The last three values provide for spacing out the grid tracks within a grid container. If you specify space-between, any unused space is allocated evenly between the grid tracks. Space is also allocated evenly between the grid tracks when you specify space-around, but half-size spaces are added before the first grid track and after the last grid track. And with space-evenly, full-size spaces are added before the first grid track and after the last grid track. Note that when you use these values, you won't typically include a gap between the grid tracks.

To align the grid items within a grid area, you can set the justify-items, justify-self, align-items, and align-self properties. These properties can have the values start, end, center, and stretch. The justify-items and align-items properties align all the grid items within a grid area, and the justify-self and align-self properties override the alignment of specific grid items within a grid area.

Properties for horizontal alignment

Property	Description
justify-content	Horizontally aligns grid tracks within a container when the tracks are less than the overall width of the container. The default is start.
justify-items	Horizontally aligns grid items within a grid area. The default is stretch.
justify-self	Overrides the grid area's horizontal alignment for an individual grid item.

Properties for vertical alignment

Property	Description
align-content	Vertically aligns grid tracks within a container when the tracks are less than the overall height of the container. The default is start.
align-items	Vertically aligns grid items within a grid area. The default is stretch.
align-self	Overrides the grid area's vertical alignment for an individual grid item.

Common values for these properties

Value	Description
start	Aligns grid items or grid tracks at the beginning of the container.
end	Aligns grid items or grid tracks at the end of the container.
center	Aligns grid items or grid tracks in the center of the container.
stretch	The grid items or grid tracks extend from the start of the container to the end of the container.
space-between	Allocates space evenly between grid tracks.
space-around	Allocates space evenly between grid tracks with half-size spaces before the first grid track and after the last grid track.
space-evenly	Allocates space evenly between grid tracks with full-size spaces before the first grid track and after the last grid track.

Description

- You can align the column tracks within a container horizontally, and you can align the row tracks vertically.
- You can also align the grid items within a grid area horizontally and vertically, and you can override a grid area's horizontal and vertical alignment for individual grid items.

Figure 14-4   The properties for aligning grid items and grid tracks

A page layout that uses alignment

To illustrate how alignment works, part 1 of figure 14-5 presents a page layout that uses both horizontal and vertical alignment. Here, the HTML for the page shows that the body contains four structural elements: a <nav> element that displays a logo and a navigation menu, a <section> element that displays an image, a <main> element that displays three products, and a <footer> element with a copyright notice.

A page layout that uses alignment

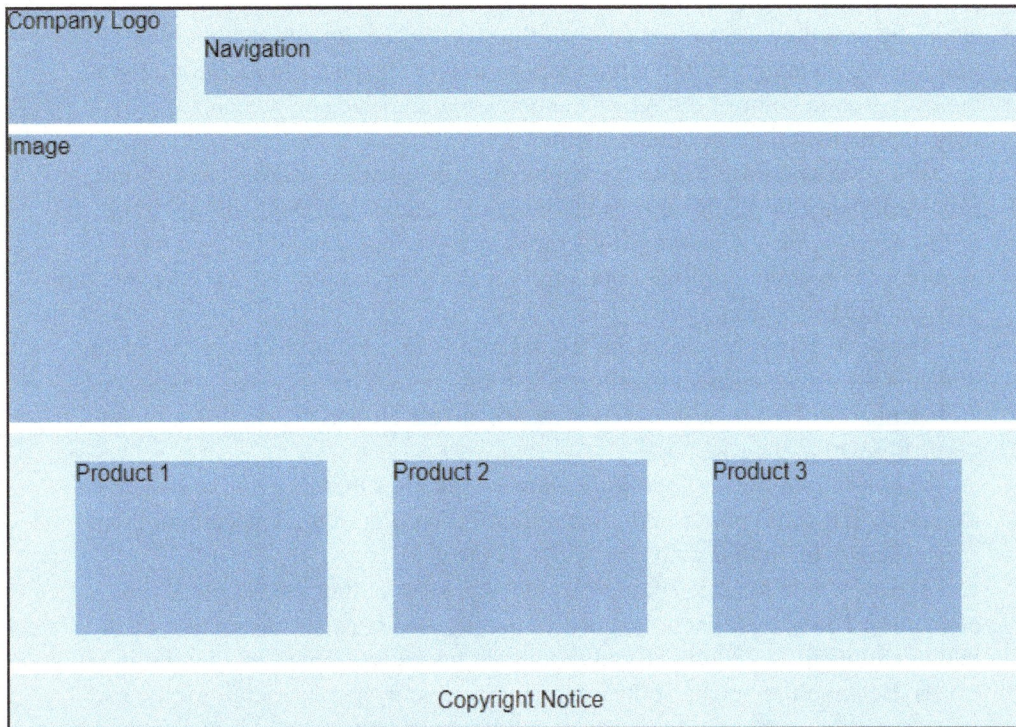

The HTML for the body of the page

```
<body>
    <nav>
        <div>Company Logo</div>
        <div>Navigation</div>
    </nav>
    <section>
        <div>Image</div>
    </section>
    <main>
        <div>Product 1</div>
        <div>Product 2</div>
        <div>Product 3</div>
    </main>
    <footer>
        <p>Copyright Notice</p>
    </footer>
</body>
```

Figure 14-5   A page layout that uses alignment (part 1)

Part 2 of figure 14-5 shows the CSS for this page with all four of these structural elements as grid containers that occupy 100% of the width of the body. This CSS sets the background color of the structural elements to light blue so you can see them against the white background of the page. In addition, it sets the background color of the <div> elements to dark blue so you can see where they appear within the structural elements.

The style rule for the <nav> element specifies that its grid consists of one row track that's 80 pixels high and two column tracks. The first column track is 120 pixels wide and displays the company logo. The second column track occupies the remainder of the container's width (1fr), except for the 20-pixel gap between the two tracks.

The next style rule selects the second <div> element in the <nav> element, which is the <div> element for the navigation menu. Then, it sets the height of that element to 40 pixels, and it uses the align-self property to align the menu vertically within the row track.

The style rule for the <section> element specifies that its grid consists of one row track that's 200 pixels high. Since there's no column track specified, this creates one column that spans the width of the grid.

The style rule for the <main> element starts by setting the height of the element to 160 pixels. Then, it specifies that its grid consists of one row track that has a height of 120 pixels and three column tracks each with a width of 180 pixels. Because the height of the container is greater than the height of the row track, the align-content property centers the row track vertically in the container. In addition, the justify-content property is set to space-evenly so when the page displays at a width that's greater than the widths of the three column tracks, the column tracks are spaced evenly.

Finally, the style rule for the <footer> element sets its height to 40 pixels. Then, the align-items and justify-items properties center the grid item (the <p> element that contains the copyright notice) within the grid area for the footer.

The CSS for the page

```css
* {
    margin: 0;
    padding: 0;
}
body {
    width: 90%;
    margin: auto;
    font-family: sans-serif;
}
div {
    background: #6FA8CF;            /* dark blue */
}
nav, section, main, footer {
    display: grid;
    width: 100%;
    margin-bottom: 7px;
    background: #C8DFEE;            /* light blue */
}
nav {
    grid-template: 80px / 120px 1fr;
    gap: 20px;
}
nav div:nth-of-type(2) {
    height: 40px;
    align-self: center;            /* vertically aligns the menu */
}
section {
    grid-template-rows: 200px;
}
main {
    height: 160px;
    grid-template: 120px / repeat(3, 180px);
    align-content: center;         /* vertically centers the row track */
    justify-content: space-evenly; /* horizontally spaces items evenly */
}
footer {
    height: 40px;
    align-items: center;           /* vertically centers the content */
    justify-items: center;         /* horizontally centers the content */
}
```

Figure 14-5   A page layout that uses alignment (part 2)

How to define grid areas

Now that you know the basics of working with grid layout, you're ready to learn how to define the grid areas that you use to display elements. To do that, you can use three different techniques: numbered lines, named lines, and template areas. In addition, you can use numbered lines with a 12-column grid.

How to use numbered lines

When you define a grid using the grid-template-rows, grid-template-columns, and grid-template properties, each line in the grid is assigned a numeric value as shown by the diagram in figure 14-6. Here, the grid consists of three column tracks and four row tracks. Then, the leftmost grid line is assigned the number 1, the next grid line to the right is assigned the number 2, and so on. Similarly, the grid lines for the row tracks are numbered from 1 starting at the topmost grid line.

To position grid items using these numbered lines, you set the grid-row and grid-column properties. These properties indicate the starting and ending line numbers for the rows and columns that an element occupies. This is illustrated by the example in this figure. Although it isn't shown in this figure, the HTML for the structural elements looks like this:

```
<body>
    <header>Header</header>
    <nav>Navigation</nav>
    <section id="s1">Section 1</section>
    <section id="s2">Section 2</section>
    <section id="s3">Section 3</section>
    <footer>Footer</footer>
</body>
```

In the CSS, the style rule for the <body> element sets the properties that create the grid. In addition, it sets the minimum height of the body to 600px, the width of the body to 90%, and the margin to auto. Although you don't typically need to set the height of a grid, this example sets it because the body doesn't contain any content.

The CSS for the elements within the body sets the grid-row and grid-column properties that create the grid areas. For example, the grid-row property for the <header> element indicates that the header occupies the first row track (grid row line 1 to grid row line 2), and the grid-column property indicates that the header occupies all three column tracks (grid column line 1 to grid column line 4). Similarly, the <nav> element occupies the second and third row tracks (grid row line 2 to grid row line 4) within the first column track (grid column line 1 to grid column line 2).

The definition of the grid in the style rule for the <body> element sets the gap property to 16 pixels. However, the diagram shows that the gap is only added between the grid areas. In other words, gaps aren't added between cells within the same grid area.

Properties for using numbered lines to define grid areas

Property	Description
grid-row	The starting and ending line numbers for the rows of a grid area.
grid-column	The starting and ending line numbers for the columns of a grid area.

The numbered grid lines and HTML tags for a grid container

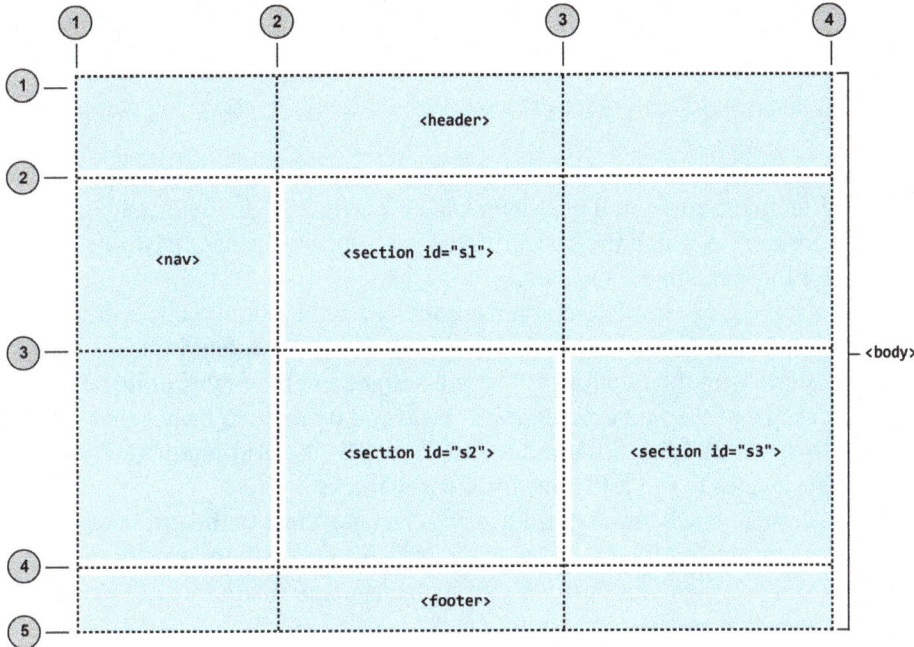

Use numbered lines to define the grid areas

```
body {
    display: grid;
    grid-template-columns: 200px 1fr 1fr;
    grid-template-rows: 80px 1fr 1fr 60px;
    gap: 16px;
    min-height: 600px;
    width: 90%;
    margin: auto;
}
header { grid-row: 1 / 2; grid-column: 1 / 4; }
nav    { grid-row: 2 / 4; grid-column: 1 / 2; }
#s1    { grid-row: 2 / 3; grid-column: 2 / 4; }
#s2    { grid-row: 3 / 4; grid-column: 2 / 3; }
#s3    { grid-row: 3 / 4; grid-column: 3 / 4; }
footer { grid-row: 4 / 5; grid-column: 1 / 4; }
```

Description

- When you define a grid, each grid line is assigned a numeric value. You can use those numbers to identify the rows and columns for a grid area that an element occupies.

Figure 14-6   How to use numbered lines to define the grid areas

How to use named lines

Figure 14-7 shows how to get similar results using named lines. Here, the diagram shows the names that have been assigned to each line in the grid. But note that multiple names have been assigned to some of the lines.

For example, the names body-start and nav-start have been assigned to the line at the left side of the first column track to indicate that it's at the start of the <body> and <nav> elements. Similarly, the names nav-end and sec-start have been assigned to the line between the first and second column tracks to indicate that it's at the end of the <nav> element and the start of the <section> element. This can make it easier to remember the names of the lines as you code your CSS.

To name a grid line, you code the name within brackets on the grid-template-columns, grid-template-rows, and grid-template properties. You code these names in the position in which they occur in the grid, and you separate two or more names for the same line with spaces.

To illustrate, the grid-template-columns property for the <body> element in this figure shows how to assign names to the grid lines for the column tracks. This property starts with the two names that are assigned to the first grid line, followed by the size of the first column track, followed by the two names that are assigned to the second grid line, and so on. Similarly, the grid-template-rows property assigns names to the grid lines for the row tracks.

After you assign names to the grid lines, you can use them in the grid-row and grid-column properties for each area in the grid. You code these properties the same way you do when you position grid items using numbered lines, except you use the names that you assigned to the lines.

In the example in this figure, the grid-row property for the header indicates that it occupies the row that starts at the grid line named row1-start and ends at the grid line named row2-start. Then, the grid-column property indicates that it occupies the columns that start at the grid line named body-start and end at the grid line named body-end. If you review the rest of the grid-row and grid-column properties, you should be able to figure out how they define the grid areas shown in this figure.

The named lines for a grid container

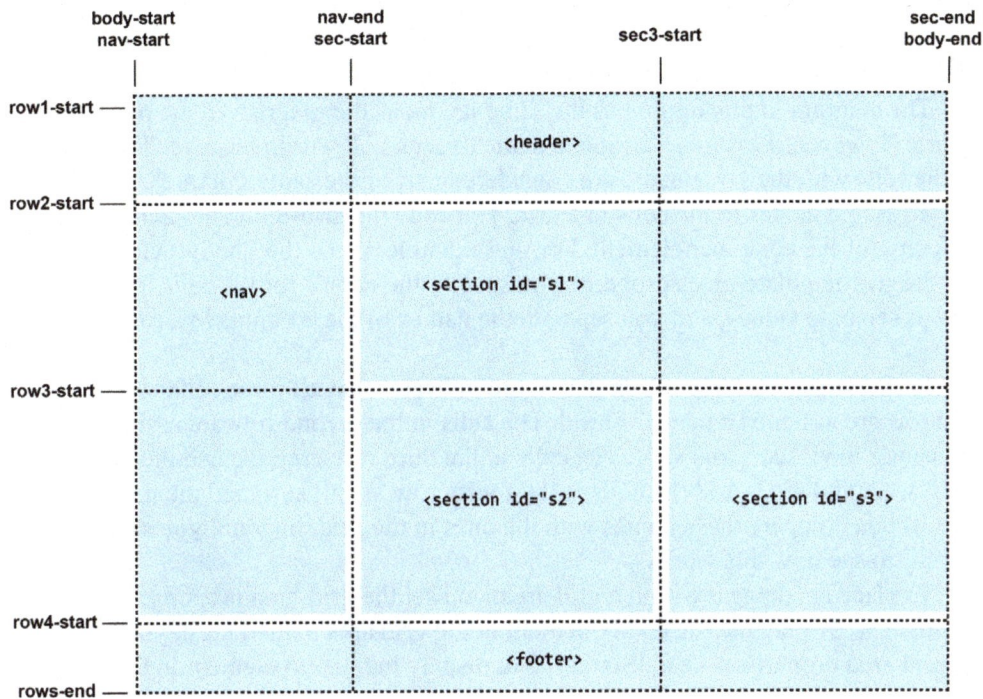

Use named lines to define the grid areas

```
body {
    display: grid;
    grid-template-columns: [body-start nav-start] 200px
        [nav-end sec-start] 1fr [sec3-start] 1fr [sec-end body-end];
    grid-template-rows: [row1-start] 80px [row2-start] 1fr
        [row3-start] 1fr [row4-start] 60px [rows-end];
    gap: 16px;
    ...
}
header { grid-row: row1-start/row2-start; grid-column: body-start/body-end; }
nav    { grid-row: row2-start/row4-start; grid-column: nav-start/nav-end; }
#s1    { grid-row: row2-start/row3-start; grid-column: sec-start/sec-end; }
#s2    { grid-row: row3-start/row4-start; grid-column: sec-start/sec3-start; }
#s3    { grid-row: row3-start/row4-start; grid-column: sec3-start/sec-end; }
footer { grid-row: row4-start/rows-end; grid-column: body-start/body-end; }
```

Description

- To name the lines in a grid, you code the name or names of each line in brackets on the grid-template-columns and grid-template-rows properties in the position they appear in the grid.

- To identify the rows and columns for a grid area that an element occupies, you code the line names on the grid-row and grid-column properties.

Figure 14-7   How to use named lines to define the grid areas

How to use template areas

Because numbered and named lines can be tedious to use, it's often easier to use template areas to position grid items. Figure 14-8 shows how this works.

The diagram in this figure has the same layout as the diagram in the previous figure. However, it assigns a template name to each cell within the grid. Then, all of the cells with the same name are considered part of the same grid area.

To assign names to the cells in a grid, you code the grid-template-areas property for the container element. For this example, that's the <body> element. On the grid-template-areas property, you specify the names for the cells in each row as separate values, and you separate the names of the columns in a row with spaces.

For example, the property in this figure indicates that all three cells in the first row are assigned a name of head. The cells in the second row are assigned the names navi, sec1, and sec1. The cells in the third row are assigned the names navi, sec2, and sec3. And the cells in the fourth row are all assigned the name foot. If you compare these names with the ones in the grid diagram, you should be able to see how this works.

To place an element within a grid area, you set the grid-area property. For example, to display the <header> element in the grid area named head, you set its grid-area property to head. Similarly, to display the <nav> element in the grid area named navi, you set its grid area property to navi. This makes it easier to tell where an element is positioned than when you use numbered or named lines.

When you use template areas, you should know that you don't have to assign a name to every cell in the grid. However, you do have to indicate the position of unnamed cells on the grid-template-areas property. To do that, you code a period as a placeholder for the cell. For example, if you didn't need to place a grid item in the cell in this figure named sec3, you could code the grid-template-areas property like this:

```
grid-template-areas:
    "head head head"
    "navi sec1 sec1"
    "navi sec2 ."
    "foot foot foot";
```

Properties for using template areas to define grid areas

Property	Description
grid-template-areas	Names the grid areas for the elements. A row is created for each string, and a column is created for each named cell in a string.
grid-area	Identifies the named grid area that an element occupies.

The template names for each cell within a grid container

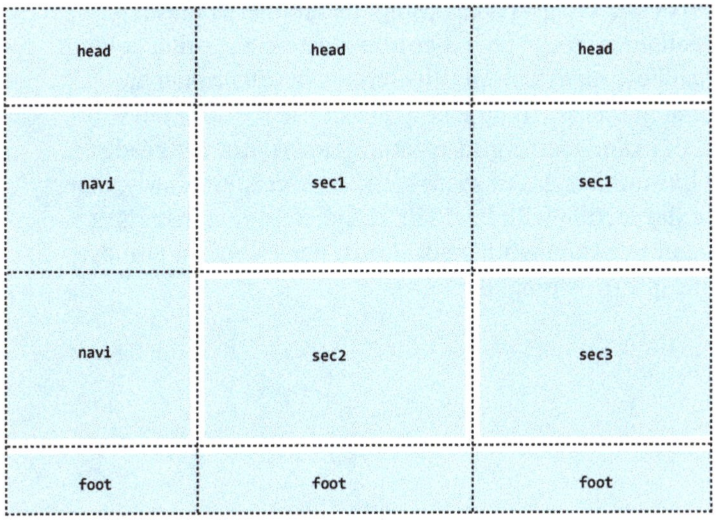

head	head	head
navi	sec1	sec1
navi	sec2	sec3
foot	foot	foot

Use template names to define the grid areas

```
body {
    display: grid;
    grid-template-columns: 200px 1fr 1fr;
    grid-template-rows: 80px 1fr 1fr 60px;
    gap: 16px;
    grid-template-areas:
        "head head head"
        "navi sec1 sec1"
        "navi sec2 sec3"
        "foot foot foot";
    ...
}
header { grid-area: head; }
nav    { grid-area: navi; }
#s1    { grid-area: sec1; }
#s2    { grid-area: sec2; }
#s3    { grid-area: sec3; }
footer { grid-area: foot; }
```

Description

- All of the cells with the same name form a grid area that spans those cells.
- If you don't want to include a cell in a grid area, you can code a period in place of a template name.

Figure 14-8   How to use template areas to define the grid areas

How to use the 12-column grid

Figure 14-6 showed how to position grid items using numbered lines. Now, figure 14-9 shows you how to use this technique to create a *12-column grid*, a popular layout that's used frequently with responsive web design.

A grid container that uses a 12-column layout consists of 12 proportionally-sized columns. To create these columns, you can use the repeat() function to create 12 columns each with a size of 1fr. Then, you can use the grid-row and grid-column properties to define the grid areas using the numbered lines.

When you use this technique, it can be easier to indicate the starting column number for a grid area and how many columns it consists of rather than the starting and ending column numbers. To do that, you can use the span keyword as shown in this figure. For example, the grid-column property for the header indicates that it starts at line number 1 and spans all 12 columns. Similarly, the grid-column property for the section with the id of s1 indicates that it starts at column 5 and spans 8 columns. Although it's not shown here, you can also use the span keyword with the grid-row property.

A 12-column grid for a container

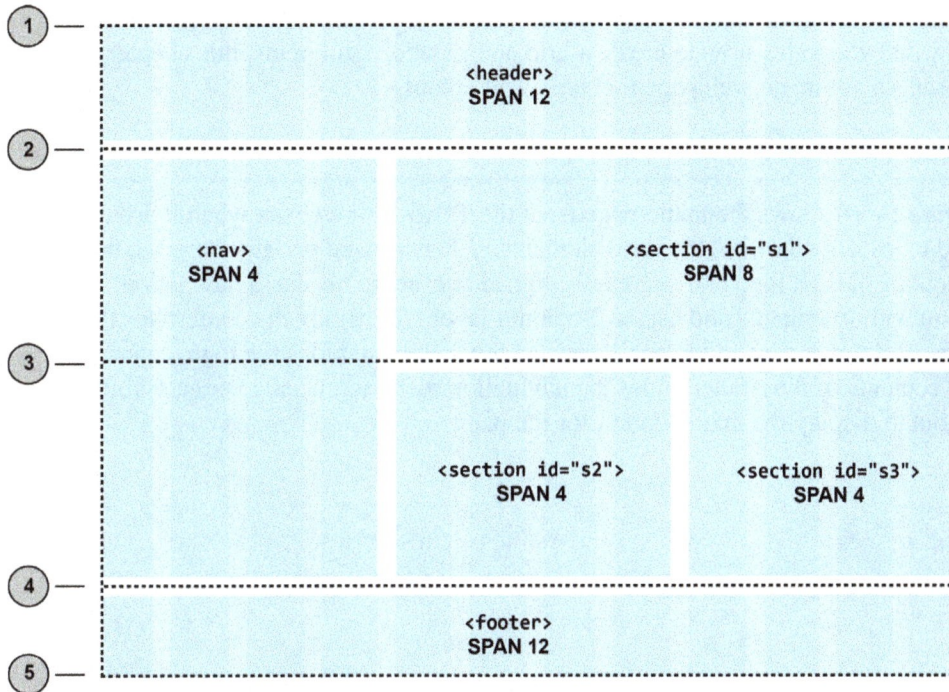

Use a 12-column grid to define the grid areas

```css
body {
    display: grid;
    grid-template-columns: repeat(12, 1fr);
    grid-template-rows: 80px 1fr 1fr 60px;
    gap: 16px;
    ...
}
header { grid-row: 1 / 2; grid-column: 1 / span 12; }
nav    { grid-row: 2 / 4; grid-column: 1 / span 4; }
#s1    { grid-row: 2 / 3; grid-column: 5 / span 8; }
#s2    { grid-row: 3 / 4; grid-column: 5 / span 4; }
#s3    { grid-row: 3 / 4; grid-column: 9 / span 4; }
footer { grid-row: 4 / 5; grid-column: 1 / span 12; }
```

Description

- The *12-column grid* is a popular layout that you can apply using numbered lines. To create a 12-column grid, you use the repeat() function to define 12 proportional columns.

- To identify the columns that an element occupies, you can use numbered lines. Instead of specifying the ending line number, you can use the span keyword to indicate how many columns an element spans.

Figure 14-9   How to use the 12-column grid

A web page that uses grid layout

Now that you know how to create a grid and position grid items, this chapter presents a complete web page that uses grid layout.

The web page

Figure 14-10 shows the home page from the Town Hall website when it uses grid to lay out all of its structural elements. When viewed on large screens, this layout displays a logo and some left-aligned text in the header, a navigation menu with five items, and uses a 2-column layout to display the content for the page. But when viewed on small screens, the web page hides the logo, centers the contents of the header, hides the navigation menu items, and uses a 1-column layout to display the main content for the page.

A web page that uses grid layout

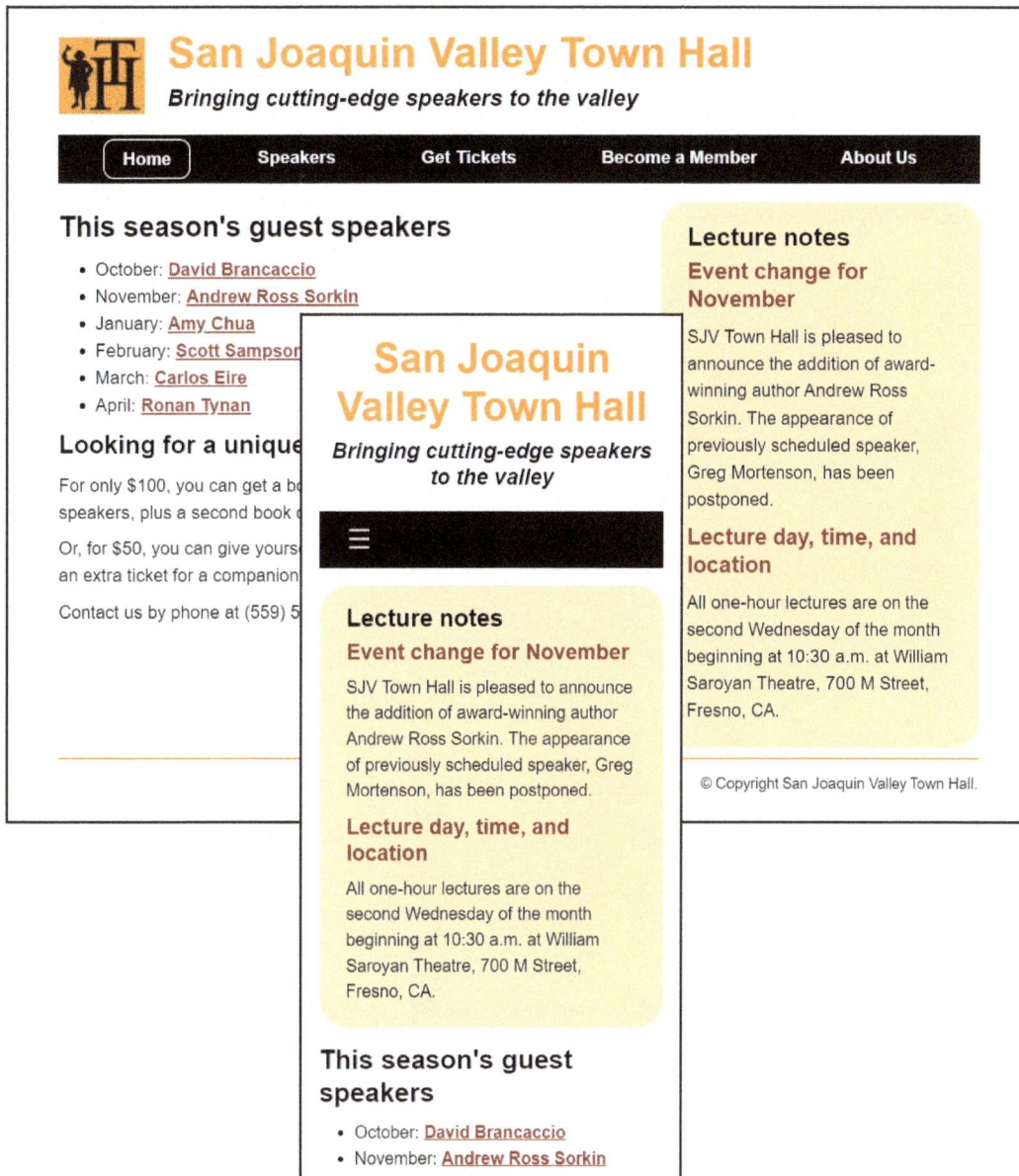

Description

- This web page uses grid layout to lay out all of the structural elements on the page.
- When viewed on small screens, the web page hides the logo, centers the contents of the header and footer, hides the navigation menu items, and uses a single column to display the main content for the page.

Figure 14-10   A web page that uses grid layout

The HTML for the structural elements

Figure 14-11 presents the HTML for the structural elements of the web page. This HTML is almost identical to the code for this page in chapter 8. However, to keep the focus on the use of grid layout, this page doesn't include submenus. As a result, the navigation menu consists of just five <li> elements.

The HTML for the structural elements

```
<body>
    <header>
        <img src="images/logo.gif" alt="Town Hall Logo">
        <h2>San Joaquin Valley Town Hall</h2>
        <h3>Bringing cutting-edge speakers to the valley</h3>
    </header>
    <nav class="navbar">
        <!-- hamburger menu button -->
        <input type="checkbox" id="checkbox-toggle" />
        <label for="checkbox-toggle" class="menu-btn"></label>

        <ul class="menu">
            <li><a class="current">Home</a></li>
            <li><a href="#">Speakers</a></li>
            <li><a href="#">Get Tickets</a></li>
            <li><a href="#">Become a Member</a></li>
            <li><a href="#">About Us</a></li>
        </ul>
    </nav>
    <main>
        <section>
            <h1>This season's guest speakers</h1>
                .
                .
            <h2>Looking for a unique gift?</h2>
                .
                .
        </section>
        <aside>
            <h2>Lecture notes</h2>
            <h3>Event change for November</h3>
                .
                .
            <h3>Lecture day, time, and location</h3>
                .
                .
        </aside>
    </main>
    <footer>
        <p>&copy; Copyright San Joaquin Valley Town Hall.</p>
    </footer>
</body>
```

Figure 14-11   The HTML for the structural elements

The mobile-first CSS for small screens

Figure 14-12 shows the mobile-first CSS for the grid template areas on the web page. To start, the style rule for the body includes the properties that lay out the main structural elements of the page using grid. That includes a display property that's set to grid and a grid-template property that defines a grid with four rows and one column. This code sets the height for all four rows to auto so the browser always makes them tall enough to contain their content. Then, the code specifies the width of the single column as 1fr so it takes up the full width of the <body> element.

The style rule for the body also includes a grid-template-areas property. This property assigns a name to each of the cells defined by the grid-template property. Here, the single cell in the first row is named head, the cell in the second row is named navi, and so on.

In the style rule for the <header> element, the grid-area property places the header in the grid area that's defined by the cell named head. Similarly, the grid-area property for the <footer> element places the footer in the grid area named foot, and the grid-area property for the navbar class places the navigation menu in the grid area named navi.

The style rule for the <main> element places the element in the grid area named main. Then, it uses the display property to make the <main> element a nested grid. Next, it uses the grid-template property to define two rows and one column within the nested grid. Here again, the code sets the height of the rows to auto so they'll be tall enough to contain their content. And it sets the width of the column to 1fr so it takes up the full width of the <main> element.

The style rule for the <main> element also includes a grid-template-areas property that defines template areas for the two cells in the column and names them side and sect. Then, the style rules for the <section> and <aside> elements use the grid-area property to place themselves in the grid areas named sect and side.

The mobile-first CSS for small screens

```
body {
    ...
    display: grid;                          /* make the body a grid */
    grid-template: repeat(4, auto) / 1fr;   /* define 4 rows and 1 column */
    grid-template-areas:                    /* define the template areas */
        "head"
        "navi"
        "main"
        "foot";
}
header {
    grid-area: head;             /* place the header in a grid area */
    ...
}
main {
    grid-area: main;             /* place the <main> element in a grid area */
    display: grid;               /* make the <main> element a nested grid */
    grid-template: auto auto / 1fr;   /* define 2 rows and 1 column */
    grid-template-areas:         /* define the template areas */
        "side"
        "sect";
}
section {
    grid-area: sect;         /* place the section element in a grid area */
    ...
}
aside {
    grid-area: side;         /* place the aside element in a grid area */
    ...
}
footer {
    grid-area: foot;         /* place the footer in a grid area*/
    ...
}
.navbar {
    grid-area: navi;         /* place the navigation bar in a grid area*/
    ...
}
```

Figure 14-12   The mobile-first CSS for small screens

The media queries for larger screens

Figure 14-13 shows the media queries for larger screens. The first one applies to screens that are 745 pixels wide or larger, and the second applies to screens that are 768 pixels wide or larger.

The first media query starts by displaying the header image that's hidden in the mobile-first CSS and changing the text alignment from center to left. Then, it changes the text alignment for the footer to right.

After adjusting the header and footer, the media query shows the menu and hides the menu button. To show the menu, it sets the display property to grid and sets the grid-template property to define a grid with one row and five columns. This code sets the height of the row and the width of the columns to auto so they are tall enough and wide enough to contain their content, and it uses the justify-content property to evenly space the items in the navigation menu.

The last style rule in this media query handles a potential bug that was described in chapter 8. With this bug, the #checkbox-toggle:checked ~ .menu selector sets the display property to block so the menu drop-down displays when the checkbox-toggle is checked. But if the page is resized with the checkbox still checked, the menu displays vertically instead of horizontally. To fix that, the media query needs to change the display property when the checkbox is checked. In chapter 8, it changed the value to flex, because that web page used flexbox for formatting. Here, it changes the display property to grid because this web page uses grid layout.

The second media query changes the layout of the grid in the <main> element from one column to two columns. To do that, it sets the grid-template property to create one row with two columns. This row automatically adjusts its height to fit the content, and the columns have widths of 1 fr and 35%. As a result, the second column is 35% as wide as the container element, and the first column uses the remaining width, not including the width of the column gap, which is 5% of the container.

The media queries for larger screens

```
@media only screen and (min-width: 745px) {
    /* change text alignment, display header image */
    header {
        text-align: left;
        & img {
            display: block;
            float: left;
            margin: .5em 1.25em .5em 0;
            width: clamp(40px, 10%, 80px);
        }
    }
    footer p { text-align: right; }

    /* show menu, adjust spacing, hide menu button */
    .menu {
        position: relative;
        display: grid;                            /* make the menu a grid */
        grid-template: auto / repeat(5, auto);    /* define 1 row and 5 columns */
        justify-content: space-evenly;            /* space columns evenly */
    }
    .menu-btn {
        display: none;
    }

    /* make sure menu displays horizontally if menu is expanded when resized*/
    #checkbox-toggle:checked ~ .menu {
        display: grid;
    }
}

@media only screen and (min-width: 768px) {
    /* make two columns */
    main {
        grid-template: auto / 1fr 35%;            /* define 1 row and 2 columns */
        grid-template-areas: "sect side";         /* define the template areas */
        gap: 5%;                                  /* define the gap between grid areas */
    }
}
```

Figure 14-13   The media queries for larger screens

Common page layouts that use grid

Figure 14-5 presented a common page layout that uses grid. Now, this chapter finishes by presenting three more common layouts that you can implement using grid. Although the code for these layouts isn't presented here, it's included in the download for this book along with two other common page layouts.

The headline and gallery layout

Part 1 of figure 14-14 presents a page layout called the *headline and gallery layout*. Like most layouts, it includes a company logo and navigation menu at the top of the page and a footer at the bottom of the page. In addition, it includes a headline and some text that describe the gallery of items that follows.

The entire body of this page is implemented as a grid with four rows and four columns. The first row includes the company logo and navigation menu, the second row includes the headline and text, the third row includes the gallery of items, and the fourth row includes the footer.

The first column is for the company logo, the second column is for the space between the company logo and the navigation menu, and the third and fourth columns are for the navigation menu. Four columns are used here so the width of the logo plus the space that follows are the same width as the navigation menu. Additional grids are used for the navigation menu and for the items within the gallery.

This type of layout can be used for any page that displays a series of items. For example, it can be used to display a gallery of images on a photographer's or museum's website. Or, it can be used on an ecommerce website where each item depicts a product or service the company offers.

The fixed sidebar layout

Part 1 of figure 14-14 also presents a layout called the *fixed sidebar layout*. Here, the sidebar contains a company logo and a navigation menu that is fixed at the left side of the window. Because of that, the sidebar is always visible even when the user scrolls down the page.

Unlike the headline and gallery layout, the body of a page that uses the fixed sidebar layout can't be implemented as a grid. That's because the sidebar uses fixed positioning and the rest of the page doesn't. Instead, the contents of the sidebar and the rest of the page are implemented as two separate grids.

In this case, the grid for the sidebar includes two rows and one column. The first row contains the company logo, and the second row contains the navigation menu. Then, the navigation menu itself is implemented as a grid with four rows and one column.

You can use the fixed sidebar layout for any page that requires a fixed sidebar. Then, you can simply swap in the page layout you want to use on the right side of the page. In this case, the layout is for a page that includes a title, a main image, three products, and a footer.

The headline and gallery layout

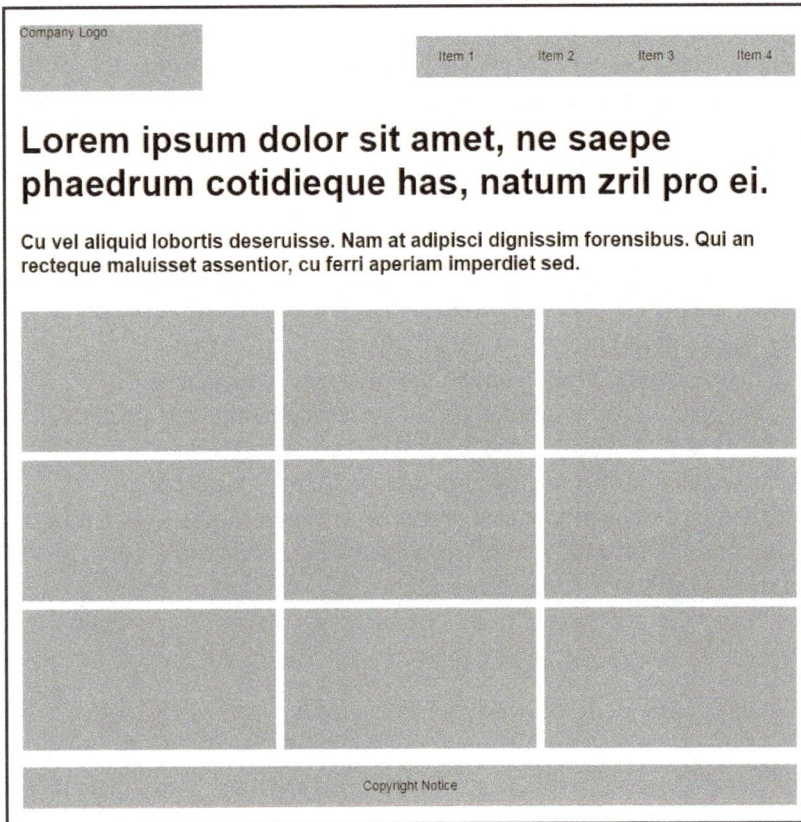

The fixed sidebar layout

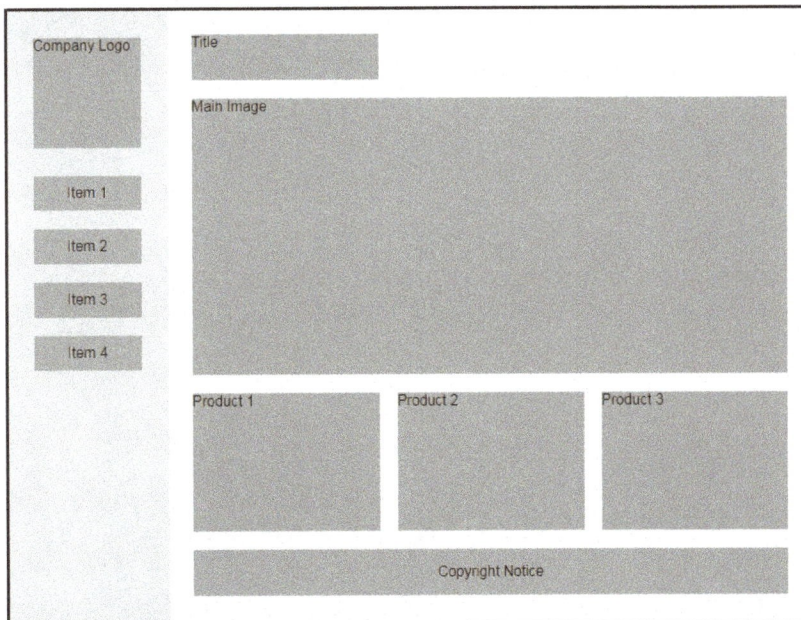

Figure 14-14   Common page layouts that use grid (part 1)

The advanced grid layout

Part 2 of figure 14-14 presents a layout called the *advanced grid layout*. This layout is more complex than the layouts presented in part 1 of this figure. Although you can use grid to develop layouts that are much more complex than what's shown here, this begins to illustrate the power of grid.

Like the headline and gallery layout, the advanced grid layout includes a company logo and a navigation menu at the top of the page, a headline and some text below that, and a footer at the bottom of the page. In addition, it includes two titles that identify two more areas of the page. The first area is for blog posts, and the second area is for featured products. However, you could use a layout like this for any content that you wanted to display in two areas.

Because all of the main structural elements for this page are at the same level, it uses only two grids. The first one is for all the elements that are children of the <body> element, and the second one is for the navigation menu.

Another way to implement this page would be to code the main content for this page (the titles, blog posts, and featured products) within a <main> element. Then, you could use another grid to lay out this content. This just shows that there's usually more than one way to lay out a page.

The advanced grid layout

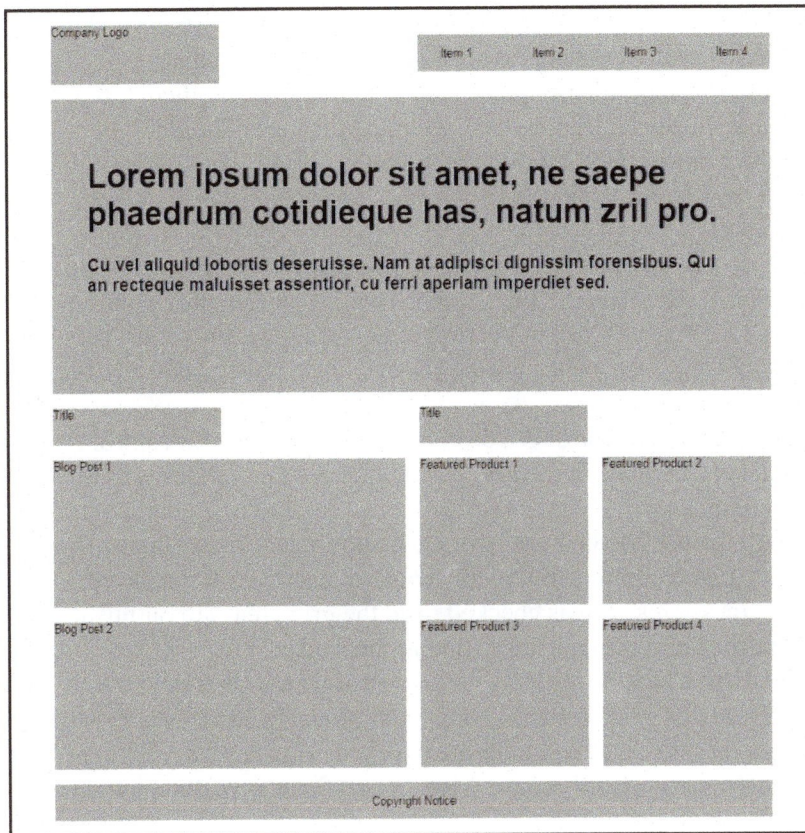

Description

- The headline and gallery layout can be used to display a series of images.
- The sidebar layout displays a vertical sidebar that's fixed to the side of the window. That way, it's accessible even when you scroll through the page. Because the sidebar uses fixed positioning, it's implemented as a separate grid.
- The advanced grid layout can be used for many purposes.

Figure 14-14   Common page layouts that use grid (part 2)

How to debug the layout for a page

No matter how carefully you plan your layouts, you will still encounter formatting problems. These problems can be hard to debug because there often isn't an error message. Instead, elements may overlap when they shouldn't or an element might have more space around it than you expected. That's why it's often helpful to use the Layout tab in the Chrome's Developer Tools to debug these types of problems.

To access the Layout tab, open the Developer Tools, select the Elements tab, and within the Elements tab select the Layout tab. Then, if you set the options as shown in the Layout tab that's shown in the first screen of figure 14-5, Chrome displays the track sizes and area names for the selected elements as shown in the second screen.

In the Layout tab, the first control is a drop-down list that lets you control how the debugger labels the lines in the grid. You can choose either line names, line numbers, or no line labels.

After the drop-down list, the first group of check boxes lets you to control how grid lines are displayed. In this figure, the length of each line is displayed because the "Show track sizes" box is checked. And the grid areas are outlined and labeled because the "Show area names" box is checked.

The second group of check boxes lets you to control which grids are displayed. In this figure, the grids for the <body> and <main> elements are shown because the boxes for these elements are checked. This is particularly useful when you have nested grids because it lets you to focus on debugging one grid at a time.

Note that the browser uses a different color for each grid element. In this figure, the browser uses one color for the grid for the <body> and another color for the grid for the <main> element.

The lines displayed for each grid can help you to debug layout issues. For example, if you have elements that display on top of each other, the grid lines can help you determine which element is overflowing its assigned grid area.

Although this figure shows how to use Chrome's Developer Tools to debug the layout for a page, other browsers also offer similar tools. For example, Firefox also provides a tool for debugging the layout of a page. However, the tools for other browsers may look and work differently.

The Layout tab in Chrome's Developer Tools

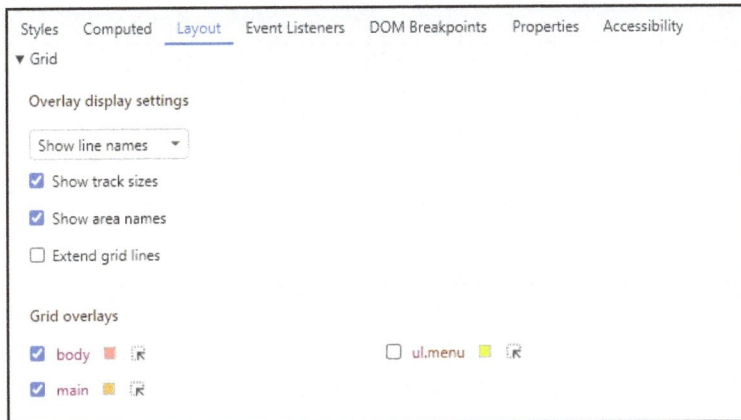

The web page in debugging mode

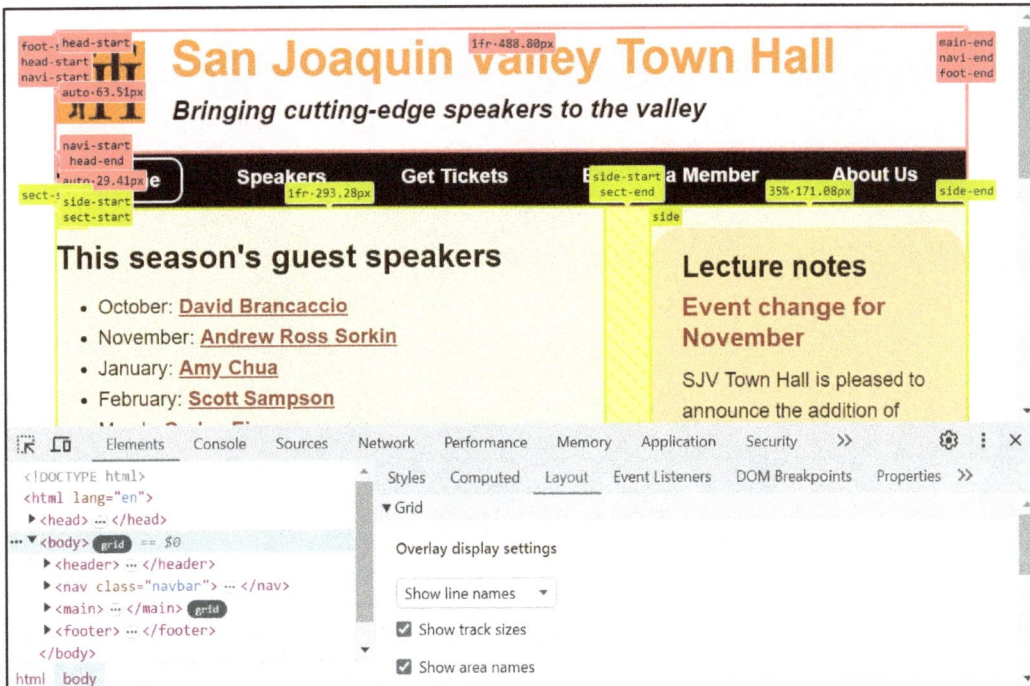

Description

- The Layout tab in a browser's Developer Tools can be used to display the boundaries of each grid area.

- Each grid area can be enabled separately and is assigned a different color.

Figure 14-15   How to debug the layout for a page

Perspective

Now that you've completed this chapter, you should be able to use grid layout to lay out the structural elements of a page in rows and columns. This provides a way to create complex page layouts. If necessary, you can even mix grid layouts with flexbox layouts. With those skills, you should be able to lay out your web pages the way you want.

Terms

grid layout	grid cell
grid	grid area
grid container	12-column grid
grid item	headline and gallery layout
grid track	fixed sidebar layout
grid line	advanced grid layout

Exercise 14-1   Convert the Town Hall home page from flexbox to grid

In this exercise, you'll update the home page of the Town Hall website to use grid template areas instead of flexbox. On larger screens, the sidebar should display to the left of the section. On smaller screens, the sidebar should display below the section.

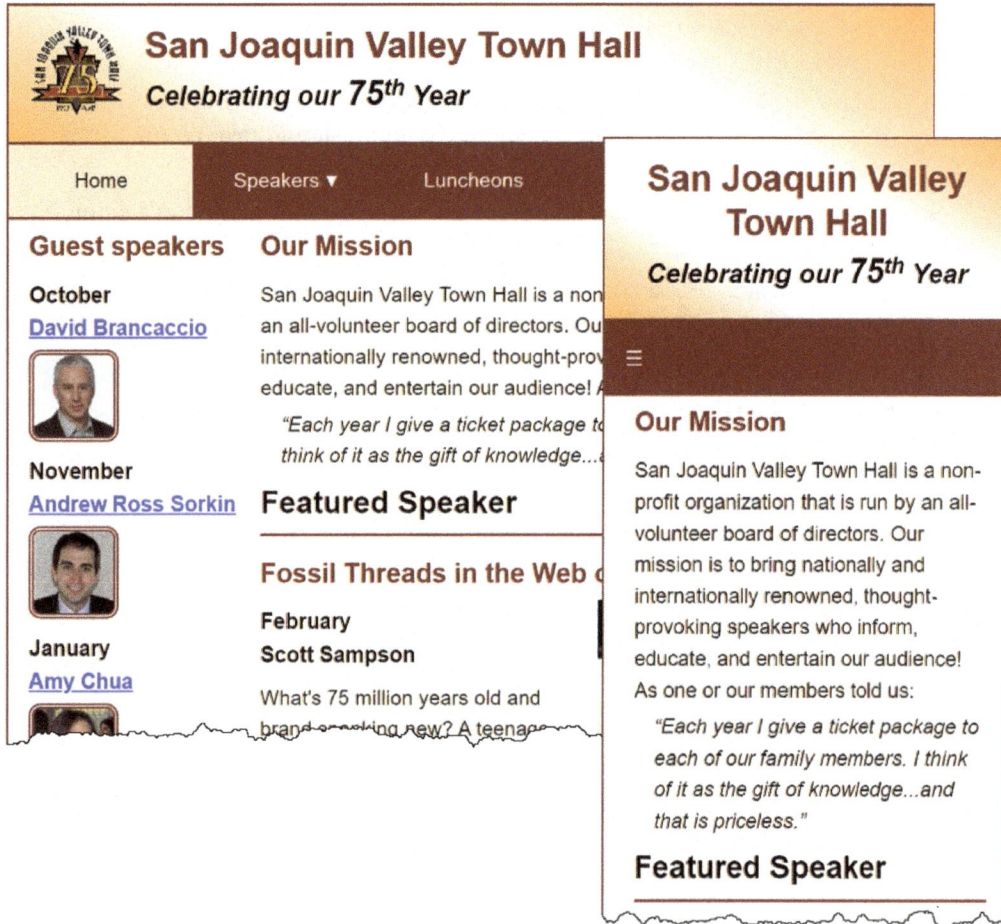

Open the HTML and CSS files for this page

1. Use your text editor to open these HTML and CSS files:

   ```
   exercises\ch14\town_hall\index.html
   exercises\ch14\town_hall\styles\main.css
   ```

2. Test the page and observe how the layout looks at different screen sizes.

Change the <body> element so it uses grid layout

3. In the CSS file, set the display property for the body to grid.

4. Define the template for the grid so it has four rows and one column. All four rows should size automatically, and the column should take up the full width of the body. No gaps are required.

5. Define grid template areas that place the header, navigation menu, main content, and footer in a single column.

6. Code grid-area properties for the <header>, <main>, and <footer> elements and the navbar class to place them in the correct grid area. Now test these changes. (This CSS is mobile first so you should test this for small screens.)

Change the <main> element so it uses grid layout

7. Change the display property for the <main> element so it uses grid layout instead of flexbox and remove the flex-direction property.

8. Define the template for the grid so it has two rows and one column. Both rows should size automatically, and the column should take up the full width of the <main> element. No gaps are required.

9. Define grid template areas that place the section and the aside in a single column.

10. Code grid-area properties for the <section> and <aside> elements to place them in the correct grid area.

11. Test these changes to make sure they work correctly on a small screen.

Fix the formatting so larger screens use a 2-column layout

12. Make the screen wider and note that the page no longer changes to a 2-column layout at wider screen sizes.

13. In the media query for screens wider than 922 pixels, find the style rule for the <main> element.

14. Remove the flex-direction property.

15. Define the grid template so it has one row and two columns. Set the row to size automatically, the first column to be 25% as wide as the container, and the second column to take the remaining width in the container.

16. Define grid template areas that place the aside and section side by side.

17. Remove the flex-basis properties from the style rules for the <section> and <aside> elements.

18. Test these changes to make sure they work correctly on all screen sizes.

Change the navigation menu so it uses grid layout on larger screens

19. In the media query for screens wider than 768 pixels, find the style rule for the menu class.

20. Set the display property to grid and remove the flex-basis property for nested <li> elements.

21. Define the grid template so it has one row that sizes automatically and five columns of equal proportion.

22. Find the style rule for the menu class next sibling of the checkbox-toggle:checked pseudo-class and change the display from flex to grid.

23. Test these changes to make sure they work correctly on all screen sizes.

Chapter 15

More CSS skills

So far, this book has already presented the essential CSS skills that you need for most websites. Now, this chapter presents a few more CSS skills that may be useful every once in a while. To start, it shows how to use CSS for transitions, transforms, animations, and filters. Then, it presents four more assorted skills. Finally, it shows how to work with the specificity algorithm that CSS uses to determine the style to apply to an element.

How to use transitions and transforms

A *transition* lets you gradually change one or more of the CSS properties for an element over a specified period of time. This lets you provide features with CSS that would otherwise require JavaScript.

A *transform* lets you rotate, scale, skew, and position HTML elements using CSS. When you combine transforms with transitions, you can create some interesting animations for your HTML elements.

Although CSS supports both 2D and 3D transforms, this chapter only covers 2D transforms. For an introduction to 3D transforms, we recommend David DeSandro's GitHub post:

http://desandro.github.io/3dtransforms

How to code transitions

Figure 15-1 summarizes the five properties that can be used for transitions. However, the transition property is the shorthand property for the other four properties. Because using this shorthand property is the easiest way to code most transitions, the examples in this chapter use that property.

The first example in this figure uses a transition to change between two properties that are applied to the <h1> element: font-size and color. In the CSS for this transition, the style rule for the h1 selector sets the font-size to 120% and the color to blue. Then, the transition property provides two sets of values that are separated by a comma.

The first set of values defines the transition for the font-size property. It says that the transition should take 2 seconds, use the ease-out *timing function* (or *speed curve*), and wait 1 second before starting. The second set of values defines the transition for the color property. It's the same as the set for the font-size property except it uses the ease-in timing function. This timing function causes the transition to start slowly and end quickly. In contrast, the ease-out timing function causes the transition to start quickly and end slowly.

The style rule for the <h1> element is followed by the style rule for the :hover pseudo-class of the <h1> element. It changes the font-size to 180% and the color to red, as well changing the cursor to a pointer instead of an arrow. So, when the user hovers the mouse over the <h1> heading, the browser gradually changes the two values in the transition. As a result, the transition happens as shown by the before and during views of the heading. When the user stops hovering, the transition returns to the original font-size and color settings.

If this example didn't include the transition property, the browser would still apply the changes in the style rule for the :hover pseudo-class. However, it would apply those changes immediately. In other words, the transition provides a gradual change that gives the appearance of animation.

The second example in this figure shows the code for a transition with just one property. The transition for this property omits the delay value. As a result, the browser uses its default value of 0, and the animation starts immediately. If you omit the timing function value, the browser uses the default value of ease.

CSS properties for working with transitions

Property	Description
transition	The shorthand property for setting the properties that follow.
transition-property	The property or properties that the transition is for. Use commas to separate multiple properties.
transition-duration	The seconds or milliseconds that the transition will take.
transition-timing-function	The speed curve of the transition. Values include ease, linear, ease-in, ease-out, ease-in-out, and cubic-bezier.
transition-delay	The seconds or milliseconds before the transition starts.

The syntax for the shorthand property

```
transition: [property] [duration] [timing-function] [delay];
```

A transition that occurs when the mouse hovers over a heading

Before the transition

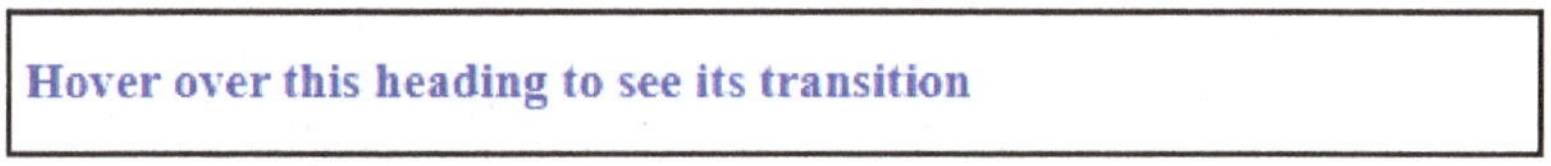

Hover over this heading to see its transition

During the transition

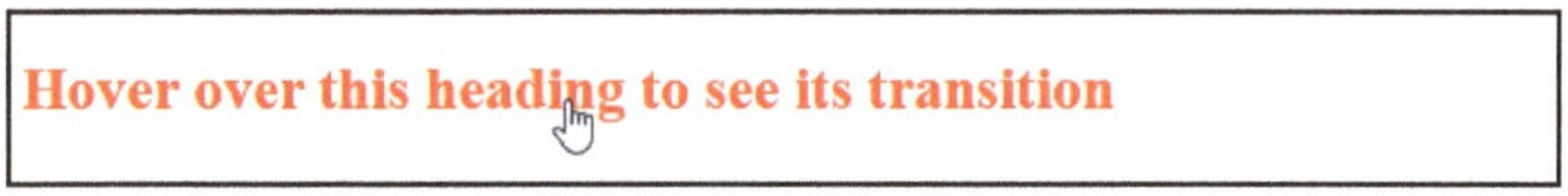

Hover over this heading to see its transition

The HTML for the heading

```
<h1>Hover over this heading to see its transition</h1>
```

The CSS for a two-property transition

```
h1 {
    font-size: 120%;
    color: blue;
    transition: font-size 2s ease-out 1s,
                color 2s ease-in 1s;
}
h1:hover {
    font-size: 180%;
    color: red;
    cursor: pointer;
}
```

A transition for one property when the mouse hovers over the heading

```
h1 {
    font-size: 120%;
    transition: font-size 2s ease-out;
}
h1:hover {
    font-size: 180%;
}
```

Description

- A *transition* provides a gradual change from one set of properties to another.

Figure 15-1   How to code transitions

And if you omit the duration value, the browser uses its default value of 0, which means that no transition takes place.

How to create an accordion using transitions

Figure 15-2 shows some of the power of CSS transitions. Here, transitions create an accordion. In web design, an *accordion* displays a list of headers for some content. Then, when the user clicks the header, the browser displays or hides the content. Accordions provide a good way to break down content into digestible chunks. They also work well for mobile-first design since they reduce how much a user has to scroll.

In this figure, the accordion has three headings and three panels that contain the contents for those headings. When the user clicks on a heading, the related panel gradually opens and any other panel that's open gradually closes.

The HTML for this accordion consists of <h3> elements for the accordion headings and <div> elements for the contents of the panels. However, the headings contain <a> elements that point to the placeholders in the <div> elements. For example, the href attribute of the first <a> element (#Q1) points to the id of the <div> element that follows (Q1).

The CSS starts with a style rule that formats the <div> elements and defines the transition that's used by this accordion. That style rule begins by setting the overflow property so it hides any content that doesn't fit in the <div> element. Then, it sets the height property to 0 to hide the <div> element. Next, it codes the transition that occurs when the height property is changed. This transition takes 1 second and uses the ease-in-out speed curve, which starts and ends slowly but progresses more quickly in between. Since this property omits the delay value, the transition starts immediately.

The second style rule makes this accordion work by using the :target pseudo-class for the <div> elements. The browser activates this pseudo-class when the user clicks on an <a> element that refers to a <div> element. Then, the browser changes the height of the <div> element to 7em, which starts the transition.

If you want, you can use this code to create an accordion without the transition property. Then, the panels for the accordion open and close immediately. However, the transition shown in this figure makes the change gradual instead of immediate, which adds an effect to the opening and closing of the panels.

An accordion created with transitions

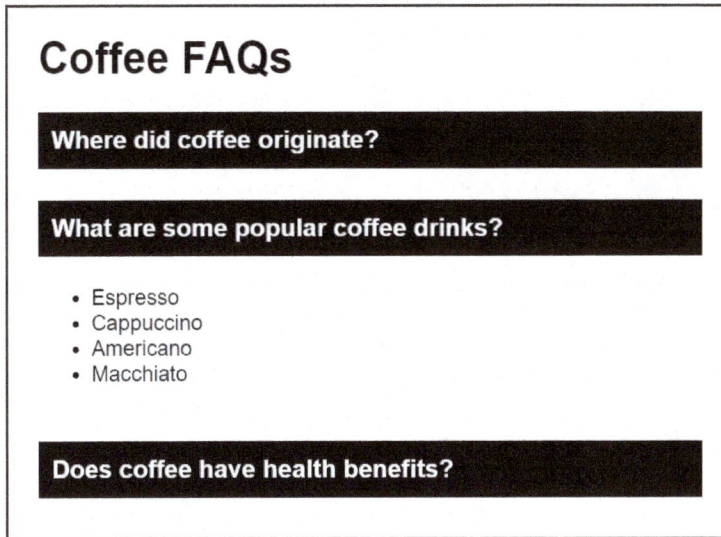

The HTML for the accordion

```
<h1>Coffee FAQs</h1>
<div id="accordion">
    <h3><a href="#Q1">Where did coffee originate?</a></h3>
    <div id="Q1"><p>Yemen</p></div>
    <h3><a href="#Q2">What are some popular coffee drinks?</a></h3>
    <div id="Q2"><ul>...</ul></div>
    <h3><a href="#Q3">Does coffee have health benefits?</a></h3>
    <div id="Q3"><p>Yes! Plain black coffee...</p></div>
</div>
```

The CSS for the transition

```
#accordion div {
    overflow: hidden;
    height: 0;
    transition: height 1s ease-in-out;
}
#accordion div:target {
    height: 7em;
}
```

Description

- In web design, an *accordion* displays a list of headers for some content. Then, when the user clicks a header, the browser displays the header's content and hides the content for any other headers.

- To create an accordion, you can use the :target pseudo-class to change the height property of a <div> element. When a user clicks on a link that targets the <div> element, you can change the height property from 0 to a height that's large enough to display its contents.

Figure 15-2   How to create an accordion using transitions

How to code 2D transforms

The first table in figure 15-3 summarizes the properties you can use to work with 2D transforms. The transform property lets you apply one or more transforms to an HTML element. The transform-origin property lets you change the origin point for the transform. The second table summarizes the functions you can use with the transform property.

The example in this figure illustrates how to use these properties and functions. Here, the HTML displays the same image twice, side by side. Then, when the mouse hovers over the image on the right, the browser rotates it by 180 degrees. As a result, it looks like a mirror image of the image on the left.

In addition, the origin point is changed so the rotation takes place from the right side of the second image. By default, the origin point is in the middle of the element being transformed. That's 50% or center on the X axis and 50% or center on the Y axis. When that's the case, the image rotates in place. Here, though, since the origin point is the right side of the image rather than the center, the image rotates away from the right side. That's why there's an empty space between the first image and the transformed image.

The HTML for this example defines two <img> elements for the same image. But, the second <img> element has a class attribute of image1.

In the CSS, the style rule for the image1 class specifies that the transition should take two seconds. Then, the style rule for the :hover pseudo-class of the image1 class defines the transform. To start, it uses the transform property to rotate the image 180 degrees on its Y axis. But it also uses the transform-origin property to change the origin point for the rotation to the right edge of the image. Without the transform-origin property, the image would rotate in place. In other words, the transform would not create space between the two images.

Properties for working with 2D transforms

Property	Description
transform	Applies one or more transform functions to the element.
transform-origin	Changes the default origin point. The parameters can be percents or keywords like left, right center, top, and bottom.

Functions for the transform property

Function	Description
rotate(angle)	Rotates an element by a specified angle.
rotateX(angle)	Rotates an element horizontally.
rotateY(angle)	Rotates an element vertically.
scaleX(value)	Scales the element's width horizontally.
scaleY(value)	Scales the element's height vertically.
scale(x-value, y-value)	Scales the element's width and height.
skewX(angle)	Skews an element along the X axis.
skewY(angle)	Skews an element along the Y axis.
skew(x-angle, y-angle)	Skews an element along the X and Y axis.
translateX(value)	Moves an element to the right or left.
translateY(value)	Moves an element up or down.
translate(x-value, y-value)	Moves an element right or left and up or down.
matrix(a, b, c, d, e, f)	Uses a matrix to rotate, scale, skew, and translate elements.

The image on the right rotated to the right when the cursor moved over it

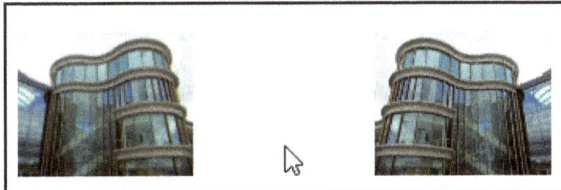

The HTML for the images

```
<img src="images/bldg.jpg">
<img src="images/bldg.jpg" class="image1">
```

The CSS for the transform

```
.image1 {
    transition: 2s;
}
.image1:hover {
    transform: rotateY(180deg);
    transform-origin: right;
}
```

Description

- *Transforms* are often combined with transitions.

Figure 15-3   How to code 2D transforms

A gallery of images with 2D transforms

Figure 15-4 shows how eight images look after the browser has applied various transforms to them. The browser applies these transforms when the user hovers over the image, and each has the default origin point. The best way to understand what these transforms do is to experiment with them on your own.

In the CSS for this figure, the matrix() function is the most difficult to understand. It lets you rotate, scale, translate, and skew elements with a single function by setting the values in a matrix. If you're mathematically inclined and are familiar with matrixes, you can learn more about this function by searching the web. Fortunately, most of the other functions are relatively easy to understand.

Images that have different transforms applied to them

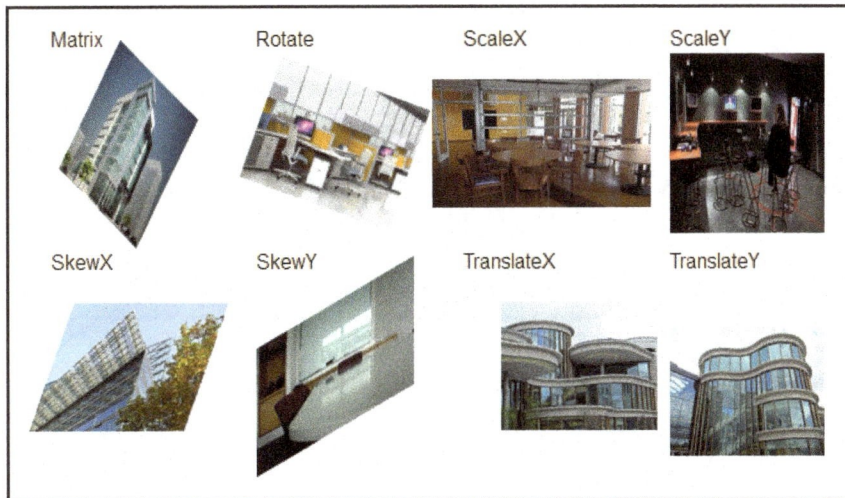

The HTML for the images

```
<ul>
    <li>Matrix      <img src="images/01.jpg" class="image1"></li>
    <li>Rotate      <img src="images/02.jpg" class="image2"></li>
    <li>ScaleX      <img src="images/03.jpg" class="image3"></li>
    <li>ScaleY      <img src="images/04.jpg" class="image4"></li>
    <li>SkewX       <img src="images/05.jpg" class="image5"></li>
    <li>SkewY       <img src="images/06.jpg" class="image6"></li>
    <li>TranslateX  <img src="images/07.jpg" class="image7"></li>
    <li>TranslateY  <img src="images/08.jpg" class="image8"></li>
</ul>
```

The CSS for the 2D transforms

```
.image1:hover { transform: matrix(0.5, 0.5, -0.5, 1, 0, 0); }
.image2:hover { transform: rotate(20deg); }
.image3:hover { transform: scaleX(1.4); }
.image4:hover { transform: scaleY(1.4); }
.image5:hover { transform: skewX(-20deg); }
.image6:hover { transform: skewY(-30deg); }
.image7:hover { transform: translateX(30px); }
.image8:hover { transform: translateY(20px); }
```

Description

- To apply transforms when the mouse hovers over an image, you can use the :hover pseudo-class.

- To learn about 3D transforms, please visit David DeSandro's excellent GitHub post:

  http://desandro.github.io/3dtransforms

Figure 15-4   A gallery of images with 2D transforms

How to use animations and filters

CSS *animations* let you animate HTML elements without using JavaScript or third-party plugins. To do that, CSS animations gradually change one or more properties of an element over time.

Filters let you change the appearance of images after they load in the browser without changing the underlying image files. For instance, you can use filters to convert an image to grayscale or to blur an image.

How to code simple animations

Figure 15-5 summarizes the primary properties for working with animations. Here, the animation property is the shorthand property for the other six properties. Because using this shorthand property is the easiest way to code most animations, the examples in this chapter use that property.

Of the six values that the animation property can include, the values for the duration, delay, and timing function work as they do for a transition. The value for iteration count sets the number of times the animation runs. And the value for direction sets the direction in which the animation runs.

The name value for the animation property sets the name you provide for the *@keyframes rule*. This rule defines the keyframes for the *animation sequence*. A *keyframe* defines what an element should look like at key points in an animation. Then, the browser uses *tweening* to fill the "in-between" frames for you. This gives the impression of motion.

The example in this figure shows how this works. This animation changes the left margin for a heading from 20% to 60%, and it changes the color from blue to red. As a result, the animation moves the heading from left to right while also changing the color of the heading from blue to red.

In the CSS for this animation, the animation property for the <h1> element uses the @keyframes rule named moveright. It also specifies that the duration of the animation should be 3 seconds, the ease-in-out speed curve should be used, the start of the animation should be delayed 1 second, and the animation should keep repeating. Finally, the animation property says that the direction should alternate. As a result, the first animation moves the heading from left to right, the second one moves the heading from right to left, and so on.

The @keyframes style rule shows how the keyframes can be defined for a simple animation. Here, the from rule sets the properties for the first keyframe, and the to rule sets the properties for the last keyframe.

Properties for working with animations

Property	Description
animation	The shorthand property for setting the properties that follow.
animation-name	The name of the @keyframes rule for the keyframe sequence.
animation-duration	The seconds or milliseconds that the animation will take.
animation-timing-function	The speed curve of the animation.
animation-delay	The seconds or milliseconds before the animation starts.
animation-iteration-count	The number of times the animation should repeat, or infinite.
animation-direction	The direction of the animation: normal, reverse, or alternate.

The syntax for the shorthand animation property

```
animation: name duration timing-function delay iteration-count direction;
```

A simple animation that moves a heading and changes its color

Starting in blue with a left margin of 20%

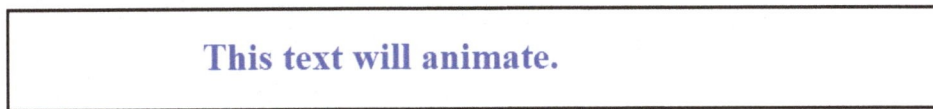

> This text will animate.

Ending in red with a left margin of 60%

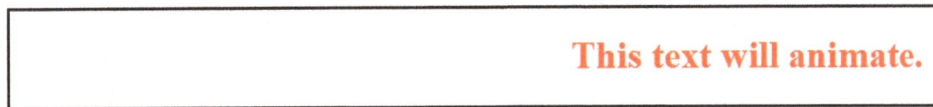

> This text will animate.

The HTML for the heading

```
<h1>This text will animate.</h1>
```

The CSS for the animation

```
h1 {
    animation: moveright 3s ease-in-out 1s infinite alternate;
}
@keyframes moveright {
    from {
        margin-left: 20%;
        color: blue;
    }
    to {
        margin-left: 60%;
        color: red;
    }
}
```

Description

- A CSS *animation* gradually changes one or more properties of an element. A *keyframe* defines what an element should look like at various points in an animation.

- For a simple animation, you can use the *@keyframes rule* to set the properties for the first and last keyframe. Then, the browser uses *tweening* to move the element between the two keyframes.

Figure 15-5   How to code simple animations

How to set the keyframes for a slide show

Figure 15-6 shows how you can set more of the keyframes for an animation and then let the browser do the tweening for the rest. This example uses CSS to create a slide show for five images and captions.

The HTML for this slide show defines an unordered list, but each list item contains an <h2> element for the caption and an <img> element for the image. To save space, this example doesn't show all the CSS for formatting the elements. But, when the browser applies that formatting, it displays all five images in a row with four of them hidden to the right of the one shown.

The CSS for the unordered list applies to the <ul> and <li> elements. Here, the <ul> element uses relative positioning. That way, the keyframes for the animation sequence for that element can also use relative positioning. In addition, the width of the <ul> element is set to 500% of its containing block. That way, it can accommodate all five <li> elements while only displaying one of them. Finally, the width of each <li> element is set to 20% of the <ul> element, which is 100% of the <ul> element's containing block (20% of 500%).

The CSS for the animation uses the animation property of the <ul> element to define the animation. This animation uses the @keyframes rule named slideshow. This rule sets eleven keyframes at different points in the animation: 0%, 10%, 20%, and so on up to 100%. Here, the starting keyframe sets the left property for relative positioning to 0%. This causes the browser to display the first list item. Then, at the 20% point in the animation, the rule changes the left property to -100%, which is the width of one list item. This causes the browser to slide the second list item into place from the right and hide the first one. This animation continues until the rule changes the left property to -400% at the 80% point in the animation. At this point, the browser slides the last item into place and stays there until the animation is finished.

Because the animation property sets the direction to alternate, the next repetition of the animation reverses the order of the slides so they go from the last to the first. If the animation property set the direction to normal, the slides would restart from the first keyframe for the next repetition.

For a demonstration of how this works, you can review and run the example for this figure that's available from the download for this book. If you do that, you can also experiment with the animation property and the keyframe settings to see how you can control the slide show.

A slide show animation with captions above the images

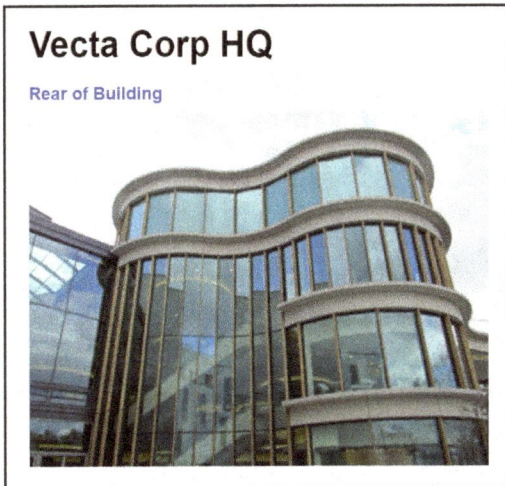

The HTML for an unordered list with five items

```
<ul>
    <li>
        <h2>Front of Building</h2>
        <img src="images/front.jpg" alt="">
    </li>
    </* four more list items go here */
</ul>
```

The CSS for the unordered list

```
ul {
    display: flex;
    list-style: none;
    width: 500%;
    position: relative;
    & li {
        width: 20%;
    }
}
```

The CSS for the animation

```
ul { animation: slideshow 15s infinite alternate; }
@keyframes slideshow {
    0%     {left:    0%;}
    10%    {left:    0%;}
    20%    {left: -100%;}
    30%    {left: -100%;}
    40%    {left: -200%;}
    50%    {left: -200%;}
    60%    {left: -300%;}
    70%    {left: -300%;}
    80%    {left: -400%;}
    90%    {left: -400%;}
    100%   {left: -400%;}
}
```

Figure 15-6   How to set the keyframes for a slide show

How to use filters

Figure 15-7 summarizes the filter property and ten filter functions that it supports. Most of these filter functions accept percent values that can either be expressed as a percent like 50% or a decimal value like .5. But some require degree values, and the drop-shadow() function requires a series of values that specify the horizontal offset, the vertical offset, the blur radius, and the color. This works much like the text-shadow property described a little later in this chapter.

The example in this figure shows how to invert an image. Here, the browser displays the same image side by side, but it inverts the colors of the one on the right when the user hovers the mouse over it. In this case, the function specifies an 80% inversion, but an inversion can range from 0 (no inversion) to 100% (full inversion).

The syntax for adding a filter to an element

```
filter: filter-function(value);
```

The filter functions

Function	Description
blur(value)	Applies a Gaussian blur. The value is in pixels.
brightness(value)	Adjusts the brightness, from 0% (black) to 100% (unchanged). Numbers higher than 100% result in a brighter image.
contrast(value)	Adjusts the contrast, from 0% (black) to 100% (unchanged).
drop-shadow(values)	Adds a drop shadow using values for horizontal offset, vertical offset, blur radius, and color.
grayscale(value)	Converts the image to grayscale where 100% is completely grayscale and 0% leaves the image unchanged.
hue-rotate(angle)	Adjusts the hue rotation of the image. The angle value is the number of degrees around the color circle.
invert(value)	Inverts the colors of the image. 100% is completely inverted, while 0% leaves the image unchanged.
opacity(value)	Applies transparency to the image. 0 results in an image that is completely transparent while 1 leaves the image unchanged.
saturate(value)	Saturates the image. 0% results in an image that is completely saturated while 100% leaves the image unchanged.
sepia(value)	Converts the image to sepia. 100% is completely sepia, while 0% leaves the image unchanged.

An image before and after its colors are inverted

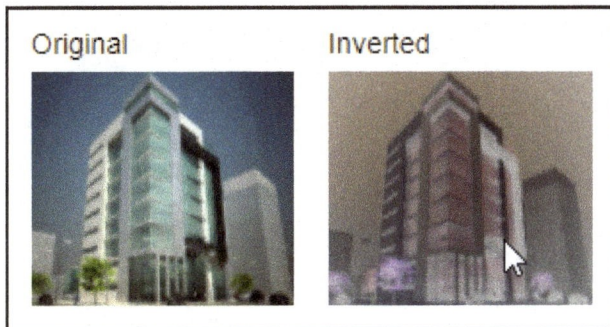

The HTML for the images

```
<li>Original <img src="images/01.jpg"></li>
<li>Inverted <img src="images/01.jpg" class="image1"></li>
```

The CSS for the filter

```
.image1:hover { filter: invert(.80); }
```

Description

- You can use *filters* to change the appearance of an image in the browser.
- Percentage values like 50% can also be expressed as decimal fractions like .5.

Figure 15-7   How to use filters

Ten filter functions applied to the same image

Figure 15-8 shows what the same image looks like when the browser applies ten filters to it. The last example shows how you can apply two filters to the same image. To do that, you code one function after the other without using commas to separate them. In this case, the filter applies the sepia() and blur() functions to the image.

Since the CSS in this figure doesn't use the :hover pseudo-class, the browser applies these functions when it loads the page. However, if you want to display these images when the mouse hovers over an image, you can modify the style rules to use the :hover pseudo-class like this:

```
.image1:hover { filter: blur(2px); }
```

Different filters applied to the same image

The HTML for the image and its filters

```
<ul>
    <li>Blur        <img src="images/01.jpg" class="image1" alt=""></li>
    <li>Brightness  <img src="images/01.jpg" class="image2" alt=""></li>
    <li>Contrast    <img src="images/01.jpg" class="image3" alt=""></li>
    <li>Drop Shadow <img src="images/01.jpg" class="image4" alt=""></li>
    <li>Grayscale   <img src="images/01.jpg" class="image5" alt=""></li>
    <li>Hue Rotate  <img src="images/01.jpg" class="image6" alt=""></li>
    <li>Invert      <img src="images/01.jpg" class="image7" alt=""></li>
    <li>Opacity     <img src="images/01.jpg" class="image8" alt=""></li>
    <li>Saturate    <img src="images/01.jpg" class="image9" alt=""></li>
    <li>Sepia/blur  <img src="images/01.jpg" class="image10" alt=""></li>
</ul>
```

The CSS for the filters

```
.image1  { filter: blur(2px); }
.image2  { filter: brightness(50%); }
.image3  { filter: contrast(50%);}
.image4  { filter: drop-shadow(2px 2px 5px #333); }
.image5  { filter: grayscale(50%); }
.image6  { filter: hue-rotate(90deg); }
.image7  { filter: invert(.8); }
.image8  { filter: opacity(.50); }
.image9  { filter: saturate(30%); }
.image10 { filter: sepia(100%) blur(1px); }
```

Description

- You can apply a filter to an image when the page loads or when the mouse hovers over an image.
- You can't use transitions with filters.

Figure 15-8   Ten filter functions applied to the same image

Four more CSS skills

The figures that follow show how to how to add shadows to text, format your page based on user preferences, use pseudo-classes that are functions, and work with logical properties that are based on the start and end of the text flow rather than the physical screen. These are useful skills to have, even if you may not use them as often as the skills presented earlier in this book.

How to add shadows to text

To add shadows to text, you can use the text-shadow property shown in figure 15-9. As the syntax shows, you can set four values for this property.

The first value specifies the horizontal offset, which is how much the shadow should be offset to the right (a positive value) or left (a negative value). The second value specifies the vertical offset, which is how much the shadow should be offset down (a positive value) or up (a negative value). The third value specifies the size of the blur radius for the shadow. And the fourth value specifies the color for the shadow.

The first example shows text with a shadow that's offset 4 pixels to the right and 4 pixels down, with no blur and with the shadow the same color as the text. By contrast, the second example shows text with a shadow that's offset to the left by 2 pixels and up by 2 pixels, with a blur radius of 4 pixels and with the shadow in red. Since the heading is in blue, this provides an interesting effect.

When you use the text-shadow property, it's important to consider accessibility for people who are visually impaired. For example, if the offsets or blur are too large, the shadow can make the text more difficult to read.

The syntax of the text-shadow property

```
text-shadow: horizontalOffset verticalOffset blurRadius shadowColor;
```

A heading with a text shadow

The HTML

```
<h1>San Joaquin Valley Town Hall</h1>
```

The CSS

```
h1 {
    color: #ef9c00;
    text-shadow: 4px 4px;
}
```

The heading in a browser

A different text shadow for the same heading

The CSS

```
h1 {
    color: blue;
    text-shadow: -2px -2px 4px red;
}
```

The heading in a browser

Accessibility guideline

- Remember the visually-impaired. Too much shadow or blur makes text harder to read.

Description

- Positive values offset the shadow to the right or down.
- Negative values offset the shadow to the left or up.
- The blur radius determines how much the shadow is blurred.

Figure 15-9   How to add shadows to text

How to work with user preferences

Users can change the settings on their devices to indicate preferences such as light or dark color schemes, reduced motion in animations, and more or less contrast. Then, you can use media queries to format your web page to accommodate these preferences.

Figure 15-10 presents some common media features for user preferences that you can use in media queries. Then, it shows examples of media queries that use these features.

The first example shows how to use the prefers-reduced-motion media feature. This example starts with a style rule for an <h1> element that uses a CSS animation like the ones presented earlier in this chapter. Then, it presents a media query that sets the animation property for that element to none when the user prefers reduced motion.

In general, you should stop an animation completely in response to this user preference unless the animation is essential to the functioning of your web page. In that case, you should change the animation to use the minimal amount of motion needed to function properly.

The second example shows how to use the prefers-contrast media feature. This example starts with a style rule for the <main> element that sets a 2 pixel dashed purple border. Then, it shows a media query that makes the border wider and solid when the user prefers more contrast, and a media query that makes the border less wide when the user prefers less contrast.

The third example shows how to use the prefers-color-scheme media feature. Since the <body> element uses a light color scheme by default (white background, black text), this example only shows a media query that defines a dark background with white text when the user prefers a dark color scheme. In addition, this media query changes the border for the <main> element to white.

You can also use the light-dark() function to format light and dark color schemes based on user preference. This function is presented in chapter 10.

To test your web page, you can use Chrome's Developer Tools to emulate what a web page looks like when these user preferences are set. To do that, follow the steps presented in this figure to display the Rendering panel. Then, scroll down until you find the drop-down lists that let you change the user preferences.

For instance, to test the prefers-reduced-motion feature, find the drop-down list labeled "Emulate CSS media feature prefers-reduced-motion". Then, select "prefers-reduced-motion: reduce". When you're done testing this feature, you can select "No emulation" to turn off the emulation.

Media features for user preferences

Media feature	Description
prefers-reduced-motion	The user prefers that motion-based animations be reduced or removed. Possible values are no-preference and reduce.
prefers-contrast	The user prefers more or less contrast. Possible values are no-preference, more, less, and custom.
prefers-color-scheme	The user prefers light or dark color schemes. Possible values are light and dark.

Reduce motion

```
h1 {
    animation: move 4s ease-in-out infinite alternate;
}
@media screen and (prefers-reduced-motion: reduce) {
    h1 {
        animation: none;
    }
}
```

Increase and decrease contrast

```
main {
    border: 2px dashed rebeccapurple;
}
@media screen and (prefers-contrast: more) {
    main {
        border-width: 3px;
        border-style: solid;
    }
}
@media screen and (prefers-contrast: less) {
    main { border-width: 1px; }
}
```

Provide for a dark color scheme

```
@media screen and (prefers-color-scheme: dark) {
    body {
        background-color: rebeccapurple;
        color: white;
    }
    main { border-color: white; }
}
```

How to emulate user preferences with Chrome's Developer Tools

1. In the Developer Tools toolbar, click on the Customize button (three vertical dots next to the close button).
2. Select "More tools" →Rendering.
3. In the Rendering panel, scroll down to the drop-down lists for user preferences, and select an appropriate item.

Description

- You can use media queries to work with the preferences that a user has set.

Figure 15-10 How to work with user preferences

How to use functional pseudo-class selectors

Chapter 4 showed how to use pseudo-class selectors, which are predefined classes like :hover that apply to specific conditions. CSS also provides several *functional pseudo-classes*. These pseudo-classes accept arguments. In other words, you code them like you code functions. Now, figure 15-11 shows how to use four functional pseudo-classes that can be used as selectors.

The :is() pseudo-class accepts a list of selectors and selects elements that match that list. For instance, the selector in the first example selects the <header> element and then uses the :is() pseudo-class to select several descendent elements of that <header> element. The statement below the example shows that you can select these elements without using the pseudo-class. However, the statement that uses the :is() pseudo-class is more concise and arguably easier to read.

The second example adds a second :is() pseudo-class to the selector. Specifically, it uses the :is() pseudo-class to select both the <header> and <nav> element. Then, it uses the :is() pseudo-class to select several descendent elements of both the <header> and the <nav> element. Again, you can select these elements without using the pseudo-class. However, the code that uses the :is() pseudo-class is much more concise.

The :where() pseudo-class works the same as the :is() pseudo-class. So, you could re-write the first two examples to use the :where() pseudo-class. The difference between the two is that the :is() pseudo-class might override other selectors, while the :where() pseudo-class doesn't override other selectors. This is due to the way that CSS calculates specificity, and there's an example of when it makes sense to use :where() instead of :is() later in this chapter.

The :not() pseudo-class accepts a list of selectors and selects elements that *don't* match that list. For instance, the selector in the third example selects <a> elements that don't have an href attribute. You can use this to select the link for the current page in a navigation menu.

The :has() pseudo-class accepts a list of selectors and selects elements that have related elements that match that list. For instance, the fourth example selects <h2> elements that have an adjacent sibling that's assigned to a class named subhead. You can use code like this to reduce or remove the padding for a heading that's followed by a subheading.

The fifth example uses the :has() pseudo-class to select <li> elements that have a descendent that's assigned to a class named submenu. As shown in chapter 7, you can use code like this in a navigation menu to select a list item that has a submenu.

The sixth example uses the :has() pseudo-class to style the <body> element based on whether it contains a checkbox that's checked. You can use code like this for a checkbox that toggles between a light and dark color scheme.

CSS functional pseudo-class selectors

Pseudo-class	Description
:is(selector-list)	Selects elements that match the list of selectors.
:where(selector-list)	Works the same as :is(), but it never overrides other selectors.
:not(selector-list)	Selects elements that do not match the list of selectors.
:has(selector-list)	Selects elements that have elements that match the list of selectors.

Select multiple descendent elements

```
header :is(h1, h2, h3, a) {
    color: rebeccapurple;
}
```

Without the pseudo-class

```
header h1, header h2, header h3, header a {
    color: rebeccapurple;
}
```

Select multiple descendent elements of multiple elements

```
:is(header, nav) :is(h1, h2, h3, a) {
    color: rebeccapurple;
}
```

Without the pseudo-class

```
header h1, header h2, header h3, header a, nav h1, nav h2, nav h3, nav a {
    color: rebeccapurple;
}
```

Select <a> elements that don't have an href attribute

```
a:not([href]) { font-weight: 900; }
```

Select <h2> elements that have an adjacent sibling assigned to a class

```
h2:has(+ .subhead) { padding-bottom: 0; }
```

Select <li> elements that have a descendent assigned to a class

```
li:has(.submenu) {
    position: relative;
    &:hover > .submenu { display: block; }
}
```

Select the <body> element when a checkbox is checked

```
body:has(#dark:checked) {
    background-color: rebeccapurple;
    color: aliceblue;
}
```

Description

- A *functional pseudo-class* selector accepts a list of selectors that it uses to select elements.

Figure 15-11   How to use functional pseudo-class selectors

How to work with logical properties

Figure 15-12 presents two versions of the CSS box model. The physical box model was presented in chapter 5. It has properties based on the physical dimensions of the screen, such as margin-left and padding-top. When your website only needs to support the English language, or other languages that flow from left to right, it makes sense to use the physical box model.

By contrast, the *logical properties* of the logical box model are based on the flow of the content rather than the dimensions of the screen. For instance, when the flow of the content is left to right, the padding-inline-start property is the same as the padding-left property. But, when the flow of the content is right to left, the padding-inline-start property is the same as the padding-right property. That's because, in that case, the padding at the start of the flow is the right padding, not the left padding. If your website needs to support languages that flow from right to left or top to bottom, it makes sense to use the logical model.

This figure also presents two CSS properties that allow you to control the flow of the content. The direction property determines whether the content flows from left to right or right to left. The default is left to right.

You should know that there's also an HTML attribute named dir that you can use to change the direction of the content, like this:

```
<p dir="rtl">...</p>
```

If possible, you should use the HTML dir attribute rather than the CSS direction property. That way, the direction is correct even if the CSS fails to load.

The writing-mode property determines whether the content flows horizontally or vertically. The default value of horizontal-tb means that the content flows horizontally from top to bottom. Whether the horizontal flow is left to right or right to left depends on the value of the dir attribute or direction property.

The writing-mode value of vertical-lr means that the content flows vertically and the next vertical line is to the left of the previous line. Conversely, the value of vertical-rl means that the content flows vertically and the next vertical line is to the right of the previous line. Whether the vertical flow is top to bottom or bottom to top depends on the value of the dir attribute or the direction property.

The examples in this figure give you an idea of how this works. The first example sets the padding-inline-start property to 1em. When the direction of the flow is left to right, this sets the left padding, since the flow starts on the left. But when the direction of the flow is right to left, this sets the right padding, since the flow starts on the right. The second example is similar to the first, except it sets an inline-end property rather than an inline-start property.

The last example sets a border for two sides at once. To do that, it omits the start or end part of the property. When the writing mode is horizontal, this sets the top and bottom borders. But when the writing mode is vertical, this sets the left and right borders. Even when you're using the physical model, this is a convenient way to set two sides at once.

The CSS box model

Physical

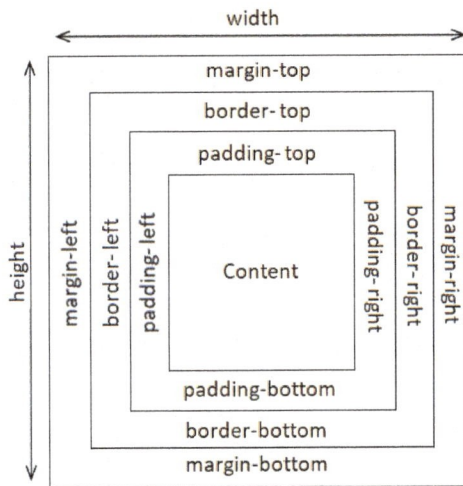

Logical (when left to right)

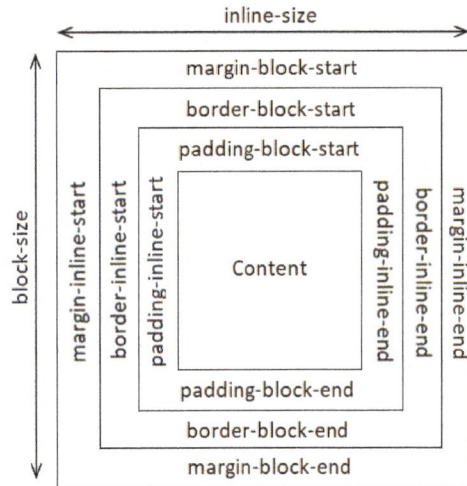

CSS properties that control the flow of the content

Property	Description
direction	Determines whether the flow is from left to right or right to left. Possible values are ltr and rtl. Default is ltr.
writing-mode	Determines whether the flow is horizontal or vertical. Possible values are horizontal-tb, vertical-lr, and vertical-rl. Default is horizontal-tb.

Set padding for one side

```
padding-inline-start: 1em;     /* Sets the left padding when the direction
                                  is left to right. Sets the right padding
                                  when the direction is right to left. */
```

Set the margin for one side

```
margin-inline-end: 1em;        /* Sets the right margin when the direction
                                  is left to right. Sets the left margin
                                  when the direction is right to left. */
```

Set the border for two sides

```
border-block: 1px solid;       /* Sets the top and bottom borders when the
                                  writing mode is horizontal. Sets the left
                                  and right borders when the writing mode is
                                  vertical. */
```

Description

- The CSS *logical properties* control layout based on the flow of the content rather than the physical dimensions of the screen. This is useful for languages that flow from right-to-left or top-to-bottom.

- The *block dimension* is perpendicular to the flow of the content, while the *inline dimension* is parallel to the flow of the content.

Figure 15-12 How to work with logical properties

How to work with specificity

Chapter 4 showed how CSS resolves conflicts when more than one style is applied to an element. In most cases, the browser applies the style rule with the highest *specificity*. Now, this chapter shows how specificity is calculated, and how you can work with specificity to get the results you want.

How specificity is calculated

The specificity algorithm calculates the *weight* of a style rule by determining the number of selectors in three *weight categories*. These weight categories correspond to the three types of selectors presented at the top of figure 15-13. Each selector in a weight category adds a value of 1 to that category.

The ID weight category counts the number of ID selectors. The class weight category counts the number of class selectors. This category includes pseudo-class selectors like :hover and :is(), and attribute selectors like [href] and [type="input"]. And the type weight category counts the number of type selectors. This category includes pseudo-element selectors like ::before and ::after.

To calculate the weight of a style rule, the specificity algorithm adds the value of each weight category to one of the columns in a three column value. It adds the ID category to the first column, the class category to the second column, and the type category to the third column.

The first example in this figure shows how this works. The first selector is an ID selector, so the algorithm adds a value of 1 to the first column for a weight value of 1-0-0. The second selector is a class selector so it adds 1 to the second column. And the third selector is a type selector so it adds 1 to the third column.

The fourth selector contains both a type selector and a class selector, so the algorithm adds 1 to the second column and 1 to the third column. The fifth selector contains two class selectors, so it adds 2 to the second column. And the sixth selector contains an ID, class, and type selector, so the algorithm adds 1 to all three columns. Since combinators don't count, the direct child combinator (>) in this selector doesn't add anything to the weight value.

Once the browser uses the algorithm to calculate weights for the selectors, it compares those weights to determine which selector to apply. First it compares the ID category, then the class category, and then the type category. If one of the categories has a higher value, the browser applies that selector and doesn't make any further comparisons. For instance, if a selector has a higher ID category value, the browser applies that selector and doesn't compare the other categories.

The algorithm doesn't include the universal selector (*) or the :where() pseudo-class. In other words, these add 0-0-0 to the weight value. However, the algorithm does include the other functional pseudo-classes.

The second example in this figure presents a specificity issue. The CSS in the media query is supposed to style both <ul> elements in the navbar class. But, it doesn't override the mobile-first CSS because that selector has a higher value in the class weight category. In the next three figures, you'll learn several ways to fix this issue.

The three weight categories

Category	Description
ID	ID selectors. For each ID in a selector, add 1-0-0 to the weight value.
Class	Class selectors, pseudo-class selectors, and attribute selectors. For each of these in a selector, add 0-1-0 to the weight value.
Type	Type selectors and pseudo-element selectors. For each of these in a selector, add 0-0-1 to the weight value.

Example selectors

```
#my-element {...}              /* 1-0-0 */
.my-class {...}                /* 0-1-0 */
div {...}                      /* 0-0-1 */
p .copyright                   /* 0-1-1 */
.footer .copyright             /* 0-2-0 */
#my-nav .menu > ul             /* 1-1-1 (combinator doesn't count) */
```

A media query that doesn't work as intended because of a specificity issue

The HTML

```
<nav class="navbar">
    <ul class="menu">
        <li>Menu item 1</li>
        <li>Menu item 2
            <ul class="submenu">
                <li>Submenu item 1</li>
                <li>Submenu item 2</li>
            </ul>
        </li>
    </ul>
</nav>
```

The mobile-first CSS

```
.navbar .menu .submenu {...}            /* 0-3-0 */
```

The media query

```
@media screen and (min-width: 600px) {
    .navbar ul {...}                    /* 0-1-1 */
}
/* media query selector doesn't override the mobile-first selector */
```

Description

- *Specificity* refers to the algorithm browsers use to determine which style rule to apply to an element. The algorithm calculates the *weight* of a style rule by determining the number of selectors in three *weight categories* that correspond to the types of selectors.

- The specificity algorithm is a three-column value that tallies the incidence of each selector type. It compares the number of ID selectors, class selectors, and type selectors.

- The specificity algorithm doesn't use the universal selector (*) or the functional :where() pseudo-class, but it does use the functional :is(), :not(), and :has() pseudo-classes. The column for these functions is based on the most specific selector in the list.

- The browser starts by comparing the values in the first column and only moves on to the next column if necessary.

Figure 15-13   How specificity is calculated

How to use classes

Figure 15-14 shows how to use classes to fix the specificity issue presented in the previous figure. In the HTML, the class names of the nested elements incorporate the class names of the parent element. Specifically, the <nav> element is assigned to the navbar class, its child <ul> element is assigned to the navbar-menu class, and the descendant <ul> element is assigned to the navbar-menu-submenu class. When you name your classes this way, you can select the nested elements with selectors that have lower specificity.

In the CSS, the mobile-first selector uses one class selector rather than the three shown in the last figure. This reduces the value of the class weight category to 1, which matches the value of this category in the media query selector. Because of that, the algorithm moves on to compare the type weight categories. Since the media query selector has a higher value for that category, it's applied and overrides the mobile-first selector, which is what you want here.

Over the years, using class names to control specificity became increasingly common, and developers created naming conventions to make this more uniform. For example, one popular naming convention is called BEM (block, element, modifier).

While the technique described in this figure is common, it does have some downsides. First, it requires you to change the HTML as well as the CSS when you format your web page. Second, the class names can become long and unwieldy.

Use class names to fix a specificity issue

The HTML

```
<nav class="navbar">
    <ul class="navbar-menu">
        <li>Menu item 1</li>
        <li>Menu item 2
            <ul class="navbar-menu-submenu">
                <li>Submenu item 1</li>
                <li>Submenu item 2</li>
            </ul>
        </li>
    </ul>
</nav>
```

The mobile-first CSS

```
.navbar-menu-submenu {...}                      /* 0-1-0 */
```

The media query

```
@media screen and (min-width: 600px) {
    .navbar ul {...}                            /* 0-1-1 */
}
/* media query selector overrides the mobile-first selector */
```

Description

- You can use class names to control specificity in your CSS.

- If the class names of the nested elements incorporate the class names of the parent element, you can select a nested element without increasing the specificity of the selector.

- Using class names to control specificity requires changes to the HTML and can result in long and unwieldy class names.

Figure 15-14   How to use classes to work with specificity

How to use the :where() pseudo-class

Figure 15-15 shows how to use the :where() pseudo-class to fix the specificity issue presented in figure 15-13. In the HTML, the class names are the same as in figure 15-13. In other words, you don't need to adjust the HTML to use this technique, which is a benefit.

In the CSS, the mobile-first selector uses the :where() pseudo-class that receives three class selectors. This reduces the weight of the mobile-first selector to 0-0-0. That's because the specificity of the :where() pseudo-class is always zero, regardless of the specificity of the selectors it receives. So, the media query selector has a higher class weight and overrides the mobile-first selector, which is what you want here.

Use the :where() pseudo-class to fix a specificity issue

The HTML

```
<nav class="navbar">
    <ul class="menu">
        <li>Menu item 1</li>
        <li>Menu item 2
            <ul class="submenu">
                <li>Submenu item 1</li>
                <li>Submenu item 2</li>
            </ul>
        </li>
    </ul>
</nav>
```

The mobile-first CSS

```
:where(.navbar .menu .submenu) {...}          /* 0-0-0 */
```

The media query

```
@media screen and (min-width: 600px) {
    .navbar ul {...}                          /* 0-1-1 */
}
/* media query selector overrides the mobile-first selector */
```

Description

- The weight value of the :where() pseudo-class is always 0-0-0.

- You can use the :where() pseudo-class to reduce the specificity of a selector that isn't working as intended.

Figure 15-15   How to use the :where() pseudo-class to work with specificity

How to use cascade layers

Chapter 4 described the cascade order, and how the important and normal declarations in the web browser, user style sheets, and your CSS interact to determine which styles to apply. Then, within each level of the cascade order, specificity is applied based on the three types of selector.

Cascade layers allow you to add more layers to the cascade order. Then, within these cascade layers, specificity applies as usual. But, specificity doesn't apply between cascade layers. In other words, a less specific selector overrides a more specific selector if it's in a layer that has higher priority.

The first example in figure 15-16 shows how to use the *@layer rule* to declare cascade layers. This code declares three cascade layers named defaults, theme, and overrides.

The first cascade layer that's coded has lowest priority and each subsequent layer has higher priority. Here, the defaults layer has the lowest priority, the theme layer has higher priority than defaults, and the overrides layer has higher priority than both. Importantly, CSS that isn't within a layer has higher priority than CSS that is within a layer. So, an unlayered declaration has higher priority than any of these layers.

Once you've declared a cascade layer, you can add style rules to it. To do that, you use the @layer rule again, as shown in the second example.

When working with cascade layers, you don't have to declare a layer before you add style rules to it. For instance, you can code the layer in the second example and add style rules to it without first coding the declaration in the first example. In that case, the second example declares the defaults layer and also adds style rules to it. In general, though, it's better to declare your cascade layers in one place and then add style rules to them later. That way, it's easy to view the priority of the layers. In addition, if you declare your layers, you can add the style rules in any order since the order of the layers has already been declared.

Another way to add style rules to a cascade layer is to import a style sheet. To do that, you use the *@import rule*, as shown in the third example. Any @import statements must be at the top of the file, after the @layer statement that declares the layers but before any other CSS.

The fourth example shows how to use layers to fix the specificity issue presented in figure 15-13. Here, the mobile-first CSS is in a cascade layer named mobile-first, while the media query CSS is not in a layer. Since unlayered CSS has priority, the media query selector overrides the mobile-first selector, even though the specificity of the mobile-first selector is higher.

When you use cascade layers, the !important rule reverses the priority of the layer. For instance, in the first example, the defaults layer has the lowest priority. But, if you code an important declaration in that layer, it overrides styles in the overrides layer, rather than the other way around. As you can imagine, this can cause a lot of confusion. That's yet another reason why it's a good practice to avoid important declarations whenever possible.

Declare three cascade layers

```
@layer defaults, theme, overrides;
/* defaults has lowest priority, overrides has highest priority */
```

Add style rules to a layer

```
@layer defaults {
    /* style rules */
}
```

Import a style sheet into a layer

```
@import url('menu.css') layer(theme);
```

Use cascade layers to fix specificity issue

The HTML

```
<nav class="navbar">
    <ul class="menu">
        <li>Menu item 1</li>
        <li>Menu item 2
            <ul class="submenu">
                <li>Submenu item 1</li>
                <li>Submenu item 2</li>
            </ul>
        </li>
    </ul>
</nav>
```

The mobile-first CSS

```
@layer mobile-first {
    .navbar .menu .submenu {...}            /* layered */
}
```

The media query

```
@media screen and (min-width: 600px) {
    .navbar ul {...}                        /* unlayered */
}
/* unlayered style overrides the layered style */
```

Description

- You can use *cascade layers* to define an order of precedence for your CSS.
- You can use the *@layer rule* to declare a cascade layer and to add styles to a layer.
- You can use the *@import rule* to import a style sheet into a layer.
- The first cascade layer you declare in your code has lowest priority, the next has higher priority, and so on.
- Unlayered styles (styles that aren't declared within a layer) have priority over layered styles.
- You can code the same @layer rule multiple times in your CSS. The order of priority is based on where it *first* appears.

Figure 15-16   How to use cascade layers

Perspective

Now that you've finished this chapter, you should be able to use transitions, transforms, animations, and filters in your own web pages. But what should you do with them? If you're looking for ideas, just search the web for examples. For instance, you can find lots of examples of transitions and animations that include accordions, slide shows, carousels, and some incredible and fun animations. You can also find plenty of examples for transforms and filters.

In addition, you should now be able to work with advanced skills like accommodating user preferences as well as working with functional pseudo-classes, logical properties, and the specificity algorithm.

Terms

transition	functional pseudo-class
timing function	logical property
speed curve	block dimension
accordion	inline dimension
transform	specificity
animation	weight
@keyframes rule	weight category
keyframe	cascade layer
animation sequence	@layer rule
tweening	@import rule
filter	

Exercise 15-1   Use transitions, transforms, and animation

In this exercise, you'll move an image in the aside when the user hovers the mouse over it, and you'll animate the image on a speaker page so it appears to pulse. When both of those features are in progress, a speaker page will look something like this:

Use a transition

1. Use your text editor to open these files:

    ```
    exercises\ch15\town_hall\speakers\sampson.html
    exercises\ch15\town_hall\styles\main.css
    ```

2. Find the style rule for the <aside> element. Then, add a transition that moves an image in the aside 30 pixels to the right when the user hovers the mouse over it. The duration of the transition should be 2 seconds.

3. View the web page in a browser and make sure this transition works correctly.

4. Comment out the transition and test this again. The movement should still happen, but it should happen immediately. Then, uncomment the transition.

Use a transition with a transform

5. In the style rule for the :hover pseudo-class for the images in the aside, comment out the CSS that moves the image 30 pixels to the right.

6. Add a transform that rotates the image by 180 degrees.

7. Change the origin of this transform so it's at the right edge of the image. Note the difference in the transform.

Use an animation with filters and a transform

8. Find the style rule for the speaker class. Then, animate the image in the <section> element that's assigned to this class. The animation should run immediately when the page loads, last 2 seconds, use the ease-in-out speed curve, run infinitely, and alternate directions.

9. In the @keyframes rule, use the to rule to set a 10 pixel drop-shadow filter and a 125% brightness filter. This should make the image appear to pulse.

10. Add a transform to this animation that increases the size of the image 1.1 times. You can use the scale() function to do that.

Add formatting for users who prefer reduced motion

11. Add a media query for the prefers-reduced-motion media feature.

12. In the media query, set the animation of the image in the article on the speaker page and the transition and transform of the images in the aside to none.

13. Use Chrome to test these changes by emulating a user who prefers reduced motion.

Use the :not() pseudo-class to replace a class

14. In Chrome, navigate to the home page and notice how the Home link in the navigation menu is styled to indicate it's the current page. Navigate to the speaker page for Scott Sampson and see how the Speaker link and the Scott Sampson link in the submenu are also styled to indicate the current page.

15. Use your text editor to open this file:

    `exercises\ch15\town_hall\index.html`

16. In the HTML for the home page, find the navigation menu and remove the class attribute for the Home link.

17. In the HMTL for the speaker page, remove the class attribute for the Speaker link and the Scott Sampson submenu link.

18. Test the website and see how the navigation menu on each page looks without the class attributes for the Home and Speaker links.

19. In the CSS, find the style rule for the navbar class. Replace the nested selector for the a.current class with a selector that uses the :not() pseudo-class to select <a> elements that don't have an href attribute.

20. Test the website again. The formatting for the current page should be restored.

Exercise 15-2   Fix a specificity issue

In this exercise, you'll fix the specificity issue in the web page shown below. When you're done, the "Contact us" heading at the bottom should be left-aligned and the same size as the other headings when displayed on small screens like this:

Find the specificity issue

1. Use your text editor to open these files:

   ```
   exercises\ch15\students\index.html
   exercises\ch15\students\styles\main.css
   ```

2. Review the HTML and CSS. Then, view the page in a browser and narrow the browser width. Note that the h2 style rule in the media query doesn't override the h2 style rule in the footer.

3. In the CSS, find the style rule for <h2> elements and calculate the weight of the selector.

4. Find the style rule for <h2> elements in the footer and calculate the weight of the selector.

5. In the media query, find the style rule for <h2> elements and calculate the weight of the selector.

6. Compare the weights of the style rules. Since the selector for the <h2> element in the media query has the same weight as the selector for the <h2> element in the style sheet, it overrides that selector because it appears later in the style sheet. However, since it has a lower value in the type category than the selector for the <h2> element that's in the footer, it doesn't override that selector.

Use the :where() pseudo-class to fix the issue

7. Pass the selector for <h2> elements in the footer to the :where() pseudo-class, like this:

    ```
    :where(footer h2) {...}
    ```

8. Test the page. This should fix the formatting on small screens. However, it doesn't format the "Contact us" heading correctly on large screens because the font-size of that heading is supposed to be small on large screens.

    Why is this happening? The weight of the :where() pseudo-class selector is always 0-0-0. Since this is less than the weight of the selector for <h2> elements that appears earlier in the stylesheet, this selector no longer overrides that one. To fix this, you need to make the weight of the selector for <h2> elements in the footer the same as the weight of the earlier selector for <h2> elements.

9. Update the :where() pseudo-class to receive only the <footer> element like this:

    ```
    :where(footer) h2 {...}
    ```

10. After you make this adjustment, test the page. It should work correctly now. That's because the type selector that's outside the :where() pseudo-class adds 1 to the type weight category, so it matches the weight of the earlier selector.

Use cascade layers to fix the issue

11. Remove the :where() pseudo-class that you added in the previous steps.

12. Test the page again. The formatting issue should occur on small screens.

13. Declare two cascade layers named defaults and page.

14. Add the first four style rules to the defaults layer, and the next two style rules to the page layer. Leave the media query unlayered.

15. Test the page. This should fix the formatting issue for small screens and large screens.

Section 3

Web design and deployment

The two chapters in this section present some useful information about web design and deployment. You can read these chapters any time after you complete the chapters in section 1, and you can read these chapters in whatever sequence you prefer.

Chapter 16

Users, usability, and web design

Section 1 presented the skills you need to use HTML and CSS to develop the pages for a website. Now, this chapter presents some information about users and usability as well as some basic guidelines for designing a website.

Users and usability

Before you design a website, you need to think about who your users are and what they expect. After all, it is your users who determine the success of your website.

An introduction to usability

What do users want when they reach a website? They want to find what they're looking for as quickly and easily as possible. And when they find it, they want to extract the information or do the task as quickly and easily as possible.

How do users interact with a web page? They don't read it in an orderly way, and they don't like to scroll. Instead, they scan the page to see if they can find what they're looking for or a link to what they're looking for. Often, they click on a link to see if it gives them what they want, and if it doesn't, they click on the Back button to return to where they were. In fact, users click on the Back button more than 30% of the time when they reach a new page.

If users can't find what they're looking for or get too frustrated, they leave the site. It's that simple. For some websites, more than 50% of first-time visitors to the home page leave without ever going to another page.

In web development terms, what users want is *usability*. This term refers to how easy it is to use a website, and it's one of the key factors that determines the effectiveness of a website. If a site is easy to use, it has a chance to be effective. If it isn't easy to use, it probably won't be effective.

Figure 16-1 presents one page of a website that has a high degree of usability, and it presents three guidelines for improving usability. First, you should try to present the critical information for a page "above the fold." This term refers to what's shown on the screen when a new page is displayed, which is analogous to the top half of a newspaper. This reduces the need for scrolling, and it gives the page a better chance for success.

Second, you should try to group related items into separate components and limit the number of components on each page. That makes the page look more manageable and helps people find what they're looking for.

Third, you should adhere to the current conventions for website usability. For instance, clickable links should look like they're clickable and items that aren't clickable shouldn't fool users by looking like they are clickable.

If you look at the website in this figure, you can see that it has implemented these guidelines. All of the critical information is presented above the fold. The page is divided into a header and other well-defined components. It's also easy to tell where to click.

It's relatively easy to make a small website usable. By contrast, building usability into a large website with dozens of product categories and hundreds of products is a serious challenge.

A small website that is easy to use

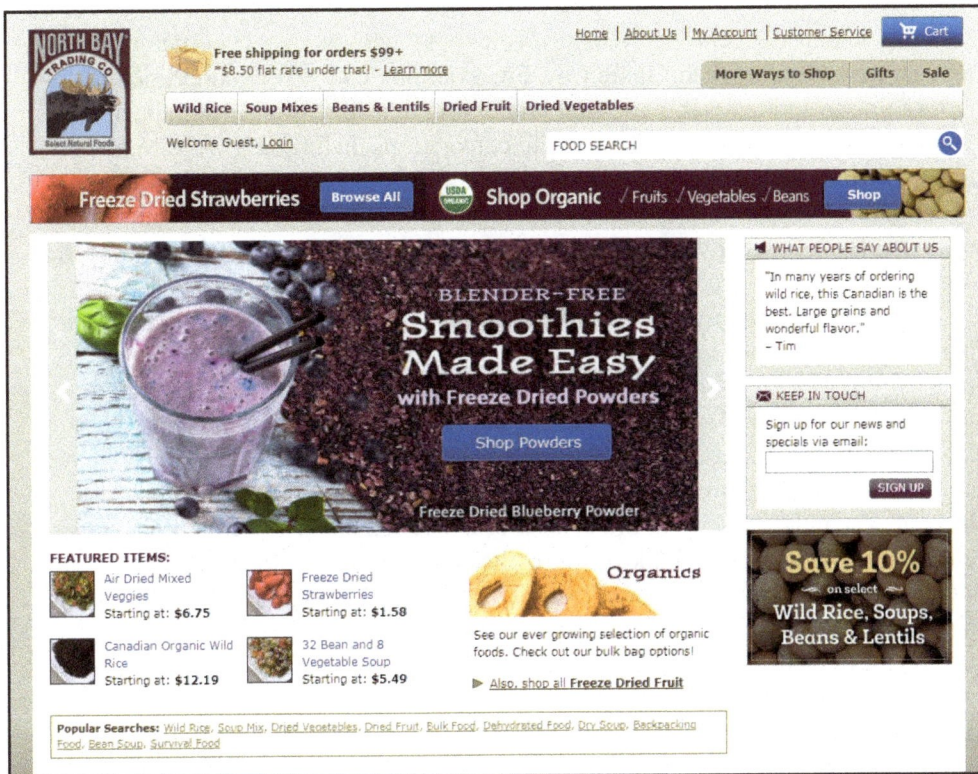

What website users want

- To find what they're looking for as quickly and easily as possible.
- To get information or perform a task as quickly and easily as possible.

How people use a web page

- They scan the page to find what they're looking for or a link to what they're looking for, and they don't like to scroll. If they get frustrated, they leave.
- They often click on links and buttons with the hope of finding what they're looking for, and they frequently click on the Back button when they don't find it.

Three guidelines for improving usability

- Present all of the critical information "above the fold" so the user doesn't have to scroll.
- Group related items into separate components, and limit the number of components on each page.
- Adhere to the current conventions for website usability (see the next figure).

Description

- *Usability* refers to how easy it is to use a website, and usability is a critical requirement for an effective website.

Figure 16-1   An introduction to usability

The current conventions for usability

If you have experience using websites, you know that you expect certain aspects of each website to work the same way. For example, you expect underlined text to be a link to another web page, and you expect that something will happen if you click on a button. These are website conventions that make a website easier to use because they work the same on almost all sites.

Figure 16-2 summarizes some of the other conventions that lead to improved usability. By following these conventions, you give your users what they expect, and that makes your website easier for them to use.

To start, a header usually contains a logo, a search function, and a navigation bar that provides links to the sections of the website. In addition, a header usually provides links to one or more *utilities*, like creating an account, signing in, or getting customer service. Some headers also include a *tag line* that identifies what's unique about the website.

The navigation conventions are also critical to the usability of a website. In brief, a search function should consist of a text box and a search button, clickable items should look like they're clickable, items that aren't clickable shouldn't look like they're clickable, and clicking on the logo in the header of a page should take you back to the home page.

If you implement all of these conventions on your site, you will be on your way to web usability. But that's just a start. The rest of this chapter presents additional ways to improve the usability of a site.

A web page that illustrates some of the website conventions

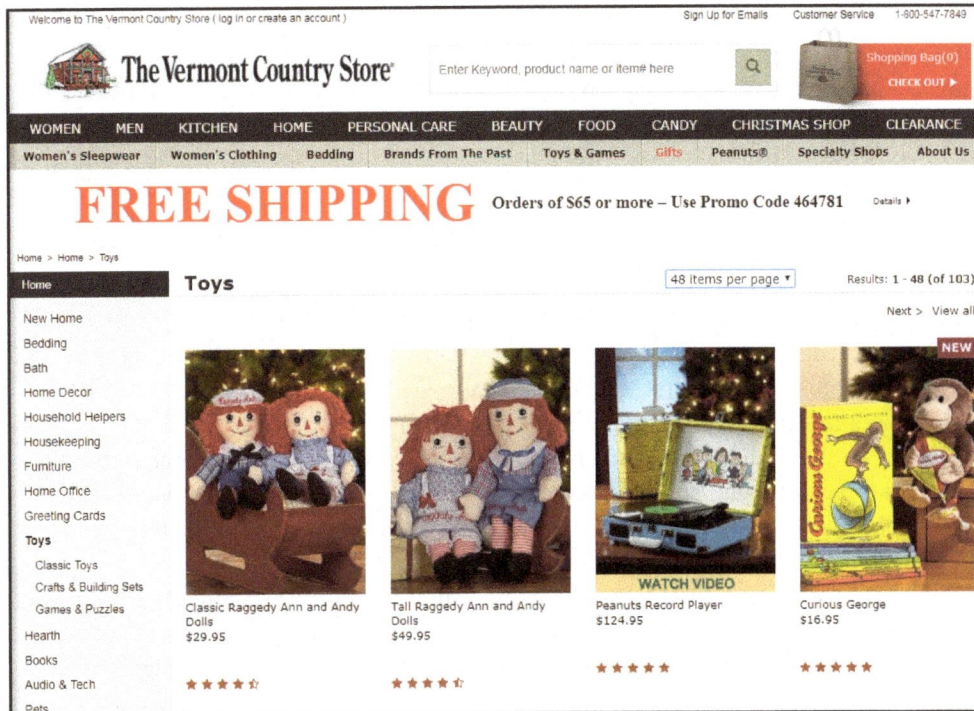

Header conventions

- The header consists of a logo, an optional tag line, utilities, and a navigation bar.
- If there's a tag line, it identifies what's unique about the website.
- The navigation bar provides links that divide the site into sections.
- The utilities consist of links to useful but not primary information.
- If your site requires a search function, it should be in the header, and it should consist of a large text box for the entry followed by a search button.

Navigation conventions

- Underlined text is always a link.
- Images that are close to short text phrases are clickable.
- Short text phrases in columns are clickable.
- If you click on a cart symbol, you go to your shopping cart.
- If you click on the logo in the header, you go to the home page.

Description

- If your website implements the current website conventions, your users can use the same techniques on your site that they use on other sites.

Figure 16-2   The current conventions for usability

Web design guidelines

Have you ever thought about what makes a good website? Well, a lot of experts have. What follows is a distillation of some of the best thinking on the subject.

Use mobile-first design

Today, more than half of the visitors to a website are likely to be using mobile devices with small screens. As a result, for most websites, it's not a good practice to design your web pages for desktop computers with large screens and then handle mobile devices with small screens as an afterthought.

Instead, many web developers consider it a best practice to design your web pages for mobile devices first, and then expand the design for other devices like desktop computers. This can be referred to as *mobile-first design*.

Another approach is to design a web site for devices that have larger screens like desktop computers, and then reduce the design so it works for devices that have smaller screens like mobile phones. This can be referred to as *desktop-first design*, and it might make sense for some websites.

The two screens in figure 16-3 show the same page for a website on a desktop computer and a mobile phone. Here, the content for both pages is pretty much the same. It's just arranged in a way that works the best for each device. To accomplish this, you can use mobile-first or desktop-first design. Since either approach can work, you can choose the approach that seems right for your users. However, since so many people use their mobile phones to access the web today, it usually makes sense to use mobile-first design.

A web page on a desktop computer and a mobile phone

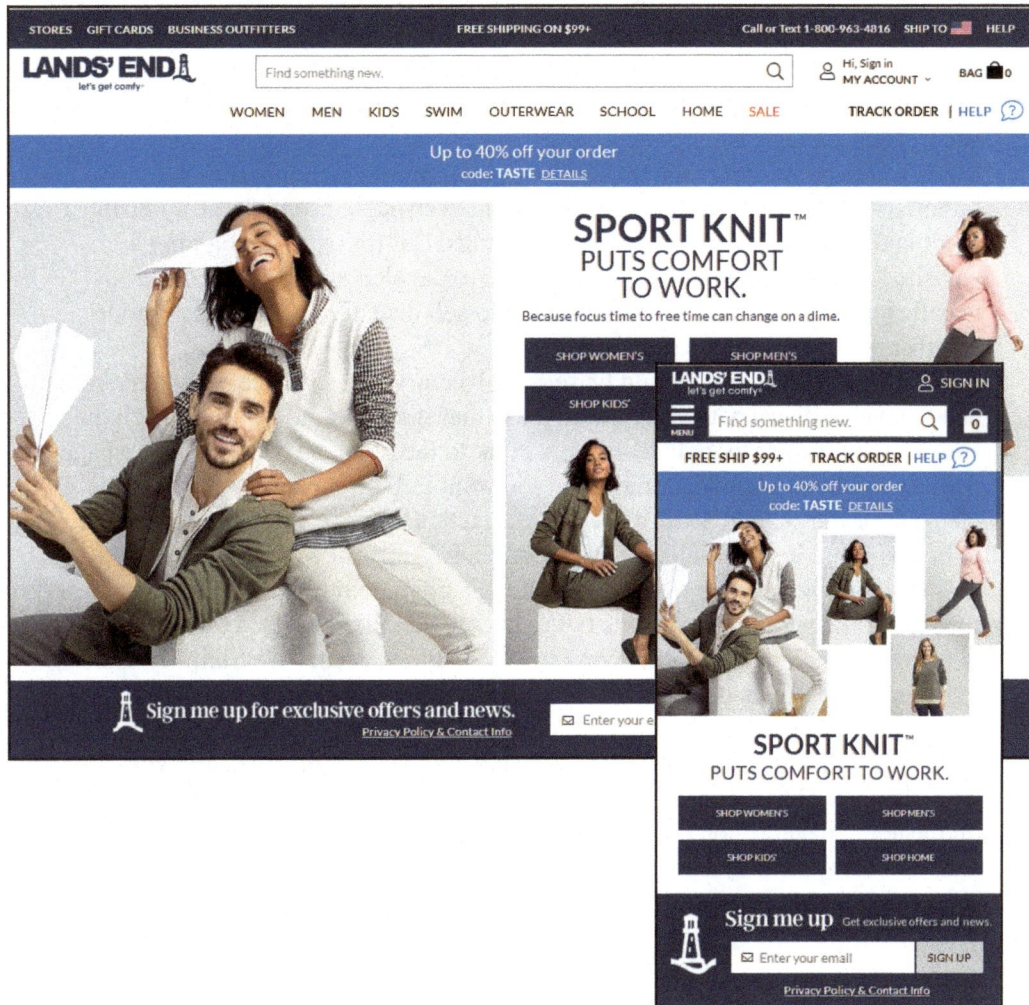

Mobile-first design

- Design for mobile devices first.
- Expand the design for other devices with larger screens.

Description

- Because so many people use mobile devices to view websites, many developers consider it a best practice to use *mobile-first design*. With mobile-first design, you start by designing your pages so they work for mobile devices and then expand them so they work for devices with larger screens.
- Another approach is to design your pages for desktop computers first, and then reduce them for devices with smaller screens. That can be referred to as *desktop-first design*.

Figure 16-3   Use mobile-first design

Use the home page to sell the site

Figure 16-4 presents 9 guidelines for the design of a home page. The first guideline is the most important. It says to emphasize what's different about your site and why it's valuable from the user's point of view. In other words, sell your site to your visitors.

If you're a well-known company with a successful site, this isn't as important. That's why the home pages for the websites of some large companies don't make any special efforts to sell their sites. But if you're developing a website for a small company that still needs to develop a customer base, it usually makes sense to use the home page to sell the site. That may determine whether or not your site is successful.

The other guidelines should be self-explanatory. For instance, you don't need to welcome users to your site because that space should be used to sell the site. You don't want to include an active link to the home page on the home page because it would just reload the page. You want to limit the length of the <title> element for the page to 8 words and 64 characters to make sure it's short enough to be displayed in the results for search engines. And you want to use a different design for the home page to make it stand out from the other pages of the site.

The home page shown in this figure tries to sell the site. Its tag line says: "love what you buy." Then, the copy in the top block adds: "We Simplify The Complex" and shows their three-step procedure for doing that. That encourages the user to at least try one of their recommendations for a product category.

If you're competing with a known brand, and most companies are, you obviously need to let people know what you do better, and you need to state that on the home page. If you don't do that, it is likely to limit the success of your site.

A home page that tries to sell the site

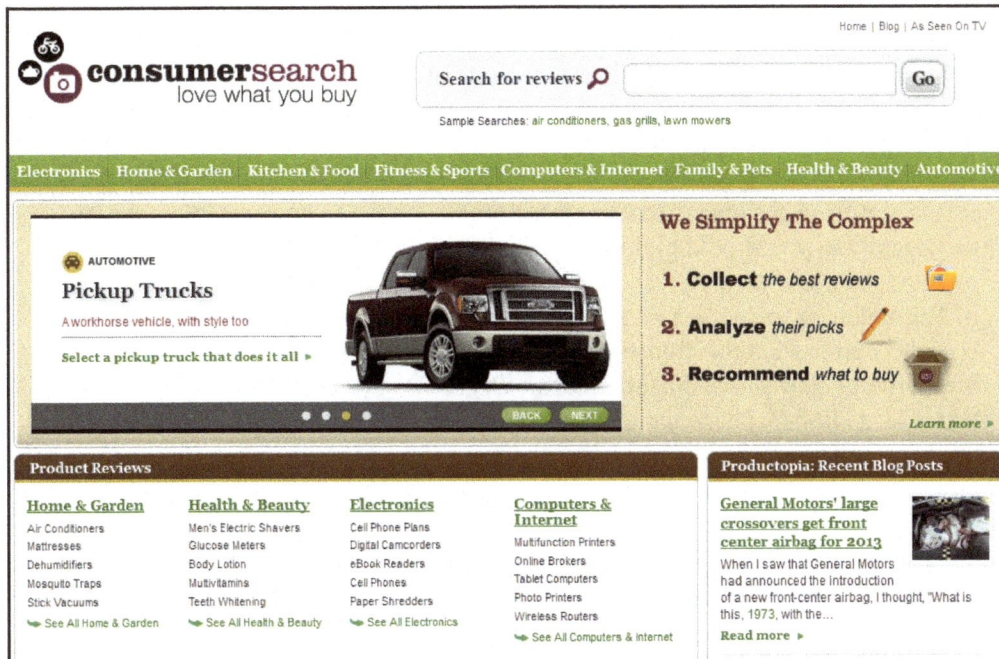

9 guidelines for developing an effective home page

1. Emphasize what your site offers that's valuable and how your site differs from competing sites.
2. Emphasize the highest priority tasks of your site so users have a clear idea of what they can do.
3. Don't welcome users to your site because that's a waste of space.
4. Group items in the navigation areas so similar items are next to each other.
5. Only use icons for navigation if the users will recognize them.
6. Design the home page so it stands out from the other pages of the site.
7. Don't include an active link to the home page on the home page.
8. Code the <title> element for the home page as the organization name, followed by a short description, and limit the title to 8 or fewer words and 64 or fewer characters.
9. If your site provides shopping, include a link to the shopping cart on your home page.

Description

- To a large extent, the home page of your website determines whether or not it gets the results that you want.

Figure 16-4   Use the home page to sell the site

Let users know where they are

As users navigate a site, they like to know where they are. That's why letting users know where they are within a site is one of the current conventions for website usability. Also, since some users may reach your site via search engines, they might not arrive at the home page. In that case, they may need to figure out where they are within your site.

As figure 16-5 shows, there are three primary ways to let users know where they are. First, you should highlight the links that led the user to the current page. Second, the heading for the page should be the same as the link that led to it. Third, you can provide *breadcrumbs* that show the path to the page.

The web page in this figure shows how breadcrumbs work. Here, the breadcrumbs start above the left sidebar: Orvis / Dogs / Traveling with Dogs. In this case, a slash (/) character separates the items in the path, but you can also use other separator characters such as the greater than (>) character. Similarly, Orvis represents the home page, but it's common to just use Home to represent that page.

Farther down the sidebar, the link that led to the page (Traveling with Dogs) is bold. In addition, the heading for the page is TRAVELING WITH DOGS, which is the same as the link that led to the page. On some sites, the link in the navigation bar is also highlighted. Here, that link would be DOGS.

A product page with breadcrumbs and highlighted links

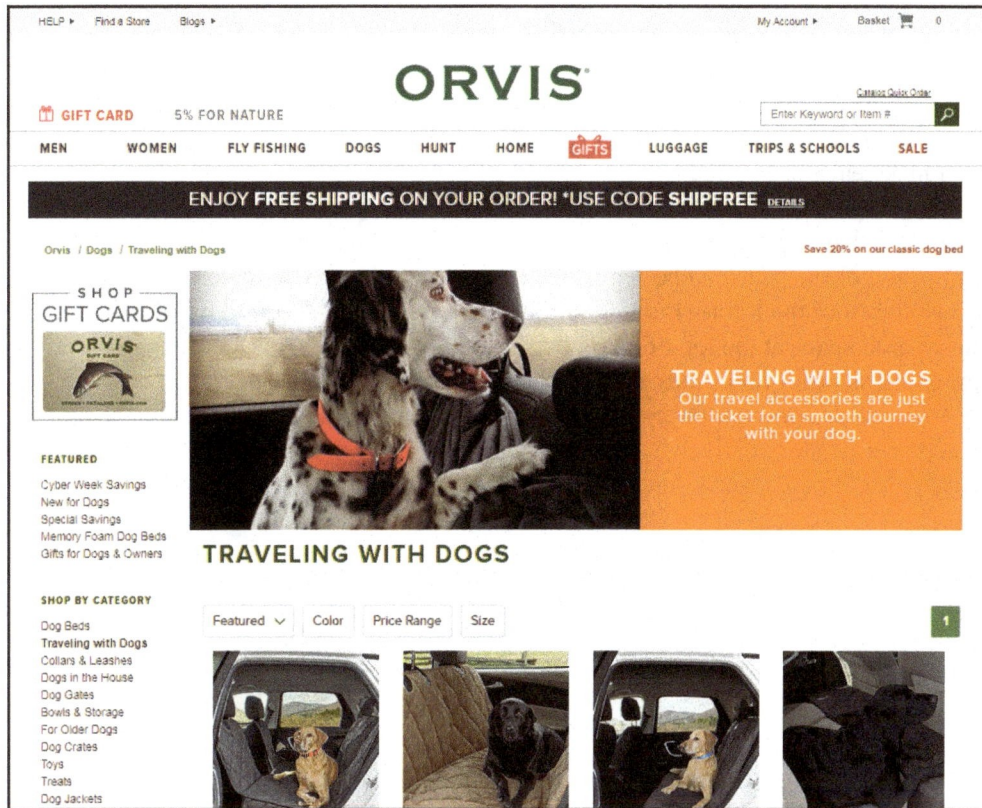

How to let the users know where they are

- Highlight the active links.
- Make the heading for the page the same as the link that led to the page.
- Provide breadcrumbs in this general format: Home > Page1 > Page2.

Description

- As your site gets more complex, users are likely to lose track of where they are in the site. But even simple sites should let users know where they are.

Figure 16-5   Let users know where they are

Make the best use of web page space

As you design a web page, remember that the part of the page that's above the fold is the most valuable space. As a result, you want to put the most important components of the page in that space. That's why many of the most successful websites have small headers. This gives them more space for the components that make each page effective.

To emphasize this point, the screen in figure 16-6 presents an example of a web page that doesn't get the most out of the space above the fold. In fact, none of the text of the page can be seen above the fold. Instead, you have to scroll down for all of the information. The photos are beautiful (if you like fishing), but they waste space at the top of every page on the website. This is a case where the graphics, as beautiful as they may be, diminish the usability of the page.

As you design your pages, try to follow the three guidelines in this figure. Keep your header relatively small. Prioritize the components that go on the page. And give the best space to the most important components.

Wasted space on a primary page

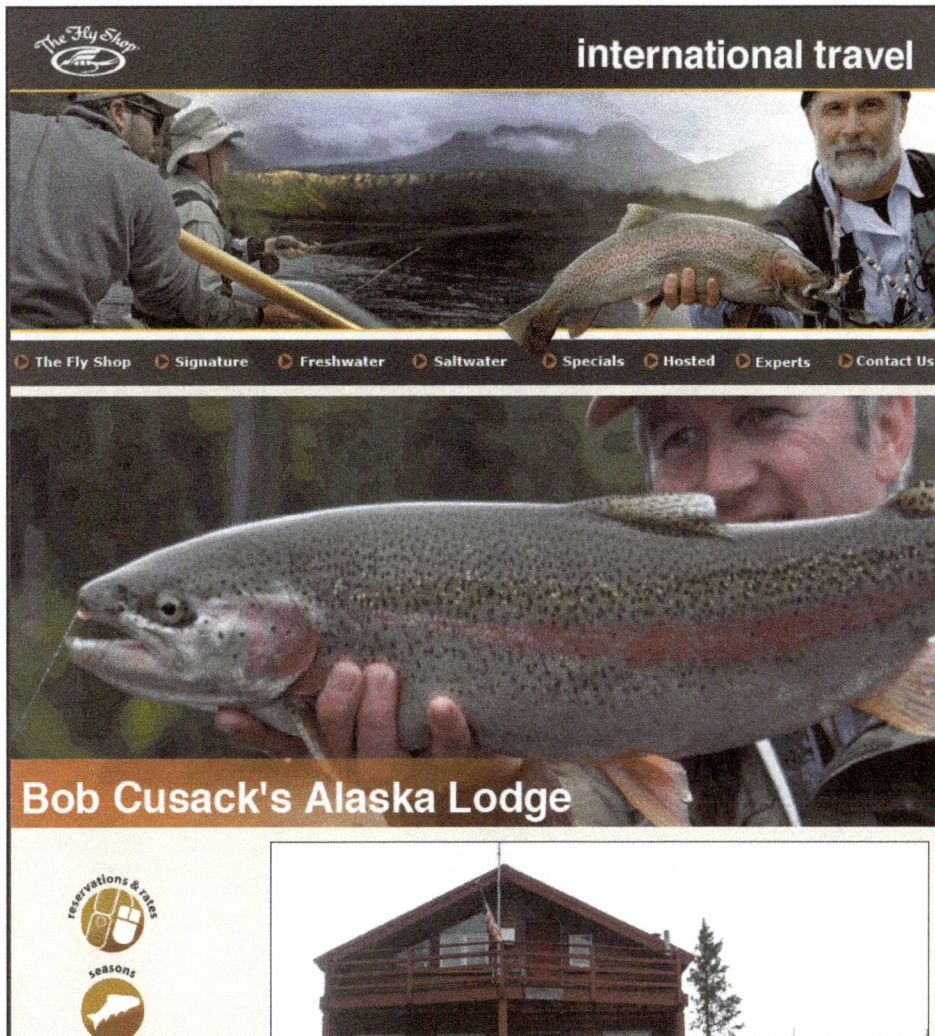

Guidelines for the effective use of space

- Keep the header relatively small.
- Prioritize the components for each page.
- Give the most important components the primary locations.

Description

- The most valuable space on each web page is the space above the fold. To get the most from it, you need to prioritize the components on each page and give the best locations to the highest-priority components.

Figure 16-6   Make the best use of web page space

Divide long pages into shorter chunks

Website users don't like to scroll. So, a general guideline for web page design is to limit the amount of scrolling to one-and-one-half or two times the height of the browser window. But what if you need to present more information than that?

The best solution is to use *chunking* to divide the content into logical pieces or topics. This is illustrated by the screen in figure 16-7. Here, a potentially long page of copy is chunked into smaller topics like "What makes Murach books the best". Then, if users are interested in the topic, they can click on the MORE button to expand that copy. Below the two topics with MORE buttons, an accordion presents "Five more reasons why our customers love our books." So in all, the copy for this page has been broken down into seven topics.

This approach lets users select the topics that they're interested in so they have more control over their website experience. This makes it easier for users to find what they're looking for. And this reduces the need for scrolling. This provides better usability than forcing users to scroll through a long page of text trying to find what they're looking for.

Incidentally, you can use JavaScript to reveal and hide portions of text as in this example. You also use JavaScript for common features like accordions and tabs that reveal and hide text. As a result, you may want to use JavaScript to implement effective chunking. If you do, we recommend *Murach's Modern JavaScript* as the perfect companion for this HTML and CSS book.

Chunking on a web page

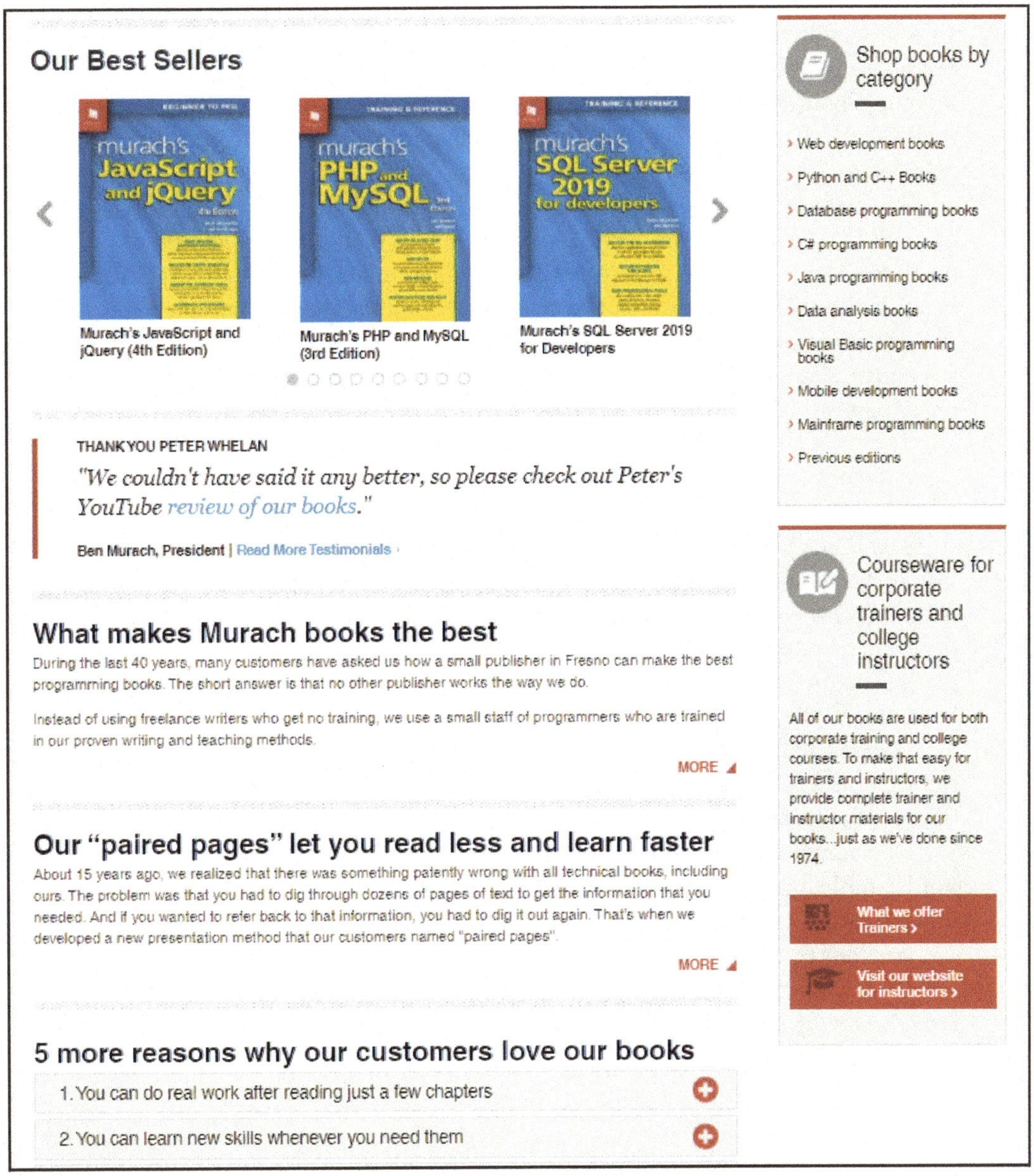

Our Best Sellers

Murach's JavaScript and jQuery (4th Edition)

Murach's PHP and MySQL (3rd Edition)

Murach's SQL Server 2019 for Developers

THANK YOU PETER WHELAN

"We couldn't have said it any better, so please check out Peter's YouTube review of our books."

Ben Murach, President | Read More Testimonials

What makes Murach books the best

During the last 40 years, many customers have asked us how a small publisher in Fresno can make the best programming books. The short answer is that no other publisher works the way we do.

Instead of using freelance writers who get no training, we use a small staff of programmers who are trained in our proven writing and teaching methods.

MORE ▲

Our "paired pages" let you read less and learn faster

About 15 years ago, we realized that there was something patently wrong with all technical books, including ours. The problem was that you had to dig through dozens of pages of text to get the information that you needed. And if you wanted to refer back to that information, you had to dig it out again. That's when we developed a new presentation method that our customers named "paired pages".

MORE ▲

5 more reasons why our customers love our books

1. You can do real work after reading just a few chapters   ⊕
2. You can learn new skills whenever you need them   ⊕

Shop books by category

› Web development books
› Python and C++ Books
› Database programming books
› C# programming books
› Java programming books
› Data analysis books
› Visual Basic programming books
› Mobile development books
› Mainframe programming books
› Previous editions

Courseware for corporate trainers and college instructors

All of our books are used for both corporate training and college courses. To make that easy for trainers and instructors, we provide complete trainer and instructor materials for our books...just as we've done since 1974.

What we offer Trainers ›

Visit our website for instructors ›

Description

- *Chunking* refers to dividing content into smaller chunks that can be presented on separate web pages or in separate components on the same page. This lets the users select the chunks of information that they're interested in.

Figure 16-7   Divide long pages into shorter chunks

Know the principles of graphic design

Even if you aren't a graphic designer, you should at least know the four basic principles of graphic design that are presented in figure 16-8. These principles are implemented by all of the best websites, so it's easy to find examples of their use. So far, all of the examples in this chapter make extensive use of these principles.

For instance, *alignment* means that aligning related items gives a sense of order to a web page. Similarly, *proximity* means that related items should be close to each other. When proximity is applied to headings, it means that a heading should be closer to the text that follows it than the text that precedes it.

The principle of *repetition* means that you repeat some elements from page to page to give the pages continuity. For instance, the header is usually the same on all pages.

The principle of *contrast* draws your eye to a component. For instance, a component with a large heading or a background color will draw attention because it stands out from the other components of the page.

The home page in this figure illustrates some of these principles with the four images carefully aligned and the description of each image right below it. But like most websites, there is always room for improvement. For instance, the principle of proximity says that the Adventure Trips heading should be closer to the text below it. In other words, there shouldn't be such a large gap between the heading and the related text.

Besides adhering to the design principles, you need to get the typography right if you want your users to read the text on your website. That's why this figure presents six guidelines for doing that.

One common problem is when lines of text are longer than 65 characters. For example, the two lines of text under the Adventure Trips heading are much longer than 65 characters. This makes the text hard to read.

Another common problem is using a background image or a background color that makes the text hard to read. And a third problem is the use of reverse type (white type on a darker background), especially for text, because that can make it harder to read.

A website that adheres to the principles of graphic design

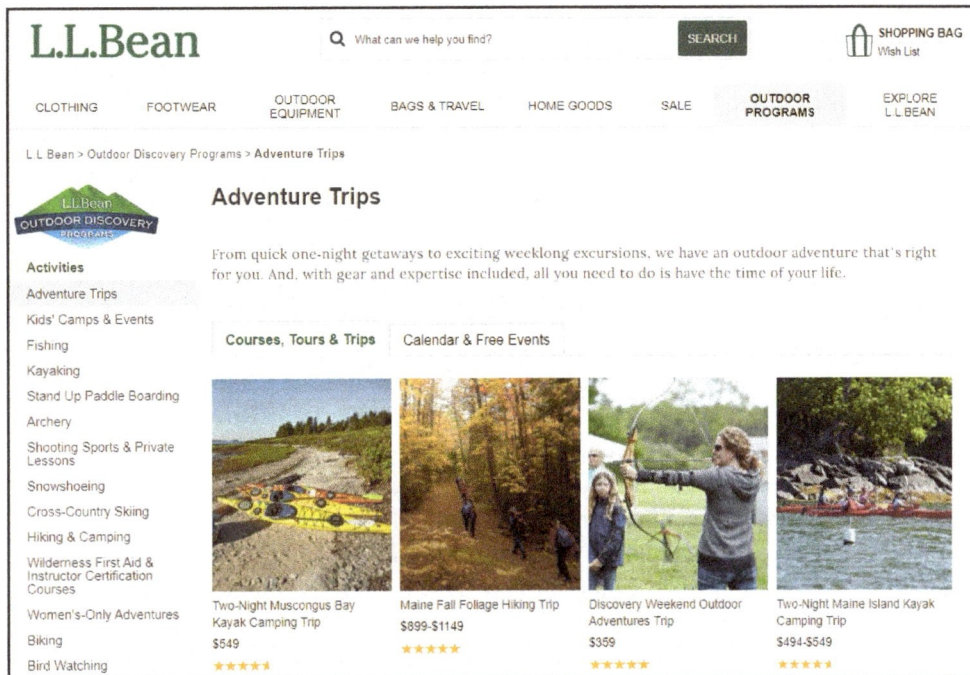

Four principles of graphic design

- *Alignment* means that related items on the page should line up with each other.
- *Proximity* means that related items should be close together.
- *Repetition* means that you should repeat some elements from page to page to give the pages continuity.
- *Contrast* is what draws your eye to some of the components on a web page.

Typographical guidelines

- Use a sans serif font in a size that's large enough for easy reading.
- Limit the line length of text to 65 characters.
- Use dark text on a light background, and don't use an image for the background.
- Don't use reverse type (white type on a colored background) for text.
- Don't center text and don't justify text.
- Avoid using all caps (all capital letters) for headings because all caps are hard to read.

Description

- If you aren't a graphic designer, you can at least implement the principles of graphic design and get the typography right.

Figure 16-8   Know the principles of graphic design

Write for the web

Instead of reading each line of text thoroughly, most web users skim and scan. They want to get the information that they're looking for as quickly as possible, and they don't want to have to dig through several paragraphs of text for it. That's why writing for a website is different from writing for the printed page.

Figure 16-9 presents a block of copy that isn't written for the web, followed by one that is. To present the same information in a way that's more appropriate for the web, the second example uses fewer words (135 to 177), which is the quickest way to improve your web writing. It also uses a numbered list to make it easy to find the flight information without having to dig through the text.

Other web writing guidelines recommend using an inverted pyramid style that presents the most important items first, using headings and subheadings to identify portions of the text, using bulleted and numbered lists to make information more accessible, and using tables to present tabular information. That way, website users can quickly get the information they're looking for.

The last guideline is to make the text for all links as explicit as possible. For instance, "Find a job" is better than "Jobs", and "Apply for unemployment compensation" is better than "Unemployment". Remember that more than 30 percent of your users' clicks are likely to be on the Back button because a link didn't take them where they wanted. The more explicit your links are, the less that should happen.

Incidentally, the first example in this figure also illustrates poor typography. In particular, most of the lines of text are more than 90 characters long, which makes the text unappealing and hard to read. To fix that problem, you can increase the font size and shorten the line width, as shown in the second example.

Writing that isn't for the web (177 words and 1041 characters)

> **The progressive air services you'll use to reach Cusack's Alaska Lodge are a wonderful reflection of your journey into the wilderness.** First, you will fly a major jet service from near your home to Anchorage, Alaska; arriving here, most itineraries will mandate an overnight stay. The next morning you will board a small plane piloted by one of the fine bush pilots of Iliamna Air Taxi (often one of their Pilatus aircraft, a high-flying, very comfortable aircraft), for the transfer between Anchorage and the little village of Iliamna. Upon arriving, Iliamna Air Taxi's Iliamna crew will switch your gear from the mid-sized plane to a smaller, float-equipped Cessna or Beaver, and after a short wait, you will be on the final leg of your adventure, touching down on the lake's surface in front of Bob's lodge a short thirty minutes later. For the remainder of your week, Bob will be your pilot, flying you into amazingly beautiful country in his two small airplanes, giving you a peek into the enormity and grandeur of his corner of Alaska.

The same copy, but written for the web (135 words and 717 characters)

> **The three-part flight to Bob Cusack's Alaska Lodge is a fascinating journey into the Alaskan wilderness**
>
> 1. You take a major jet service from your home to Anchorage, Alaska.
> 2. From Anchorage, you take the Iliamna Air Taxi to the little village of Iliamna. This flight will be piloted by one of the Air Taxi's fine bush pilots in a comfortable plane like the Pilatus.
> 3. In Iliamna, the Air Taxi's crew will switch your bags to a smaller, float plane like a Cessna or a Beaver. Then, after a short wait and a 30 minute flight, you will touch down on beautiful Lake Iliamna in front of Bob's lodge.
>
> For the remainder of your week, Bob will be your pilot as he takes you into the beauty and grandeur of his corner of Alaska.

Web writing guidelines

- Use fewer words.
- Write in inverted pyramid style with the most important information first.
- Use headings and subheadings to identify portions of the text.
- Use bulleted lists and numbered lists to make information more accessible.
- Use tables for tabular information.
- Make the text for all links as explicit as possible.

Description

- Web users skim and scan; they don't read like book readers do. So when you write for the web, you need to change the way you think about writing.

Figure 16-9   Write for the web

Perspective

If you apply what you've learned in this chapter, you should be able to design simple websites of your own. But you should also realize that designing and developing a website is a challenging process that requires a wide range of skills. That includes graphic design, HTML, CSS, JavaScript, writing for the web, and more.

Terms

usability
utility
tag line
mobile-first design
desktop-first design
breadcrumbs
chunking
alignment
proximity
repetition
contrast

Chapter 17

How to deploy a website

Once you've developed and tested your web pages on your computer, you're just a few steps away from making those pages available to anyone in the world who is connected to the internet. To do that, you just need to upload the files to a web server that's connected to the internet. This chapter shows one way to do that.

This chapter also gives you some ideas for starting your own website. That includes getting a web host and a domain name, and getting your site into search engines like Google and Bing. However, because there are so many ways to deploy a website, you have to take the ideas in this chapter and figure out the best way to use them for your website.

How to deploy a website for testing

After you're done testing your code on your local system, you're ready to *deploy* it. When you do that, you move your website to a server, such as a testing server or a production server. One way to deploy a website for testing is to use one of the many available cloud platforms. The figures that follow show how to deploy a static website to the DigitalOcean cloud platform so you can test it. To do that, you need to start by storing the files for the website on GitHub.

How to set up a GitHub account

GitHub is an online platform that developers use to create, store, manage, and share their code. GitHub uses *version control software* named Git that tracks the changes developers make to their code. GitHub is owned by Microsoft.

Figure 17-1 begins by showing the URL that you can use to sign up for a free GitHub account. After you follow the instructions and set up an account, the GitHub website displays a dashboard like the one shown in this figure. You can use this dashboard to view and create *repositories*, which contain code files and a history of the revisions made to each code file.

The GitHub website

https://github.com

The GitHub Dashboard

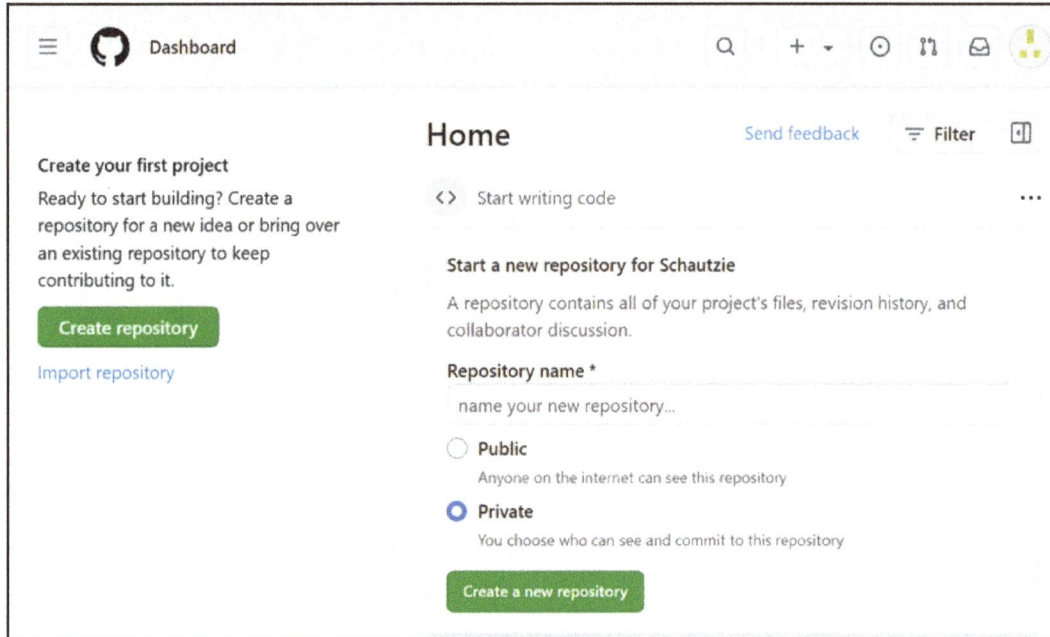

Description

- GitHub is an online platform where developers can create, store, manage, and share their code.
- GitHub uses *version control software* named Git to track the changes you make to your code files.
- When you sign up for GitHub, be sure to select the free option.

Figure 17-1   How to set up a GitHub account

How to install the GitHub Desktop app

Many developers use command line tools to work with GitHub. However, the GitHub Desktop app provides a graphical user interface that makes it easier to work with GitHub. As a result, unless you already know how to use the command line to work with GitHub, we recommend installing GitHub Desktop as described in figure 17-2.

When installing GitHub Desktop, it's important to enter your GitHub username and password. Then, you need to authorize GitHub Desktop to access your GitHub account. That way, GitHub Desktop can create and update GitHub repositories for you.

After you install GitHub Desktop, it should display a screen like the one shown in this figure. If you don't have any repositories in your GitHub account, this screen displays "Let's get started!" on its left side as shown here. Otherwise, it displays the current repository on its left side.

The URL to download the GitHub Desktop app

https://desktop.github.com/download/

How to install the GitHub Desktop app

1. Download the setup file to your computer. For Windows, this file should be an exe file. For macOS, this should be an app file that you can move into your Applications folder.
2. Double click the downloaded file.
3. In the "Welcome to GitHub Desktop" dialog, click "Sign in to Github.com".
4. Enter your GitHub username and password.
5. Select a user to authorize GitHub Desktop and click Continue.
6. Review the statement describing what you're authorizing GitHub Desktop to access. If it all looks OK to you, click Authorize Desktop.
7. If you get a dialog that the site is trying to open a file, click Open.
8. In the Configure Git dialog, accept the defaults and click Finish.

The GitHub Desktop app

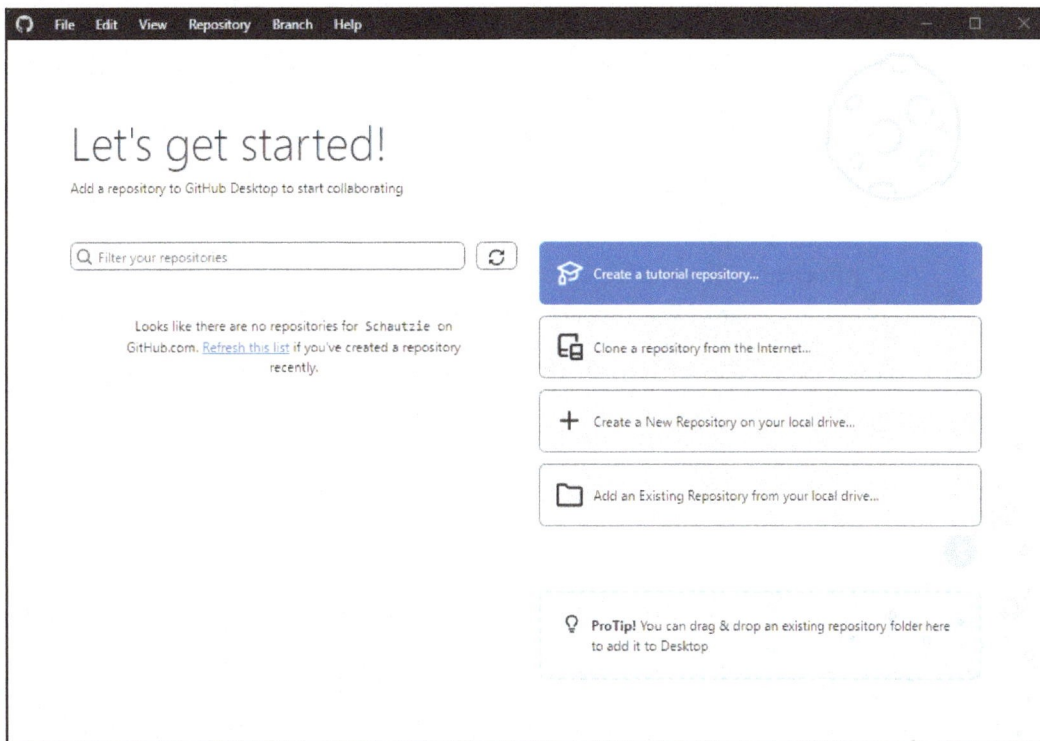

Description

- Many developers use command line tools to work with GitHub. However, the GitHub Desktop app provides a graphical user interface that makes it easier to work with GitHub.

Figure 17-2   How to install the GitHub Desktop app

How to create a GitHub repository

Figure 17-3 shows how to use GitHub Desktop to create a GitHub repository. To start, you use the first procedure in this figure to create a local repository. In other words, you create a repository that's on your computer but not yet on the GitHub website.

After you create a local repository, GitHub Desktop displays the repository as shown by the screen in this figure. Here, the top part of the left pane shows the name of the new repository as the current repository. To view a list of all local repositories, you can click the arrow to the right of the current repository name. Then, if you want to switch to another repository, you can select it from the list.

In this figure, the current repository is named town-hall. It's a common GitHub naming convention to use all lowercase letters with words separated by dashes. This is sometimes called *kebab case*, because the words separated by dashes resemble food on a skewer.

Under the name of the current repository, the left pane displays any code files that have changes. This pane also has a tab that you can use to view the change history. In this figure, the left pane has 0 changed files because this repository is new and doesn't contain any code files yet. Finally, the bottom of the left pane displays a button that allows you to commit changes.

The right pane of GitHub Desktop displays different messages and buttons depending on the state of the current repository. In this figure, it displays a "No local changes" message. That makes sense because this repository doesn't have any files yet. Below that, the right pane displays several buttons that correspond with suggestions for what to do next.

After you create a new local repository, it's a good idea to publish it to GitHub, even if it doesn't contain any files yet. To do that, you click the "Publish repository" button. In this figure, this button is highlighted. After you publish your repository to GitHub, you can click the "View on GitHub" button to view it in a browser.

How to create a local repository

1.   Start GitHub Desktop.
2.   Select File→New Repository…
3.   In the "Create a new repository" dialog, enter a name and description for your repository. You can accept the defaults for the other fields.
4.   Click Create Repository to create a local repository. GitHub Desktop should display the new repository in the left pane with 0 changed files, and the right pane should display a "No local changes" message and some suggested actions.

A new local repository with 0 changed files

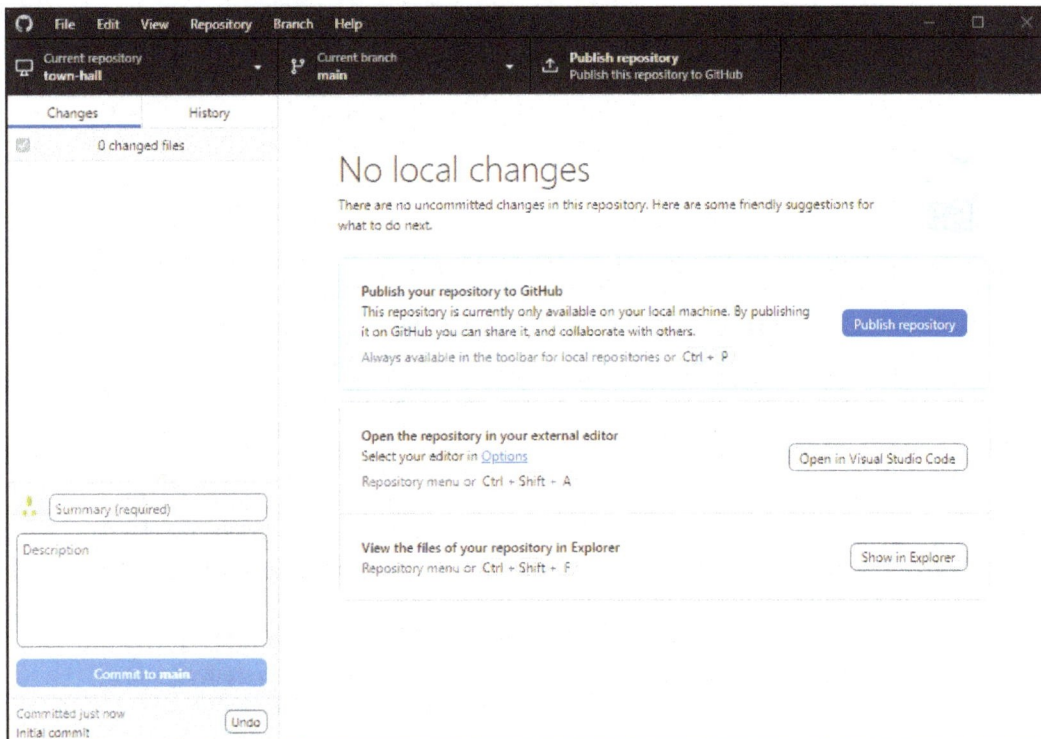

How to publish the repository to GitHub

1.   In the right pane, click "Publish repository".
2.   In the "Publish repository" dialog, accept the defaults and click "Publish repository".

How to view the repository on GitHub after you publish it

•   In the right pane, click "View on GitHub".

Figure 17-3   How to create a GitHub repository

How to add files to a repository

After you create a GitHub repository, you can add your website files to it. To do that, you can use GitHub Desktop as described in figure 17-4.

If you have an existing website that you want to add to the repository, you can use File Explorer (Windows) or Finder (macOS) to find and copy the folders and files you want to add. Then, you can use GitHub Desktop to open the folder that contains the local repository. Next, you can paste the copied files into that folder. Once you do, the left pane of GitHub Desktop displays the pasted files as changed files.

If you're creating a new website that you want to be in the repository, you can use GitHub Desktop to open the repository in VS Code. The first time you do this, VS Code might display a message that says "Git isn't found". In that case, you can use the link that VS Code provides to download and install Git for Windows or macOS. Once you've installed Git, you can switch back to VS Code.

After you open the repository in VS Code, it displays the code files of the repository, including a .gitattributes file. If it's a new repository, the .gitattributes file is the only file in the repository. Then, you can add HTML, CSS, and image files for your website, just like any other website. Once you save your changes, GitHub Desktop displays the new files in its left pane as changed files.

The screen at the bottom of part 1 of this figure shows a local repository in GitHub Desktop with 6 changed files. The left pane displays the changed files, and the right pane displays the contents of the file that's selected in the left pane. Here, the first file in the left pane is selected. Since this file is the favicon.ico file, GitHub Desktop displays the favicon image in the right pane. However, if you selected the index.html file, the right pane would display the HTML for the file. Similarly, if you selected the styles/main.css file, the right pane would display the CSS for the file.

If you're working on macOS, you may see a file named .DS_Store listed with the changed files. This file is generated by the operating system to store information about the folder, and should have no effect on your repository.

How to add existing files to a local repository

1. Open File Explorer (Windows) or Finder (macOS). Navigate to the website folder and copy the folders and files you want to add to the repository.

2. Switch to GitHub Desktop. In the right pane, click on "Show in Explorer" or "Show in Finder". This should display the local folder for the repository.

3. Paste the copied files into the folder for the repository.

4. Switch to GitHub Desktop. Its left pane should display the pasted files as changed files.

How to add new files to a local repository

1. In GitHub Desktop, in the right pane, click "Open in Visual Studio Code".

2. Switch to VS Code. If it displays a message that says that Git isn't found, click "Download Git" and download the Git file for Windows or macOS. When the file is downloaded, double-click it and follow the installation instructions.

3. In VS Code, note that your local repository contains a .gitattributes file.

4. Use VS Code to add new HTML or CSS files and save your changes.

5. Switch to GitHub Desktop. Its left pane should display the added files as changed files.

A local repository with 6 changed files

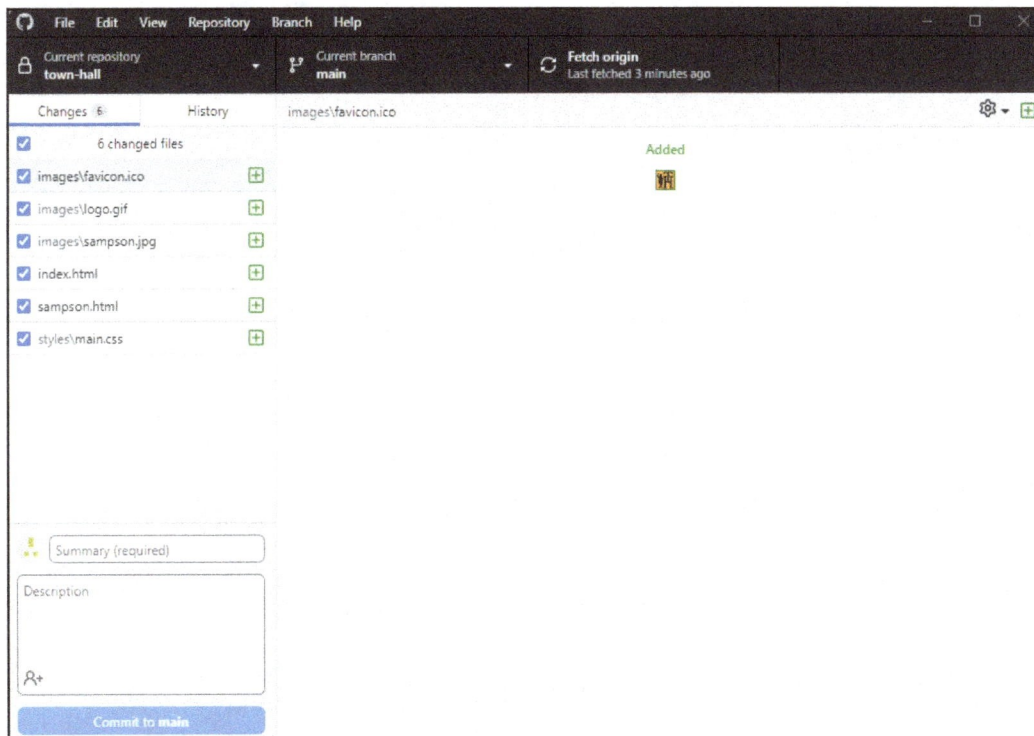

Figure 17-4   How to add files to a repository (part 1)

Once you add existing or new files to the local repository as shown in part 1 of figure 17-4, you can copy these files to your repository on GitHub as shown in part 2. First, you need to commit the changed files to the local repository. You can think of this as saving the changes that you made in part 1. To do that, you begin by entering a comment in the Summary field that briefly describes the changes. For example, you could enter "Starting files" for the changed files shown in this figure. This entry is required, and GitHub Desktop won't let you proceed with your commit without it. You can also enter a longer description of the changes in the Description field, but that is optional.

After you enter a value in the Summary field, click the "Commit to main" button to commit the changed files. Once you do that, the left pane no longer displays any files, and the right pane once again displays a "No local changes" message and some buttons for suggested actions.

At this point, you've committed your changes to the local repository, but you still need to push those committed changes to your online GitHub account. To do that, you can click the "Push origin" button. Then, you can use a browser to view those committed changes on GitHub by clicking the "View on GitHub" button.

A local repository with changed files

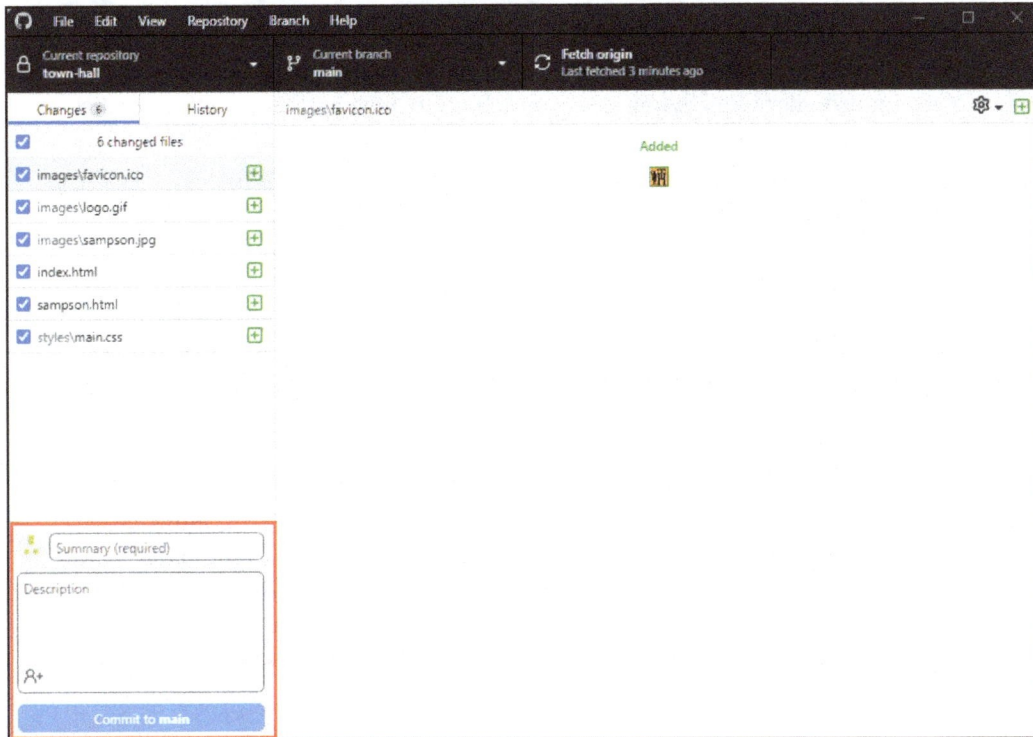

How to commit the changes in the local repository

1. At the bottom of the left pane, enter a value in the Summary field. If you want, you can also enter an optional description.

2. Click "Commit to main" to update the local repository. Then, GitHub Desktop should show 0 changed files in the left pane, and it should present some suggested actions in the right pane.

How to push the committed changes to the repository on GitHub

• In the right pane, click "Push origin".

How to view the repository on GitHub after you push changes

• In the right pane, click "View on GitHub".

Figure 17-4   How to add files to a repository (part 2)

How to set up a DigitalOcean account

At this point, you've got website files in GitHub, but they're not deployed as a website that can be viewed in a browser yet. To do that, you can use a cloud platform provider like DigitalOcean, which offers hosting for up to three static websites at no charge.

To get started, go to the DigitalOcean website presented in figure 17-5 and sign up for an account. You can sign up for DigitalOcean via email, GitHub, or Google. We recommend using the GitHub account you created earlier in this chapter as shown in this figure.

When you sign up for DigitalOcean, payment information is required. However, as long as you don't deploy more than three static websites, you won't be charged. You may see a temporary pre-authorization charge, but that should be reversed within a week.

When you finish signing up, DigitalOcean should display a Welcome page. On that page, click the "Explore our Control Panel" link to go to the control panel shown here. This control panel displays a project named first-project, which is a default project that you can store your websites in. In the future, you may want to use other projects to group your resources, but you don't need to worry about that now. From the control panel, you can use the App Platform portal indicated with the red arrow here to create a static website as shown in the next figure.

The DigitalOcean website

https://cloud.digitalocean.com

How to sign up for DigitalOcean

1. Go to the DigitalOcean website and click Sign Up.

2. Select "Sign up with GitHub".

3. Enter your GitHub username and password.

4. Click "Authorize digitalocean".

5. Fill out and submit the form on the Welcome screen. Choose "Hobbyist or Student" for the role or business type.

6. Payment information is required at signup. However, you won't be charged. You may see a pre-authorization charge, but that's temporary and should be reversed within a week.

7. In the "Welcome to DigitalOcean" page that displays, click "Explore our Control Panel".

The DigitalOcean control panel

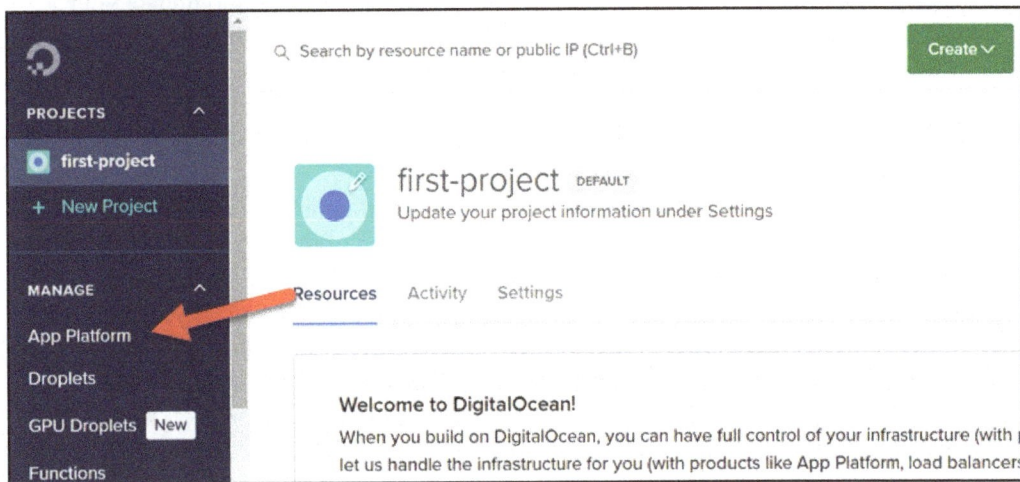

Description

- DigitalOcean is a cloud platform provider that offers free hosting for up to three static websites.

- You can use the App Platform portal to deploy a static website.

Figure 17-5   How to set up a DigitalOcean account

How to deploy a website

Figure 17-6 shows how to deploy a static website on DigitalOcean. To start, you go to the DigitalOcean control panel and click the App Platform link. This displays the Apps page.

Once you select GitHub as the source provider and click Create App, DigitalOcean displays the Resources page, which is the first of four pages that you can use to create your website. On this page, make sure to give DigitalOcean permission to access your GitHub repositories. That way, it can use your code files on GitHub to create the static website.

To give DigitalOcean permission to access your GitHub files, scroll down to the Manage Access button and click on it. Then, you can authorize DigitalOcean to access all the repositories you have on GitHub, or you can limit the authorization to select repositories.

Once you've authorized DigitalOcean to access your GitHub repositories, you can accept the defaults for the rest of the Resources page. One default value to note is the Autodeploy check box, which is checked by default. As a result, your DigitalOcean website redeploys whenever it detects changes to the code files in your GitHub repository. In other words, if you want to update your DigitalOcean static website, all you need to do is update your GitHub repository. You'll see how this works a little later in this chapter.

After you click Next on the Resources page, the Environmental Variables page is displayed. For the purposes of this chapter, you don't need any environmental variables, so you can just click Next here.

Next is the Info page, which presents the name of your app and the name of the project that contains the app. By default, the name of the app becomes part of the URL for your static website. You can accept the ocean-related name that DigitalOcean generates, or you can click Edit to change the name. If you change it, the name you choose must contain only lowercase letters and dashes.

After you click Next on the Info page, the Review page is displayed. You should review this page carefully to make sure your entries are correct. In particular, you should pay attention to the Billing section to make sure that you're creating a free website.

When you're satisfied with the Review page, click Create Resources to create your website. When you do, it may take DigitalOcean a while to build and deploy your website, but when it's done it should display a page like the one shown in this figure. To view the deployed website, you can click on the link below the app name (the one after the project name).

Note that the URL DigitalOcean generates for your deployed website includes the name from the Info page and ends with "ondigitalocean.app". However, if you have a custom domain name, you can change the static website to use that domain name instead. For more information about doing that, you can view the documentation that's at the URL listed at the bottom of this figure. In addition, you can learn more about getting a custom domain name a little later in this chapter.

How to use the DigitalOcean App Platform to deploy a static website

1. In the Control Panel, click App Platform.

2. Select GitHub as the Source Provider and click Create App.

3. On the Resources page, scroll down and click Manage Access to give DigitalOcean permission to access your GitHub repository. You can authorize all repositories, or only the repository that contains the static website you are deploying. Then, click Install & Authorize.

4. Select the repository you want to deploy. Accept the defaults and click Next.

5. Review the resources to make sure you're creating a static site. Then, click Next.

6. On the Environmental Variables page, click Next.

7. On the Info page, you can change the app name that was generated by DigitalOcean. To do that, click Edit, change the name, click Save, and click Save again. When you're done, click Next.

8. On the Review page, review all the information. Pay particular attention to the Billing section to make sure your app is free. Then, click Create Resources.

9. It may take DigitalOcean a few moments to deploy your website. When it's done, click on the link under the app name to view your website.

The Apps page after a new app is created

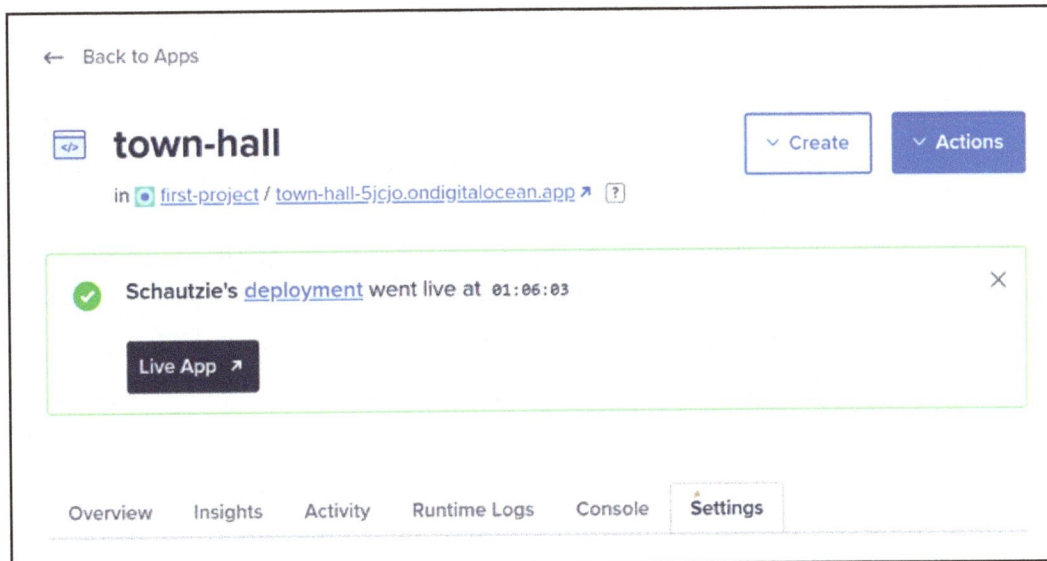

Documentation on how to use a custom domain with your static website

https://docs.digitalocean.com/products/app-platform/how-to/manage-domains

Figure 17-6   How to deploy a website

How to test a deployed website

Once you have deployed your website, you should test it. Figure 17-7 gives you some ideas for testing a deployed web page. To start, you need to make sure that the page displays all of the content correctly on different devices. Then, you need to check all the links that go to and from the web page.

For instance, a web page often links to one or more HTML, CSS, and image files. Then, if you forget to upload a supporting file, one of the links won't work correctly. In that case, you can solve the problem by pushing the supporting file to your GitHub repository.

To test an entire deployed website, you need to methodically review each of the pages on the site, including all of the links to and from each page. The larger the site, the more difficult this is, and the more methodical you need to be.

The screen in this example shows the Town Hall app from chapter 8 after it's deployed to the DigitalOcean app platform. From here, you can test the site like you would on your local computer.

The Town Hall static website deployed to DigitalOcean

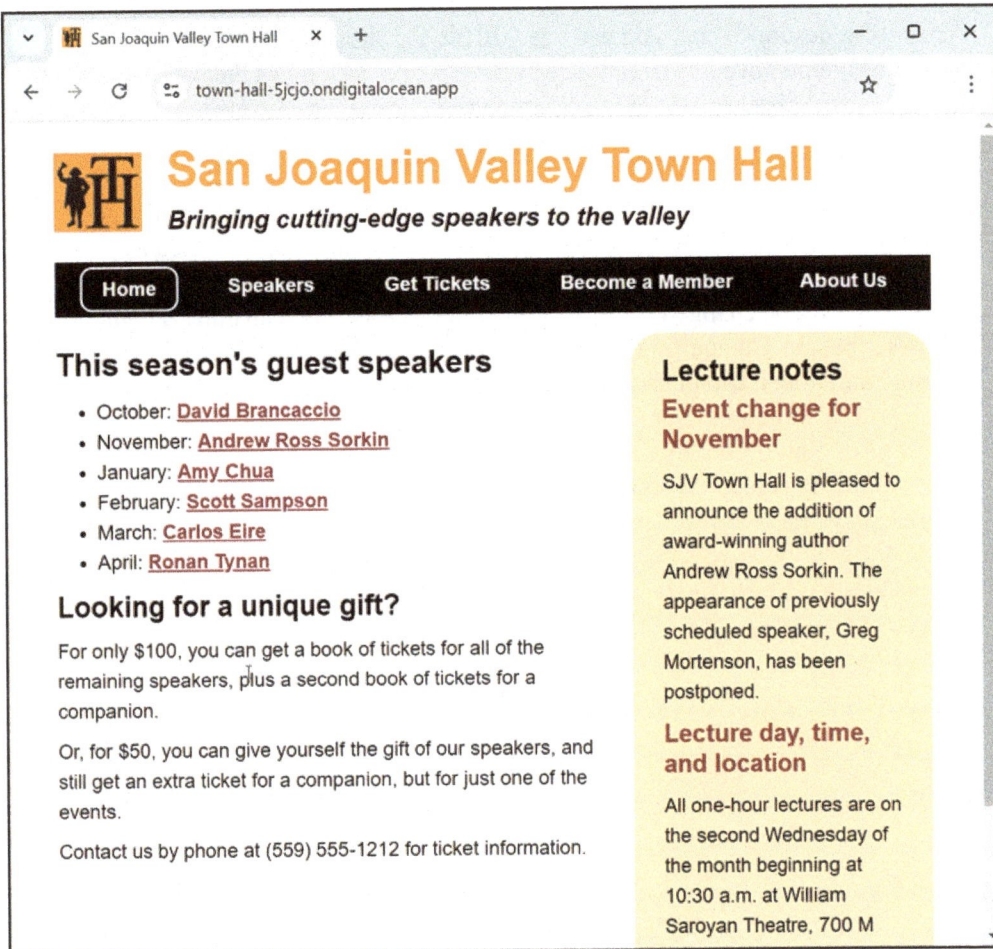

How to test a web page

1. Use Chrome to navigate to the deployed page that you want to test.
2. Review the content for the page, and make sure it's all there and it all works.
3. Test all the links on the page to make sure they work correctly.
4. Test the page in other browsers including tablet and mobile browsers.

Description

- Before you upload files to the cloud, you should test them on your own computer.

Figure 17-7   How to test a deployed website

How to make changes to a deployed website

Figure 17-8 describes how you can use GitHub Desktop to make changes to a static website deployed to DigitalOcean. This works as long as the Autodeploy check box was checked when you created your DigitalOcean website.

First, you can use GitHub Desktop to open the repository in VS Code. Once your repository is open in VS Code, you can update existing HTML and CSS files or add new files, just like any other website. Once you save your changes, the left pane of GitHub Desktop displays the files you updated or added as changed files.

To commit the changed files to the local repository, you can enter a comment that describes the changes in the Summary field and click the "Commit to main" button. Finally, you can push the committed changes to GitHub by clicking the "Push origin" button in the right pane.

At this point, if your DigitalOcean static website is configured to autodeploy, it automatically detects the changes you pushed to your GitHub repository and redeploys the website. This can take a few minutes, though, so don't be alarmed if your deployed website doesn't display the changes right away. Once the changes have been deployed, you can test them to make sure they work correctly.

How to make changes to a DigitalOcean static website

1. Open GitHub Desktop.
2. In the right pane, click "Open in Visual Studio Code".
3. Switch to VS Code. If it displays a message that says that Git isn't found, click "Download Git" and install Git.
4. In VS Code, add or update an HTML or CSS file and save your changes.
5. Switch to GitHub Desktop. It should display the file you added or changed in the left pane as a changed file.
6. At the bottom of the left pane, enter a value in the Summary field and click "Commit to main".
7. In the right pane, click "Push origin" to update your repository on GitHub.
8. Since your DigitalOcean static website is configured to automatically redeploy when there are changes to the GitHub repository, you don't need to do anything else.

The repository files open in VS Code

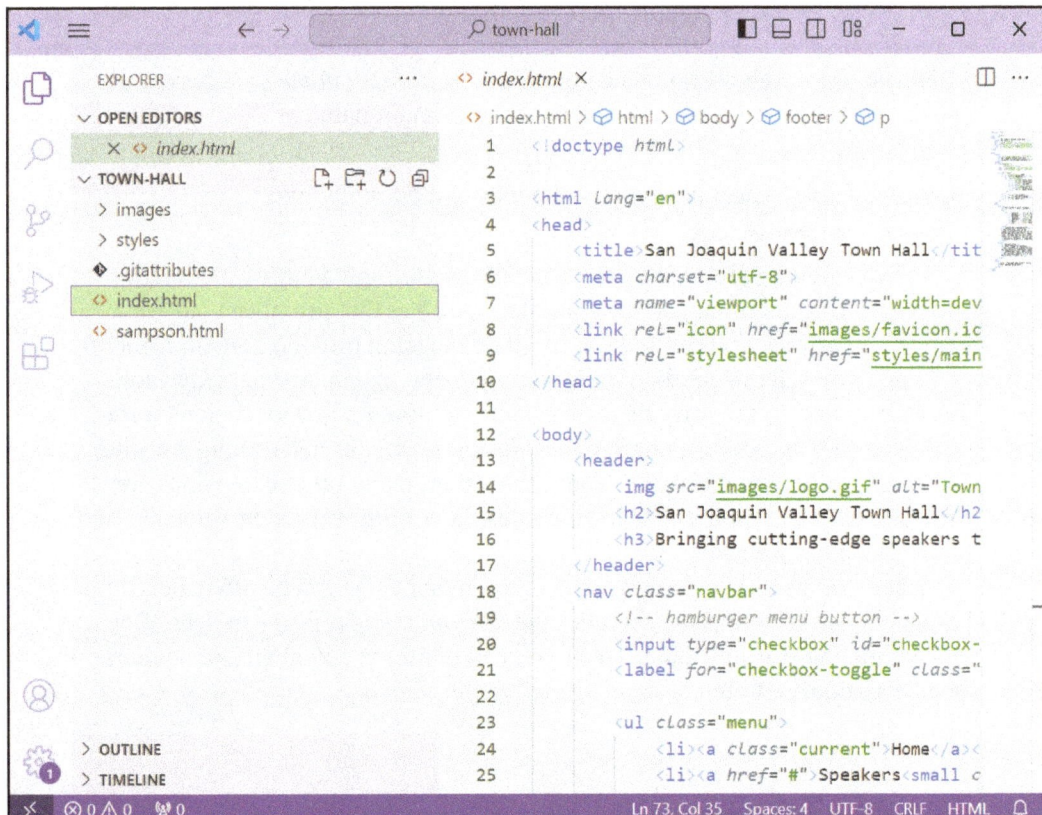

Figure 17-8   How to make changes to a deployed website

More skills for deploying a website

If you're developing a new website on your own, you need to get a web host that provides access to the internet. You need to get a domain name for your website. And you need to get your website's pages into search engines so people can find them.

How to get a web host

Figure 17-9 shows the web page for one of the many sites that provides *web hosting services*. For small websites, you only need a small amount of disk space. For larger websites, you may need more disk space, access to a database server, and a server-side programming language such as PHP.

If you search the internet, you'll find a wide range of services and prices for web hosting. And if you already pay an *internet service provider* (*ISP*) to connect to the internet, the ISP may provide free web hosting as part of its monthly fees.

How to get a domain name

When you use a web host, you can often use its *domain name* to access your website. If, for example, your web host has a domain name of

```
bluehost.com
```

you may be able to access your website with a *subdomain* like

```
murach.bluehost.com
```

For a professional website, you typically want to get your own domain name. And you can often get one from the same site that provides your web hosting. Otherwise, you can search for other sites that provide domain names.

When you select your domain name, you need to decide what extension you want to use. In the early days of the internet, only a few extensions were available, such as .com for commercial websites, .net for networking websites, and .org for other organizations. Today, however, many other extensions are available, such as .mil for military websites, .gov for government websites, and .biz for business websites.

When you use a site that provides domain names, you typically enter a domain name. Then, the site searches the *domain name registry* to determine whether the name is available. Then, you can purchase any available name for a specific amount of time.

A website for web hosting

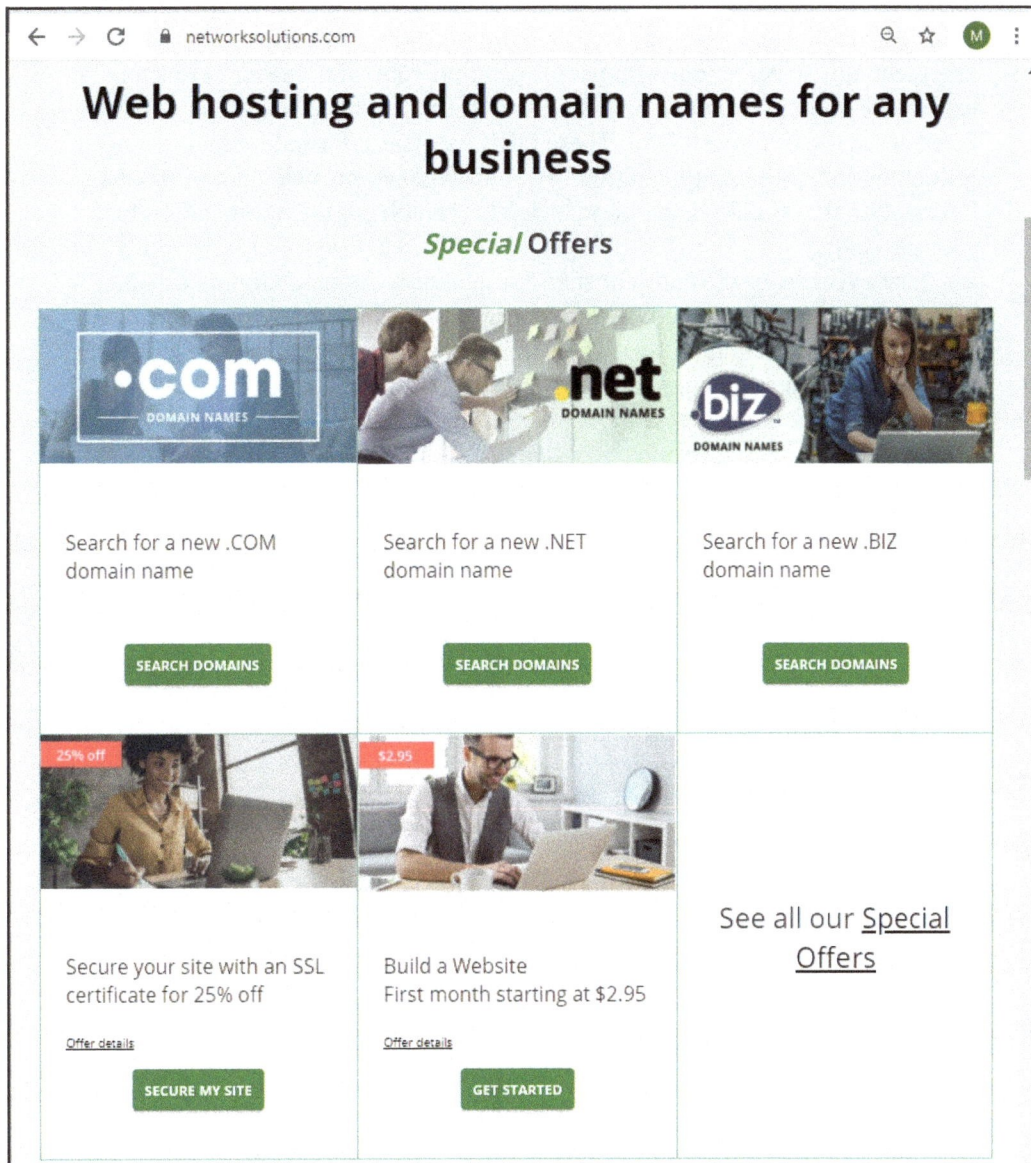

Description

- A *web host*, or *web hosting service*, provides space on a server computer that's connected to the internet, usually for a monthly fee.
- A *domain name* is a user-friendly name that's used to locate a resource that's connected to the internet.
- The .com, .net, and .org extensions are popular endings for domain names. They are intended to be used for commercial websites (.com), networking websites (.net), and other types of organizations (.org).
- The *domain name registry* is a database of all registered domain names.

Figure 17-9   How to get a web host and a domain name

How to get your website into search engines

After you deploy and test your website, you typically want to get your pages into the major search engines so people can find your site. To start, you can learn more about the Google and Bing search engines by going to the URLs listed at the top of figure 17-10. These URLs present information that you need for getting into these search engines. In addition, they provide a lot of other information that should help you improve the search engine optimization for your pages.

When you get to the page for submitting your website to Google or Bing, you only need to submit the URL of your home page. Then, if your site is linked properly, the search engine's *robot*, or *spider*, "crawls" through the rest of your pages by following the links in your site. As it crawls, the robot indexes your pages and scores them. Those scores determine how high your pages come up in the searches, and you typically want them to come up as high as possible.

Unfortunately, each search engine uses a different algorithm for determining the scores of your pages. For instance, some search engines improve the score of a page if it has links to other sites. Others improve the score if the pages of other websites link to the page. In addition, the search engines change their algorithms from time to time without any notice.

Once you've submitted your website, you don't have to do it again, even if you've made significant changes to the site. That's because the robot for a search engine periodically crawls through all of the sites and indexes the pages again, sometimes with a new algorithm for scoring.

Two search engines

Google	https://developers.google.com/search
Bing	https://www.bing.com/webmasters/about

The Google Search Central page

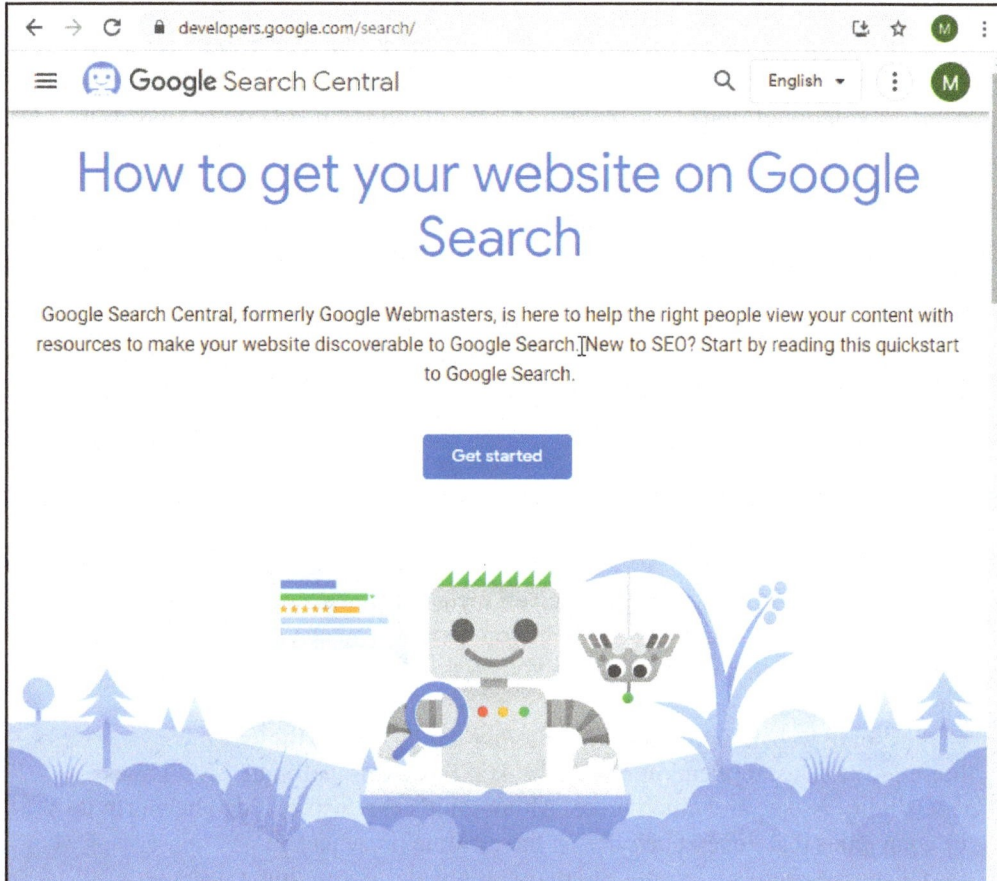

Description

- To submit your site to search engines, you can go to the website for the search engine and review the instructions.

- When you submit your site to a search engine, you provide the URL for your home page. Then, the *robot* (or *spider*) for that engine "crawls" through the pages of your site and indexes them.

- After your site is indexed the first time, the robot periodically crawls through your site again to update the index based on the changes that you've made to your site.

- At the website for each search engine, you can get a lot of information about improving the SEO for your pages.

Figure 17-10   How to get your website into search engines

How to set up, maintain, and improve a website

For many websites, there are pages and maybe even entire folders that you don't want indexed. For instance, you usually don't want your shopping cart pages indexed. So even before you submit your website to the search engines, you can use one of the two methods shown in figure 17-11 to stop or *exclude* pages from being indexed.

First, you can code a <meta> element with its http-equiv attribute set to "robots" and its content attribute set to "noindex" and "nofollow". This means that robots shouldn't index the page or follow any of the links on the page as it crawls through the pages of the site.

Second, you can add a robots.txt file to the root folder of your website. Here, the * for user-agent means that it applies to all search engines. As these examples show, you can use a robots.txt file to exclude one or more folders or files from being indexed.

In some cases, you may also want to set up a *site map* (a sitemap.xml file) that tells the search engines which pages to *include* in their index. For a small site, you can use a website like the one that's listed in this figure to create the map for you.

Once you have submitted your site to the search engines, you enter the maintenance and improvement phase. For instance, whenever you add new pages and delete old ones, you also need to update the robots.txt file and site map.

Then, to improve your pages, you can focus on web accessibility, search engine optimization, and more. To help you do that, you can use the two free tools listed in this figure to grade your website and provide ideas for improvements.

For some websites, you may want to use a more powerful tool like SiteImprove. This application reports broken links, misspellings, SEO issues, accessibility issues, and much more. Although it's expensive, it can be worth it for companies that depend heavily on their online presence.

Of course, there's a lot more to maintaining and improving a website than what's in this figure. That's why larger websites have a webmaster that's responsible for all these issues. If you're developing your own website, though, you're the webmaster, which means that you're the one who needs to master these skills. In that case, you can start by visiting Google Search Central.

However, it's also good to get your site graded by Website Grader or Nibbler. If you've adhered to the best practices and guidelines in this book, you may be pleasantly surprised by the results.

How to exclude a page from being indexed or followed

- Add a <meta> element to the page that tells the search engine robots what to do.

Don't index or follow a page

```
<meta http-equiv="robots" content="noindex, nofollow">
```

How to exclude folders and files from being indexed

- Add a robots.txt file to the root directory that tells search engine robots what to do.

Don't index the files in a folder

```
User-agent: *
Disallow: /cart/
```

Don't index the files in two folders

```
User-agent: *
Disallow: /cart/
Disallow: /private/
```

Don't index one folder and one file

```
User-agent: *
Disallow: /cart/
Disallow: /backlist/private.html
```

How to make sure web pages are included by search engines

- Add a sitemap.xml file to the root directory of the site. This *site map* tells search engines which URLs are available for indexing, how often these pages should be crawled, the last time each page was modified, and more.

- To create a free site map for small websites, you can go to: www.xml-sitemaps.com

Two free tools for grading your website

Hubspot's Website Grader	website.grader.com
Nibbler	nibbler.silktide.com

A commercial tool for grading and maintaining your website

SiteImprove	www.siteimprove.com

Description

- You can use a <meta> element or a robots.txt file to *exclude* some of the folders and files from being indexed.

- You can also use a sitemap.xml file to *include* specific pages of your site in the search engines.

- Once your site is up and running, you need to maintain and monitor it so it stays up-to-date and does well in the search engines.

Figure 17-11   How to set up, maintain, and improve a website

Perspective

Now that you've finished this chapter, you should be able to deploy a website for testing. In particular, you should be able to use GitHub to deploy a static website to DigitalOcean's app platform. Of course, there's a lot more that you can do with GitHub, especially when you're working with other developers on a project. In addition, there are many other types of projects that you can deploy to DigitalOcean. There are many good tutorials online where you can learn more about both platforms.

If you register a domain for a website and submit that website to the search engines, the job of maintenance and improvement begins. That's when you need to take the ideas presented at the end of this chapter and determine the best way to implement them for your website. To do that, you will need to do some research on your own. As you go down this road, things can get complicated, but if you stick with it, you should be able to improve and optimize your deployed website.

Terms

deploy
version control software
repository
kebab case
web hosting service
web host
internet service provider (ISP)
domain name
domain name registry
subdomain
robot
spider
site map

Appendix A

How to set up your computer for this book

This appendix shows how to install the software that we recommend for editing and testing the web pages presented in this book. That includes Visual Studio Code (VS Code) as the text editor and Chrome as the primary browser. But first, this appendix shows how to download the files that contain the code that's presented in this book.

How to download the files for this book

Figure A-1 shows how to download the files for this book. This includes the code for the complete web pages and short examples presented throughout this book. In addition, it includes the exercises at the end of each chapter and the solutions to those exercises.

The Murach website

www.murach.com

The folder that contains the source code

Documents\murach\html_css

The subfolders

Folder	Description
book_apps	The complete web pages presented throughout this book.
book_examples	The short examples presented throughout this book.
exercises	The starting points for the exercises at the end of each chapter.
solutions	The solutions to the exercises.

How to download and install the source code for this book

1. Go to www.murach.com.
2. Find the page for *HTML and CSS (6th Edition)*.
3. Scroll down to the "FREE Downloads" tab and click it.
4. Click on the DOWNLOAD NOW button for the zip file. This should download a zip file.
5. Double-click on the zip file to extract the files for this book into a folder named html_css.
6. Use File Explorer (Windows) or Finder (macOS) to create the murach folder in your Documents folder.
7. Use File Explorer or Finder to move the html_css folder into the murach folder.

Description

- We recommend storing the downloadable files in the folders shown above, but you can store them anywhere you like.

Figure A-1   How to download the files for this book

How to install Visual Studio Code

Figure A-2 shows how to install Visual Studio Code (VS Code) on both Windows and macOS. We recommend using this text editor because it's free, it has many excellent features, and it runs on Windows, macOS, and Linux.

This figure also shows how to make sure your system is set up correctly. In other words, it shows how to make sure that VS Code and the files for this book are installed correctly. To do that, you can use VS Code to open the html_css folder described in the previous figure. Then, you can use the Explorer window in VS Code to expand and collapse the folders that contain the HTML and CSS files for this book. If you can use VS Code to view these files, your system is set up correctly for this book.

The Visual Studio Code website

https://code.visualstudio.com

How to install on Windows

1. Go to the Visual Studio Code (VS Code) website. To do that, you can enter the URL shown above or search the web for "VS Code".
2. Click the button for downloading the Windows version and respond to any dialog boxes. This should download the exe file for the Setup program.
3. When the exe file finishes downloading, double-click on it to start the installation.
4. If you get a dialog box that indicates that this app isn't a verified app from the Microsoft Store, click Install Anyway.
5. If you're asked if you want to allow the program to make changes to your computer, click the Yes button.

How to install on macOS

1. Go to the Visual Studio Code (VS Code) website. To do that, you can enter the URL shown above or search the web for "VS Code".
2. Click the button for downloading the macOS version and respond to any dialog boxes. This should download the application file for VS Code.
3. Move the application file for VS Code from the Downloads folder to the Applications folder.

How to make sure your system is set up correctly

1. Start VS Code.
2. Select File→Open Folder from the menu system, and use the resulting dialog box to select this folder:

    `Documents\murach\html_css`

3. This should open the folder that contains all of the files for this book in the Explorer window that's displayed on the left side of the main VS Code window.

Description

- *Visual Studio Code*, also known as *VS Code*, runs on Windows, macOS, and Linux.
- Chapter 2 presents a tutorial that shows how to get started with VS Code.

Figure A-2   How to install Visual Studio Code

How to install Chrome and other browsers

If you use Windows, you already have the Edge browser installed on your computer. And if you use macOS, you already have the Safari browser installed on your computer. However, Chrome is the most-used browser. As a result, we recommend that you use it as the primary browser for testing your web pages.

If you haven't already installed Chrome, figure A-3 shows how. As the last step in this procedure suggests, we recommend that you make Chrome your default browser.

You may also want to install Firefox and Opera so you can test your web pages with those browsers too. To do that, you can use the website addresses shown in this figure.

The Chrome website

https://www.google.com/chrome

How to install Chrome

1. Go to the Chrome website and click on the link for downloading Chrome.
2. When the download finishes, run it.
3. As you step through the wizard that follows, you can accept all of the default settings.
4. When you're asked whether you want to make Chrome your default browser, we recommend doing that.

The Firefox website

https://www.mozilla.com

The Opera website

https://www.opera.com

Description

- Because Chrome is the most popular browser today, we recommend testing all the exercises for this book on that browser.
- If you use Windows, Edge is installed by default.
- If you use macOS, Safari is installed by default.
- Because Firefox and Opera are also popular browsers, you may want to install those browsers too.

Figure A-3   How to install Chrome and other browsers

Index

100% Guarantee

When you order directly from us, you must be satisfied. Try our books for 30 days or our eBooks for 14 days. They must work better than any other programming training you've ever used, or you can return them for a prompt refund. No questions asked!

Mike Murach, Publisher

Ben Murach, President

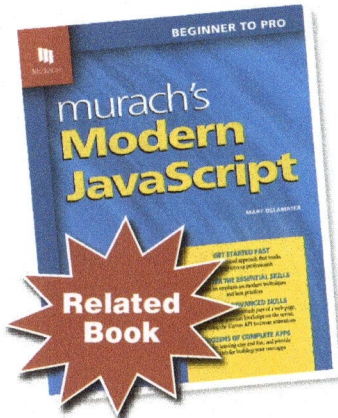

Related Book

Ready to learn JavaScript?

Murach's Modern JavaScript skips confusing old features and jumps straight to the best practices of modern JavaScript. This makes it easier than ever for you to take the next step in your web development career.

Web development

HTML and CSS (6th Ed.)	$64.50
Modern JavaScript	59.50
JavaScript and jQuery (4th Ed.)	59.50
PHP and MySQL (4th Ed.)	59.50
ASP.NET Core MVC (2nd Ed.)	59.50
Java Servlets and JSP (3rd Ed.)	57.50

Databases

MySQL (4th Ed.)	$59.50
Oracle SQL and PL/SQL (3rd Ed)	59.50
SQL Server 2022 for Developers	59.50

Programming languages

Python Programming (2nd Ed.)	$59.50
C# (8th Ed.)	59.50
C++ Programming (2nd Ed.)	59.50
Java Programming (6th Ed.)	59.50

Data science

Python for Data Science (2nd Ed.)	$59.50
R for Data Analysis	59.50

Prices and availability are subject to change. Please visit our website or call for current information.

We want to hear from you

Do you have any comments, questions, or compliments to pass on to us? It would be great to hear from you! Please share your feedback in whatever way works best.

www.murach.com

1-800-221-5528
(Weekdays, 8 am to 4 pm Pacific Time)

murachbooks@murach.com

twitter.com/murachbooks

facebook.com/murachbooks

linkedin.com/company/
mike-murach-&-associates

instagram.com/murachbooks

What software you need for this book

- **Any text editor.** We recommend Visual Studio Code (VS Code) because it provides many excellent features.
- **At least one web browser.** We recommend Google Chrome because it provides excellent developer tools.

This software can be downloaded from the internet for free.

To view step-by-step instructions for installing it, please see appendix A.

What the downloadable files include

- The complete web pages presented in this book
- The short examples presented in this book
- The starting points for the exercises at the end of each chapter
- The solutions to those exercises

How to download the files for this book

1. Go to www.murach.com.
2. Navigate to the page for *HTML and CSS (6th Edition)*.
3. Scroll down to the "FREE downloads" tab and click it.
4. Click the Download Now button for the zip file. This should download a zip file.
5. Double-click the zip file to extract the files for this book into a folder named html_css.
6. Create the murach folder within your Documents folder.
7. Copy the html_css folder into the murach folder.

For more details, please see appendix A.